Springer Series in Wood Science

Editor: T.E. Timell

M.H. Zimmermann
Xylem Structure and the Ascent of Sap (1983)

J.F. Siau
Transport Processes in Wood (1984)

R.R. Archer
Growth Stresses and Strains in Trees (1986)

W.E. Hillis
Heartwood and Tree Exudates (1987)

S. Carlquist
Comparative Wood Anatomy (1988)

L.W. Roberts/P.B. Gahan/R. Aloni
Vascular Differentiation and Plant Growth Regulators (1988)

C. Skaar
Wood-Water Relations (1988)

J.M. Harris
Spiral Grain and Wave Phenomena in Wood Formation (1989)

B.J. Zobel/J.P. van Buijtenen
Wood Variation (1989)

Bruce J. Zobel · Johannes P. van Buijtenen

Wood Variation

Its Causes and Control

With 98 Figures

Springer-Verlag
Berlin Heidelberg New York
London Paris Tokyo

Prof. Dr. BRUCE J. ZOBEL
Zobel Forestry Associates, Inc.
P.O. Box 37398
Raleigh, NC 27627, USA

Dr. JOHANNES P. VAN BUIJTENEN
Head, Reforestation Department
Texas Forest Service and Professor
Forest Science Department, Texas Agricultural Experiment Station
Texas A&M University, College Station
Texas 77843, USA

Series Editor:

T.E. TIMELL
State University of New York
College of Environmental
Science and Forestry
Syracuse, NY 13210, USA

Cover: Transverse section of *Pinus lambertiana* wood. Courtesy of Dr. Carl de Zeeuw, SUNY College of Environmental Science and Forestry, Syracuse, New York

ISBN-13: 978-3-642-74071-8 e-ISBN-13: 978-3-642-74069-5
DOI: 10.1007/978-3-642-74069-5

Softcover reprint of the hardcover 1st edition 1989

Typesetting: K+V Fotosatz GmbH, Beerfelden

2131/3145-543210 – Printed on acid-free paper

This book is dedicated to Dr. H. E. Dadswell

Much of the best early work on wood variation was done in Australia by Dr. H. E. Dadswell, who was a leader in stimulating research on a world basis and who had a very strong influence on the concepts and ideas expressed in this book. Although not formally a teacher, he taught many persons about the importance and recognition of wood variability and the necessity to use this knowledge. Dr. Dadswell had a major influence on the philosophies, ideas, and knowledge of the authors about wood variation, as well as on the whole field of wood utilization.

Preface

Wood is the usual end product of a forestry operation. Because of its importance, numerous studies have been made relative to wood properties, the causes of wood variation, and how best to develop wood for desired products. There is voluminous literature related to these subjects, but it is neither well known nor appreciated by foresters because the publications are often not available or are not well understood by the forester or by those who use the wood. Frequently, the literature is confusing and contradictory, making it difficult for the nonspecialist to use what information is available.

In order to produce and use wood efficiently, the variation patterns within trees, among trees within species, and among species must be understood. This also requires some knowledge of the causes of variation and the effects of different wood properties upon utilization. The information about variation patterns, their causes, and control and effect upon the product must be known by the tree grower, the tree breeder, and the tree harvester as well as by those who ultimately convert wood into a final, salable product.

It is the objective of this book to take the vast amount of information available on wood variation, its causes, and control and sort it out, categorize it, and summarize the main concepts and trends for the reader. We will attempt to separate the important from the unimportant and to emphasize those concepts that are based upon the best and most reliable information available to us. Unfortunately, many studies relating to variation in wood are of limited value, and sometimes even leave the wrong impression, because the studies were too limited, were incomplete, violated biological principles, or were poorly designed or analyzed.

The book is much more than a literature review in that it contains frequent summaries of the literature plus the stated attitudes and experiences of both of us whose entire professional careers have dealt with wood quality, wood variation, the genetic control of wood, and utilization of wood. Special emphasis will be placed on the wood of tree species that are grown as exotics; this is done because exotic forestry is rapidly increasing in importance and wood of the exotic species sometimes has marginal utility when the trees are grown in the new environments. Additionally, some especially troublesome problems relative to wood have arisen when forest trees are grown in exotic environments that differ from those of the indigenous range of the species. These shortcomings must be recognized and corrected as much as possible through silvicultural manipulation, through the application of genetics, and through improved and innovative manufacturing technology.

Wood is a very variable substance, with differences occurring among species and genera, among geographic sources within a species, among trees within a geographic source as well as within each individual tree. Tree to tree variability is especially large, with differences within a species often being strongly genetically influenced. The book will concentrate on the possible methods of producing more uniform and usable wood. The greater the uniformity of wood available for manufacturing, the greater will be the efficiency of producing a specific product, and the quality of the final product will be much improved.

When forests are well managed, wood is truly a renewable resource. As explained in the text, when the best forest management practices and genetic manipulation currently available are used, it would appear that wood production can be easily quadrupled on a world basis. It is our opinion that it should be possible to again double the wood produced in the world as improved biology and better forest management, along with better utilization methods, are developed.

Although not the major objective of this book, the subject of the effect of differing wood qualities on manufactured products will be briefly covered throughout the text. This is done to illustrate the importance of wood variation and its control. No attempt is made to cover completely the effects of different woods on manufactured products.

The book should be of value as a reference to the practicing forester and wood utilization specialist. It summarizes results of methods to produce more valuable wood in forestry practice. It helps the user to understand the magnitude and importance of wood variation so that manufacturing can be more efficient. It is written in such a way that it can be used as a textbook both in forestry and in wood science; both solid wood and fiber products are included in the discussions.

January 1989

Bruce J. Zobel
J.P. van Buijtenen

Acknowledgments

It is only through the generous help of many others that this book was made possible. The authors are most grateful to those persons who made suggestions and corrections, edited chapters, and supplied data, graphs, and photographs. The authors thank Dean Eric Ellwood, School of Forest Resources, North Carolina State University, Raleigh North Carolina for financial help in preparation of the book.

Following is an alphabetical listing of those who contributed to the development of the book:

Name	*Organization*	*Location*
Jeanne Adams	Office Options, Inc.	Cary, North Carolina
Pieter Baas	Rijksherbarium	Leiden, Netherlands
Leopoldo Brandão	Aracruz Florestal	Rio de Janeiro, Brazil
Edgard Campinhos	Aracruz Florestal	Aracruz, Brazil
J. A. Cisconetti	Aracruz Celulosa S. A.	Vitoria, Brazil
Ergilio Claudio-da Silva Jr.	Aracruz Celulosa S. A.	Vitoria, Brazil
Dean Einspahr	The Institute of Paper Chemistry	Appleton, Wisconsin
J. Maddern Harris	Forest Research Institute	Rotorua, New Zealand
Ted Hillis	CSIRO	Melbourne, Australia
Jara Ikemori	Aracruz Florestal	Aracruz, Brazil
Wolfgang Knigge	Universität Göttingen	Göttingen, Germany
Phillip Larson	U. S. Forest Service	Rhinelander, Wisconsin
William Ladrach	Zobel Forestry Associates, Inc.	Cary, North Carolina
Siegfried Lewark	Institute für Forstnutzung	Universität Göttingen, Germany
W. J. Lowe	Texas Forest Service	College Station, Texas
Hal Martin	Omni Engineering	Raleigh, North Carolina

Name	*Organization*	*Location*
Craig McKinley	Texas Forest Service	College Station, Texas
Ted Miller	Professor Emeritus	North Carolina State University, Raleigh, North Carolina
Gerard Nepveu	Station de Recherches sur la Qualité des Bois	Nancy, France
Ted Palmer	Tropical Products Inst.	London, England
Hubert Polge	Centre Nat'l de Recherches Foresteres	Nancy, France
Beatriz Vera Pozzi Redko	Companhia Florestal Monte Dourado	Monte Dourado, Brazil
Tony Shelbourne	Forest Research Institute	Rotorua, New Zealand
Hank van der Sijde	South African Forestry Research Institute	Pretoria, South Africa
John Talbert	University of Washington	Seattle, Washington
Fred Taylor	Mississippi State University	Starkville, Mississippi
Tore Timell	State University of New York	Syracuse, New York
Gerrit van Wyk	South African Forest Research Institute	Pretoria, South Africa
Michael Veal	Weyerhaeuser Co.	Tacoma, Washington
Alvin Yanchuk	University of Alberta	Edmonton, Alberta, Canada
Barbara Zobel	Zobel Forestry Associates	Cary, North Carolina

Contents

Chapter 1

Wood Variation and Wood Properties

1.1 What Is Wood?

Wood, a most useful substance, varies greatly. Wood properties are changing throughout the world (Zobel et al. 1983) and the magnitude and causes of the changes must be known. The importance of wood properties to the final product is becoming better known; one example is the relationship of the morphology of loblolly pine (*Pinus taeda*) tracheids and the paper made from them, as discussed by Barefoot et al. (1970). The subject of cell morphology and usefulness for the final product was covered in a general way by Keating (1983). Before the causes and control of wood variation are discussed, it is necessary to know just what wood is.

There are numerous opinions as to what constitutes wood. No specific definition of wood is totally satisfactory but the one by Webster and McKetchnie (1980) that wood is "the hard fibrous substance beneath the bark in the stems and branches of trees and shrubs: xylem" is as good as any. This, of course, includes the roots. Put slightly differently, Larson (1969b) uses the definition "... wood is the xylem of a tree. It is a product of the cambium and it consists of cells or wood elements that have passed through various stages of development. All the developmental phases of cellular division, differentiation, and maturation taken together constitute wood formation." It is important to understand that wood is complex and variable, consisting of numerous substances that are organized in differing ways, with the result that wood is a very nonuniform material.

Regardless of its definition, wood is a remarkable material with a variability and flexibility that makes it useful for many kinds of products. Its composition of cellulose, hemicelluloses, lignin, numerous types of extracts, sugars, and other organic and inorganic substances produces a raw material that can be used for many things, including papers, building materials, chemicals, energy, and even food. This subject was covered by Doimo (1984) in "What is Wood?" where he reviewed the major organic components of wood.

Although the variability of wood gives it great utility, it is also a major drawback to its efficient use as a raw material. In his discussion of wood property needs for the user, Keating (1982) states that "the variability of wood is not only one of its attractions but is also the reason why we will never be able to precisely catalogue and predict its performance." Except under unusual conditions, the uniformity so greatly desired by the manufacturer is lacking in wood, resulting in variation in quality and thus in production inefficiency. It is clear that "a major industrial need is to have greater wood uniformity" (Zobel et al. 1983). Burdon and Thulin (1965) emphasize that wood users strongly demand more uniformity both within and among trees; Larson (1969b) states that the greatest wood quality problems facing all wood-using industries is lack of uniformity.

The great need for research into the variations in wood properties, their control, and their effect on the quality of the end product is emphasized by nearly every author. Greater uniformity will help avoid inefficiencies; for example, a structure made from lumber must be designed for the weakest boards that will be used in it. Since strength properties vary considerably from board to board, inefficient utilization results because most boards are actually stronger than needed to meet the desired standard. Numerous efforts have been made to grow trees with wood having a greater uniformity and thus greater utility (Zobel et al. 1983).

It seems nearly impossible to find a satisfactory meaning for wood quality. Wood properties are easily defined but their utilization value, i.e. quality, varies according to product and utilization standards. For example, to say that juvenile wood has poor quality is correct for solid wood products, bags, and boxes, but it is not correct for writing papers, newsprint, or some tissues. Perhaps the best definition of quality is the rather cumbersome one by Gibson (1980) who states that quality is "the totality of the attributes of a product which contributes to the satisfaction of needs." Many useless arguments occur about wood quality, because wood quality can only have a meaning when the final product is known.

Although usage varies, many wood scientists refer to *wood properties* as the cellular, anatomical, and chemical characteristics of the wood within and among trees. The term *wood quality* frequently refers to the cumulative effect of the wood properties on some specified product or products. These concepts are not always observed and sometimes the terms wood property and wood quality are used interchangeably. Larson (1969b) relates the effects of growth and the physiology of wood formation to the concept of wood quality. We have seen references where wood properties referred to the strength of the wood rather than to its anatomy or chemistry. Because of differing uses, one therefore must read carefully to assess how the writer uses the terms wood properties and wood qualities. In this book, we use the terms wood properties and wood qualities as defined by the words italicized in the beginning of this paragraph.

There are many differences in terminology related to wood that can cause confusion; these will generally be discussed and defined as they are encountered in the text. However, an immediate definition is needed for the terms *earlywood* and *latewood*. From the literature it appears that earlywood (that portion of the annual ring produced early in the year) and latewood (which is produced later in the year) are the favored terms. Some authors, such as Larson (1969b), often refer to a third category called *transition wood*, which has characteristics intermediate between earlywood and latewood. Earlywood contains a high proportion of large, thin-walled cells compared to latewood, which has thicker walls and sometimes somewhat narrower and frequently flattened cells (Fig. 1.1).

A major advantage of wood resides in its low energy requirements for manufacture (Boyd et al. 1976). Items made from nonrenewable resources require appreciably more energy per ton of usable product than their wooden counterparts. The efficiency of forest production facilities is aided by the energy obtained from by-products and wood residues, which are not available in the utilization of most other products. Another saving in using wood results from the low capital investments necessary for conversion of trees into solid wood products. All these factors combined have made solid wood products and different types of particle

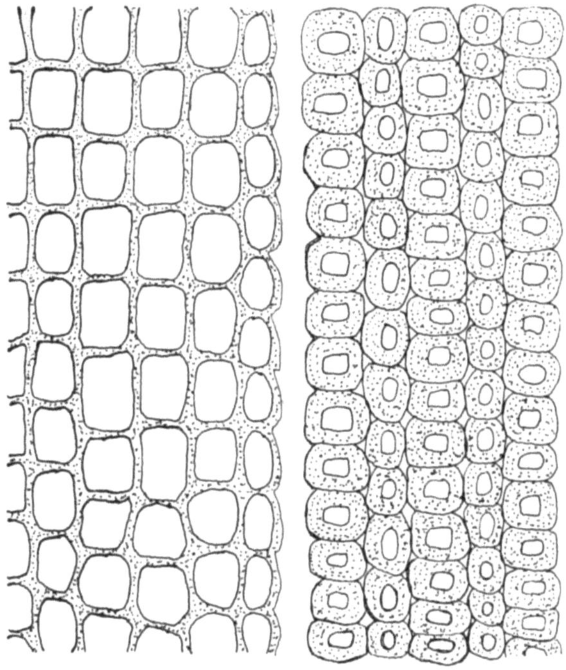

Fig. 1.1. Earlywood (*left*) and latewood cells (*right*) of loblolly pine (*Pinus taeda*) are illustrated schematically to show the differences in cell wall thickness and cell dimensions

board very competitive in construction compared to other raw materials (Bethel 1977).

1.2 Kinds of Trees and the Wood Produced by Them

There are two major categories of trees: *gymnosperms* and *angiosperms*. These are so named because of their reproductive structures; the gymnosperms do not have true flowers and their ovules and seeds are borne "naked" often subtended by a scale, while the angiosperms have flowers and the seeds and ovules are borne within ovaries. Although angiosperms include both monocotyledonous and dicotyledonous trees, almost all commercially important angiospermous trees are dicotyledonous. Dicotyledons are also commonly called *hardwoods*, and the gymnosperms are referred to as *softwoods*. This terminology can be misleading because some of the hardwoods have wood that is much softer than that of some softwoods and vice versa.

The softwoods are usually evergreens and have needlelike leaves but a few species are deciduous (i.e., they drop their leaves for part of the year). The hardwoods have the more typical flat, often broad leaf shape; some are deciduous and some evergreen. Most of the deciduous hardwoods grow in cooler climates. The softwoods are divided into smaller groups, the most important of which is the conifer group. Most of the softwoods used for forest products are conifers. There-

Table 1.1. Some books on wood that deal with wood composition and variation patterns

Year	Author	Book	Publisher	Pages
1940	Brown and Panshin	Commercial Timbers of the United States	McGraw-Hill Book Co. New York	554
1944	Tiemann	Wood Technology	Pitman Publishing Corp. New York	328
1964	Zimmermann	The Formation of Wood in Forest Trees	Academic Press New York	562
1964	Panshin, de Zeeuw and Brown	Textbook of Wood Technology	McGraw-Hill Book Co. New York	643
1965	Rydholm	Pulping Processes	Interscience Pub. (John Wiley) New York	1269
1966	Knigge and Schultz	Grundriß der Forstbenutzung	Paul Parey Hamburg	584
1969b	Larson	Wood Formation and the Concept of Wood Quality	Yale Univ. Press New Haven Connecticut	54
1969	MacDonald and Franklin	Pulp and Paper Manufacture 2nd Ed. Vol I: The Pulping of Wood	McGraw-Hill New York	769
1970	Britt	Handbook of Pulp and Paper Technology – Second Edition	Van Nostrand Reinhold Co. New York	723
1972	Dadswell	The Anatomy of Eucalypt Wood	C.S.I.R.O. Melbourne	36
1972	Koch	Utilization of the Southern Pines Vol. I – The Raw Material; Vol. II – Processing	Agr. Handbook #420, US For. Service, Washington DC Supt. of Documents	1663
1972	Zobel et al.	Wood Density of the Southern Pines	N.C. Agr. Expt. Stat. Tech. Bull. 208 Raleigh North Carolina	56
1978	Hillis and Brown	Eucalypts For Wood Production	C.S.I.R.O. Griffin Press Adelaide Aust.	434
1978	McMillin	Complete Tree Utilization of the Southern Pines	New Orleans, Louisiana For. Prod. Res. Soc.	484
1982b	Baas	New Perspectives in Wood Anatomy	W. Junk. Publ. The Hague	252
1983	Bamber and Burley	The Wood Properties of Radiata Pine	Trop. For. Paper 17 Commonw. Agr. Bureau England	84
1983	Plumptre	Pinus caribaea Vol. II Wood Properties	Commonw. For. Inst. England	145
1984	Chudnoff	Tropical Timbers of the World	US For. Service Agr. Handb. No. 607	464
1985	Koch	Utilization of Hardwoods Growing on Southern Pine Sites	U.S. For. Service Agr. Handb. No. 605	3 Vols; 3710
1985	Megraw	Wood Quality Factors in Loblolly Pine	TAPPI Press Atlanta, Georgia	89

fore, for the convenience of the reader and to relate more closely to the literature, in this book the terms *hardwoods* and *conifers* will be used to distinguish the two major categories of trees even though this usage is not absolutely technically correct in the botanical sense.

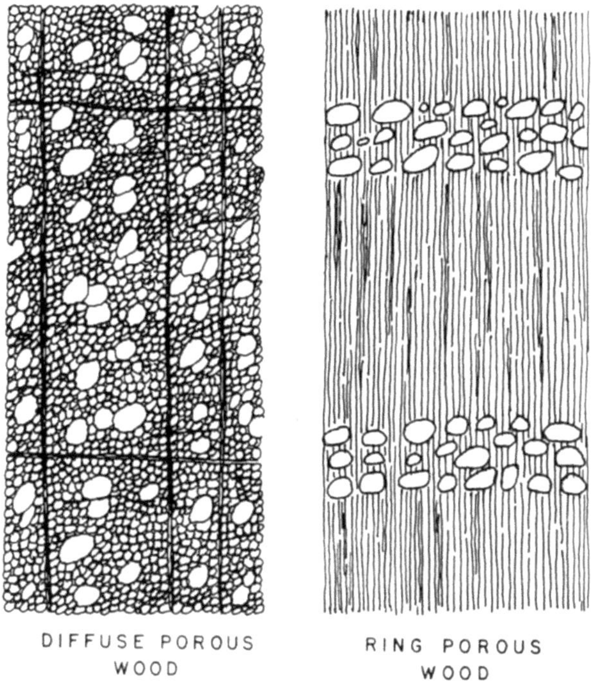

Fig. 1.2. The general cell distribution, especially of the vessels, is shown for ring-porous versus diffuse-porous wood of deciduous trees. The specific gravity values may be similar but the effect of differing cell distributions on product quality can be very different

The woods of conifers and hardwoods differ greatly from each other. Conifer wood is relatively simple, consisting of vertical tracheids, horizontal ray cells, and specialized resin canals and associated cells. Conifers usually do not have fibers. Wood cell types are defined in nearly all texts on wood technology and botany and need not be defined in this book. (See Table 1.1 for a limited list of books in which cell types are discussed.)

Hardwoods possess more complex wood than conifers, having very short and large diameter vessel elements, fibers (fiber tracheids), longitudinal parenchyma, and rays with differing types of cells (MacDonald and Franklin 1969). A key difference between conifers and hardwoods is in the length of the tracheids. The average length of the former is about 3.0 mm, while hardwood fiber length ranges from about 1.5 mm to less than 1 mm; vessel elements average about 0.5 mm in length. A general rule for conifers is that tracheid length is about 100 times tracheid width (Rydholm 1965; MacDonald and Franklin 1969).

The size and distribution of rays and vessels in the hardwoods have a major effect on wood quality and utility. For example, some genera such as the oaks (*Quercus*) produce a large number of vessels, nearly side by side, early in the growing season, with the bulk of the fibers produced later in the year. Such trees are said to be *ring-porous*. Other genera such as the poplars (*Populus*) produce vessels and fibers somewhat equally distributed throughout the growing season; these are referred to as *diffuse-porous* (MacDonald and Franklin 1969; Fig. 1.2). The patterns of vessel production are of great importance in determining the utility of hardwoods for a given product (Guiher 1965; Fig. 1.3).

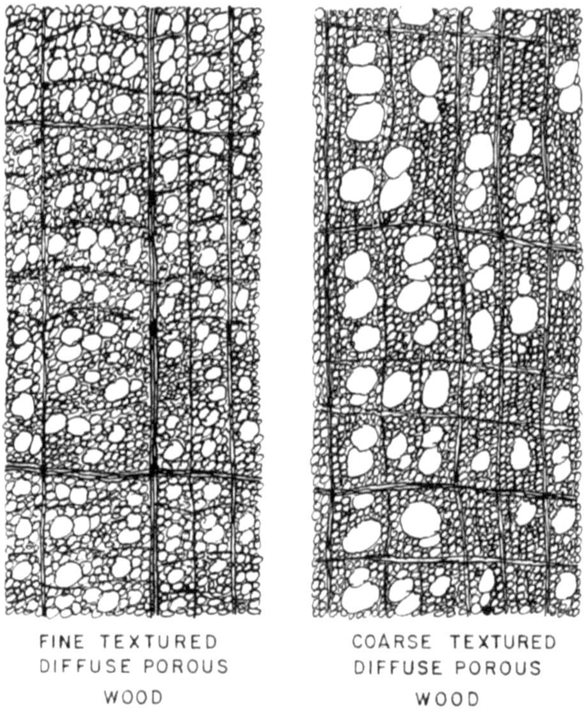

Fig. 1.3. Fine-textured versus coarse-textured diffuse-porous woods are shown schematically. The size of vessels has a major effect on product quality

Many foresters use the term "fiber" for both the wood fibers of hardwoods and the tracheids of conifers (MacDonald and Franklin 1969). Although this is botanically incorrect, the common usage of the term fiber must be recognized and numerous publications refer to the fiber characteristics of conifers as well as those of hardwoods. For example, one frequently reads or hears reference to "pine fiber length."

1.3 Important Wood Properties

There are many wood properties that could be discussed in a book on wood variation and its control. Several of these will be briefly mentioned in this introductory chapter, but the most important will be dealt with in greater depth and detail later. One objective of this first chapter is to briefly introduce the important wood properties and to describe them so that they will be familiar to the reader as they are encountered in discussions in later chapters. Fuller treatments on the description of wood properties can be obtained from any number of books and publications (see Table 1.1); one good example related to wood anatomy is the publication of Bamber and Burley (1983). Although their book specifically deals with *Pinus radiata*, their wood property descriptions are very complete. No matter what the wood properties, a reduction in variability is of key importance and is now widely emphasized as a forestry objective.

The main discussion on wood properties will deal with specific gravity (or wood density), which is by far the most important. For example, according to Burdon and Thulin (1965), the one wood property in most need of improvement in species such as *Pinus radiata* (radiata pine) is specific gravity. The importance of specific gravity has been emphasized by many investigators, e.g. van Buijtenen (1982) and Bamber and Burley (1983), who state "of all the wood properties density is the most significant in determining end use." In *Pinus elliottii* (slash pine), De Guth (1980) summarizes as follows: "wood density is strongly correlated with strength properties of wood and with pulp yield and pulp quality. . . ." Wood density is treated in depth because of its effect on the yield, strength, and general quality of most of the products produced from wood.

Most pulp and paper properties are directly related to wood specific gravity (Barefoot et al. 1965; Artuz-Siegel et al. 1968). The strength and quality of solid wood products, which are most important in sawn wood, relate to specific gravity (Lewark 1979; Pearson and Gilmore 1980). Similarly, the energy yield from wood (Boulding 1977; Goldstein 1980a) is dependent on specific gravity. Many of the characteristics of the individual cells are directly correlated to overall wood specific gravity (Zobel and Talbert 1984). Although wood density does not always directly relate to the variation of strength properties within and among trees, it still is very useful in predicting strength (Littleford 1961).

In hardwoods, the influence of the greatly differing cell types have a major effect on the quality of the final product. Despite this, Mottet (1965) and others feel that, of all the physical properties in tropical woods, density is by far the most important and most well known. As Watson (1965) emphasizes, there is only very limited information on the relationship of wood quality to fiber characteristics of tropical and semitropical hardwoods. Basic density has been found to be a much better guide to strength properties of wood than other cell characteristics per se. Tropical hardwoods are stronger than temperate zone hardwoods of the same basic density, and they show less shrinkage.

Hardwood pulps sometimes produce paper with poor printing characteristics because the vessels are "picked" from the paper during some kinds of printing, resulting in a nonuniform paper surface which affects the distribution of ink (Brown and Panshin 1940; Colley 1973). The importance of the varied cell types of hardwoods is becoming better known as hardwoods are used more generally in the pulpmill furnish and are employed less as "filler pulps" to extend the use of conifer fibers. Many mills are now using a high proportion of hardwoods; for example, Higgins and Puri (1976) and Foelkel and Zvinakevicius (1980) report that many different kinds of paper are being made using 100% *Eucalyptus* fibers.

Another variation that is of importance is the density range within an annual growth ring. As described in Chapter 3, earlywood and latewood of the conifers are usually very different. For example, Ifju (1969) showed that the specific gravity of earlywood of southern pines varies from 0.30 to 0.46, while the specific gravity of latewood varies from 0.65 to 0.92. The amount, type, and abruptness of latewood cell formation have a major effect on the quality of conifer wood and the products produced from it. Harris (1981) cites radiata or Monterey pine (*Pinus radiata*) as an even-textured softwood which has a density range within an annual ring of about 1.8 to 1. He cites loblolly pine (*Pinus taeda*) as having a range of 2.3 to 1 and some species such as Douglas-fir (*Pseudotsuga menziesii*) are coarse-textured with a ratio as high as 5 to 1. Within-ring variability in wood density can cause major difficulties in utilization.

In this book, wood characteristics will not be defined or described in detail. The emphasis here will be on the variation patterns in wood and what can be done to reduce the variability. Methods of sampling will not be outlined, but proper sampling is essential for obtaining meaningful results related to variation. For example, one rule is never to take wood samples from near the stump. In a zone approximately 1 m from the ground line, wood properties are very variable and are not representative of the rest of the tree. Another rule is that because of the usual magnitude of the tree to tree variation, about 30 trees must be sampled within a given stand for obtaining a reasonable estimate of the true mean value of the wood properties of the stand.

1.4 Wood Specific Gravity (Wood Density)

Specific gravity and wood density are expressions of how much wood substance is present in a given volume of wood. They are dealt with only briefly here because they have been covered by many authors (e.g., Bamber and Burley 1983). The literature dealing with the density of conifers has been reviewed by Elliott (1970) and for hardwoods by Koch (1985). *Wood specific gravity* is the ratio of the weight of a given volume of wood to the weight of an equal volume of water at 4 °C (grams of wood/grams of cubic centimeter of water) and is therefore a unitless measure (such as 0.50).

Wood density also is a ratio of the dry weight of wood to its volume but is measured in units such as kilograms per cubic meter or pounds per cubic foot; wood density is mostly used by industrialists and mill personnel. Examples are 450 kg/m^3, or 25.0 lbs/ft^3. These ratios can be calculated from one another; thus, specific gravity$\times$62.4 (the weight in pounds of 1 ft^3 of water) is equal to the pounds per cubic foot. For example, the wood density of a tree with a specific gravity of 0.45 = 62.4$\times$0.45 = approximately 28 lbs/ft^3 (Dadswell 1972). In the metric system, density is divided by 1000 = specific gravity. Thus, 450 kg/m^3 equals a specific gravity of 0.45.

The terminology related to wood density can become complicated but the above definitions are most often used. In some areas the terms basic or bulk density are used interchangeably with specific gravity, although basic density refers to the ratio of dry weight per unit volume of green wood (Rydholm 1965). The terms specific gravity and basic density are commonly used by reseachers. Specific gravity of wood ranges from about 0.10 to 1.40 (cork has a specific gravity of 0.24). Most commerical woods in the temperate regions have specific gravities between 0.35 and 0.65 (Brown and Panshin 1940).

For ease of understanding:

1. $\text{Specific gravity} = \dfrac{\text{dry wt. of a given volume of wood}}{\text{wt. of an equal volume of water}}$

2. Wood density (lbs/ft³) $= \frac{\text{dry wt. of wood in pounds}}{\text{green volume of the wood sampled in ft}^3}$

3. Wood density (kg/m³) $= \frac{\text{dry wt. of wood in kilograms}}{\text{green volume of the wood sampled in m}^3}$

4. Basic density (g/cm³) $= \frac{\text{dry weight in grams}}{\text{green volume in cm}^3}$.

1.4.1 What Determines Wood Specific Gravity?

Many different factors affect specific gravity, mostly through their effect on growth patterns which in turn affect cell morphology and chemistry. One growth pattern sometimes mentioned relates to height growth. It has been claimed that in some conifers the tree will deposit true latewood cells only when the terminal bud of the tree is dormant. Therefore, as reported later, growing conditions that are so ideal that growth is continuous will result in the formation of a preponderance of thin-walled cells (Schmidt and Smith 1961). Laboratory studies by Wodzicki and Witkowska (1961) on *Picea abies* showed that short-day conditions caused thick-walled cells to be formed. Factors such as site quality, fertilization, stocking, growth rate, and provenance can all be of importance in determining specific gravity under certain conditions. Each of these are covered in some of the following chapters. As Hildebrandt (1960) states "the thickness of any cell-wall within the late and early wood and also the percentage of latewood are modified by external growth conditions."

The factor most discussed that can affect specific gravity is growth rate; abundant and controversial literature has accumulated on growth influences (see Chap. 3). It is often attributed much more than its true effects; as Mitchell (1963) points out "growth rate is only one of the many factors that influence specific gravity and is definitely not a major one in the Southern pines." The true situation is that all factors interact, and it becomes quite impossible to find a distinct causal effect for each.

Specific gravity is not a simple characteristic but is determined by several characteristics of wood such as cell size and wall thickness, the ratio of earlywood to latewood, the amount of ray cells, the size and amount of vessel elements, and other factors. As Besley (1964) stated "wood density is a rather gross measurement, giving average figures which do not fully reveal the distribution of the different kinds of cells." In addition to cell characteristics, factors such as chemical deposits within and between the cells can drastically affect specific gravity. When such deposits are present, wood specific gravity must be categorized as "extracted" or "unextracted". Rather greatly inflated specific gravity values can be obtained if there are unusual deposits of extractives, such as the resins in the heartwood of some conifers, around knots or near wounds resulting from insect attack, frost cracks, or ring shake.

Many of the resinous and phenolic deposits and sugars in wood can be extracted with alcohol-benzene, and with hot or cold water (Fig. 1.4). Although they

Fig. 1.4. Resin deposits in wood, such as those shown in loblolly pine when grown in New Zealand, greatly affect specific gravity values. To compare among trees when resins are present, "extracted" specific gravity values must be used

have an effect on specific gravity, the deposited substances are often not an integral part of the cell wall and thus cause "confusion" when trying to relate strength or quality of the final product to the specific gravity of the wood. Removal of extractives is therefore necessary to assure a meaningful relationship. In *Pinus echinata* (shortleaf pine) Posey et al. (1970) indicated that extractives mask the differences between provenances. They recommend that specific gravity comparisons should be based only on extractive-free wood. Differences can be considerable; Taras and Saucier (1970) found the unextracted specific gravity of *Pinus serotina* (pond pine) to be 0.492 and that of extracted samples 0.454. They state that specific gravity is overestimated by 6.0% to 7.5% for the southern pines when use is made of unextracted increment cores. Usually it is less. For example, Megraw (1985) found that sapwood of *Pinus taeda* is 2% to 3% less with extracted compared to unextracted values when determined on an oven-dry weight basis.

For ponderosa pine (*P. ponderosa*), Barger and Ffolliott (1971) warn that the presence of extractives results in too high specific gravities; the specific gravity is reduced 12% (from 0.421 to 0.371) by removal of alcohol-benzene extractives. They present an equation to calculate extracted specific gravity from unextracted values. In *Pinus clausa* (sand pine), Clark and Taras (1969) found a correlation of r = 0.88 between extracted and unextracted wood specific gravity; unextracted was 7.9% higher than extracted. For *Pinus radiata*, Harris (1965) reported that resin extraction did not improve correlations with specific gravity in young trees. Cown et al. (1983) found resin contents of only 2% to 3% in young *Pinus caribaea* trees but much higher values in trees 25-years-old.

It is clear from the examples cited above that care must be taken to assess the extractive content when determining the actual specific gravity, but a good estimation of the extracted values can be obtained from unextracted wood samples. For operational studies on young trees, extractive removal is usually not done; unextracted specific gravity values will be higher but all values will be inflated by about the same amount, and useful comparative values are obtained with the unextracted wood. For example, in young loblolly pine (*Pinus taeda*) the difference in specific gravity between extracted and unextracted wood is usually about 0.02, but in young, fast-grown *Eucalyptus grandis*, there is no significant effect on the specific gravity of alcohol-benzene extraction.

In older trees that have differing amounts of heartwood or trees with pockets of extractives, the extraction process is necessary to obtain a reasonable estimate of the actual wood substance present in a given volume of wood. For example, Harris (1981) reports that the resin content of the sapwood of *Pinus radiata* is usually less than 2%, while heartwood of old trees can have a resin content as high as 12%.

Some tree species have chemical or crystalline products deposited in, on, or between the cell walls. This is especially so in the tropical hardwoods. There are a number of species, such as *Dicorynia guianensis* in French Guyana, with a rather high silica content in the wood (Chudnoff 1980). In Guyana, Palmer and Gibbs (1978) found much silica in *Licania venosa* and *Eschweilera sagotiana*, enough to interfere with sawing and chipping. Although they are not a part of the cell wall, these deposits, frequently in the form of crystals, greatly affect both the specific gravity of the wood and the quality of the final product. When such deposits are present, a good estimate cannot be made of the effect of specific gravity on the final product.

The characteristics that contribute to specific gravity variation in the conifers were described using diagrams by Zobel and Talbert (1984); a somewhat similar pattern was shown by Larson (1969b). These are summarized as follows:

1. *The percentage of latewood.* Some trees start producing latewood early in the growth cycle of a given year, while others begin later; all other things being equal, those starting late will have lower specific gravities. Some trees continue to produce latewood later in the season and thus have higher specific gravity wood. The growth patterns which affect latewood percent are influenced both by environmental and genetic factors. In general, for the southern pines (*P. taeda* and *P. elliottii*) the correlation between latewood percentage and specific gravity is approximately r = 0.60 (Gilmore et al. 1966). The percentage of latewood is very important in the determination of pulp quality. According to Wilson (1964), a variation in 10% in the earlywood content of any one species gave greater differences in pulp quality of the southern pines than were exhibited among the four major species. Veal et al. (1986) state that "specific gravity is a reliable index of wood quality for loblolly pine because it is highly correlated with earlywood/latewood ratio and average fiber coarseness."

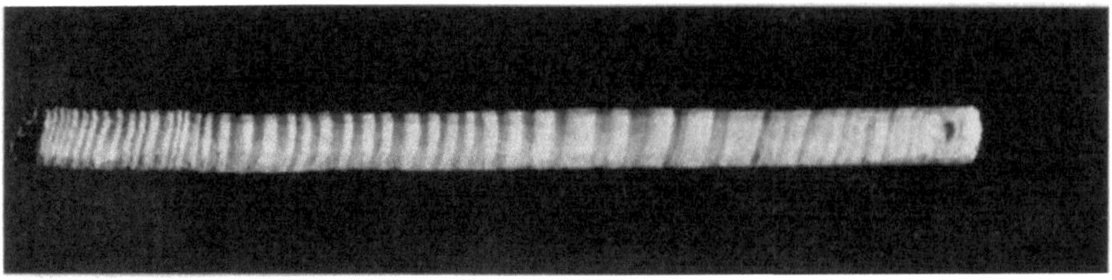

Fig. 1.5. An increment core through the center of a pine. The juvenile wood to the *right* has dark rings that are called latewood but technically their cell morphology does not fit the definition for latewood (See the text for a discussion of this discrepancy)

2. *The Cell Wall Thickness.* According to Mork (1928), cells are classified as latewood when the double wall thickness is greater than the lumen size. Some trees have very thick walls in the latewood, sometimes appearing to be almost rods with very small openings, while others barely fulfill Mork's definition. In fact, the brownish latewood produced in the juvenile wood of many pines, commonly referred to as latewood, is often actually not latewood according to Mork's definition. This was emphasized by Harris (1961) who showed that Mork's criterion could not be used for low specific gravity radiata pine because the ratio of cell wall thickness to lumen diameter was not above $\frac{1}{2}$. The latewood was easily visible even though it technically fell short of Mork's definition (Fig. 1.5).

Because of the large differences in wall thickness of the latewood cells among trees, considerable differences in specific gravity can be found in trees that have essentially the same latewood percentage. The wall thickness of fibers in the hardwoods can vary greatly, even within a species. This is illustrated for *Eucalyptus* in Fig. 1.6. These differences are very important in the utilization of the wood. Although there are minor variations, the actual cell wall material has an almost constant specific gravity (Stamm 1964). The specific gravity of the cell wall substance is 1.54 by water displacement and 1.46 by helium displacement.

3. Often trees that have the same cell wall thickness have *larger diameter cells* than others. The trees with small cells will have a higher specific gravity because of the smaller lumens.

Each of these factors are fairly strongly controlled genetically, can be manipulated by genetic selection, and can also be altered by changing growing conditions. In combination, they jointly determine what is defined as wood specific gravity, a most useful, important, and meaningful wood concept even though it is determined by a complex of characteristics. A complicating factor is that cell wall density varies directly with wood density. For example, earlywood cell walls of conifers have a density of about 0.75 g/ml, while those of latewood are about 0.85 g/ml (Wilson 1964). This is the result of anatomical differences in the cell walls.

In most conifers, especially the hard pines, the major product qualities are determined more by the characteristics of the latewood, while much less importance is attributed to the influence of the characteristics of earlywood cells. Differences in earlywood wall thickness can affect specific gravity, and occasionally a tree is found that has earlywood with walls as thick as the latewood of another

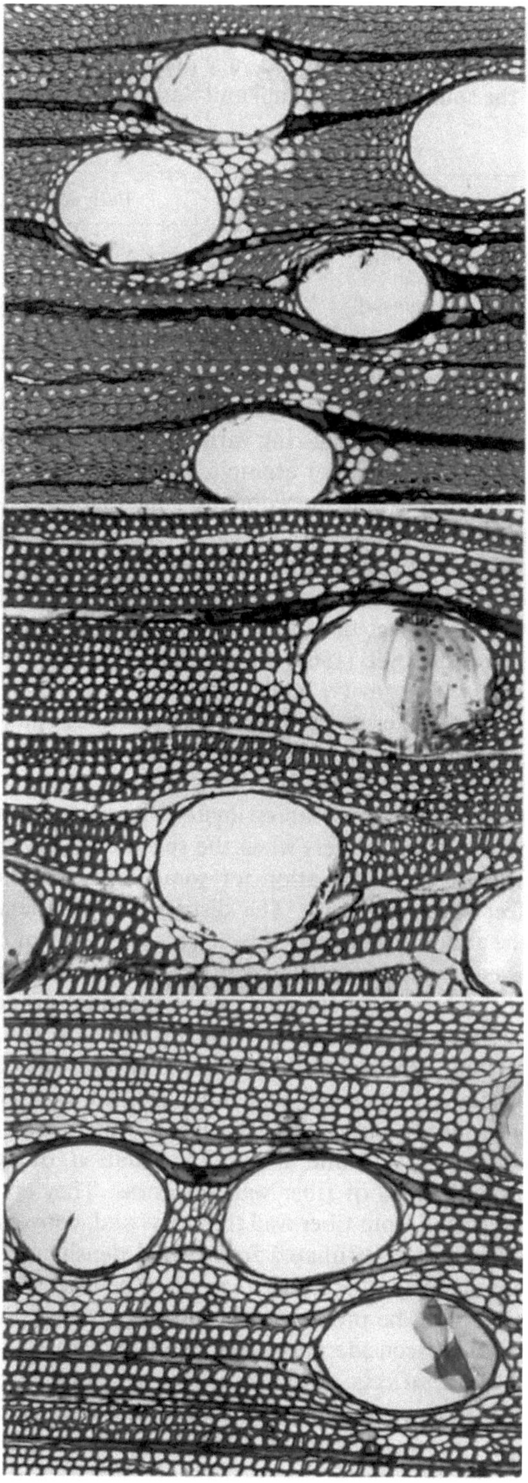

Fig. 1.6. Fiber wall thickness can vary greatly among trees of the same species as demonstrated by the wood of the *Eucalyptus grandis* hybrids shown. To the *left* are thin-walled fibers of low specific gravity wood (0.418); in the center, the fiber walls are of medium thickness in a tree with a specific gravity of 0.560; to the *right* are the thick-walled cells of a tree with a specific gravity of 0.660. Note the very small lumens (Courtesy of J. A. Cisconetti, Aracruz Florestal, Brazil)

tree of the same species. Generally, however, *wood and product quality are primarily determined by the characteristics of the cell walls in the latewood.*

The properties of earlywood vary greatly from those of latewood. In an early study of the southern pines, Paul and Smith (1950) reported the following specific gravities:

Species	Earlywood	Latewood
Longleaf pine (*P. palustris*)	0.280	0.690
Loblolly pine (*P. taeda*)	0.310	0.625
Shortleaf pine (*P. echinata*)	0.265	0.600
Slash pine (*P. elliottii*)	0.275	0.570

The subject of within-ring variation in specific gravity is covered in more detail in Chapter 3. Within-ring variation in other properties also has a major effect upon wood quality. For example, Biblis (1969) showed that specific stress and stiffness of loblolly pine latewood are 50% and 63% higher than for earlywood. Moisture affects the specific strength of latewood more than earlywood, but specific stiffness of both kinds of wood is similarly affected by moisture.

There are several factors in addition to the wall thickness of the fibers and the extractives that affect the specific gravity of the wood of hardwoods. As stated by Kaeiser and Boyce (1964) for *Populus*, the fibers are the most important in affecting specific gravity, while vessels are second in importance. The specific gravity is partly determined by the relative amount and size of the vessel elements (Brown and Panshin 1940). Another factor is the proportion and characteristics of the ray cells (Taylor 1969c).

Most wood scientists can estimate reasonably accurately the qualities of the final product from conifers when the specific gravity of the wood is known. This is not necessarily the situation for some hardwoods with their varied cell types that affect specific gravity. The distribution of the vessel elements and fibers within the annual ring can greatly affect the quality of the final product (Dadswell 1960). Even if vessel volume in diffuse-porous and ring-porous wood is the same, they can produce different qualities in the final product.

Regardless of specific gravity, a difference in cell distribution can result in quite different types of wood in the hardwoods, especially in the tropical hardwoods where cell types are frequently grouped together. In some of the soft, diffuse-porous hardwoods this is not always true. In a study on poplars, Scaramuzzi and Ferrari (1963) found a good correlation of specific gravity to various measures and ratios of fiber wall thickness. They concluded that wood density can be estimated from fiber wall thickness and, conversely, that wall thickness can be fairly accurately estimated from wood density.

In summary, specific gravity can often be used to help assess the kind of final product that will be produced. In pulping, the fact that a unit volume of heavy wood produces considerably more pulp than the same volume of low specific gravity wood affects the efficiency in wood handling and manufacture. Sometimes differences in the proportion and distribution of various tissues or the presence of chemical deposits make the use of specific gravity inefficient as an estimator of the quality of the final product. But in general, knowledge of the density variations and how to control them is of prime value to the wood user, and specific gravity is the ideal characteristic for the forester to manipulate.

1.4.2 The Importance of Specific Gravity

Specific gravity is of key importance in forest product manufacture because it has a major effect on both yield and quality of fibrous and solid wood products (Davis 1961; Barefoot et al. 1970; Lewark 1979) and because it can be changed by silvicultural manipulation (Williams and Hamilton 1961) and genetic manipulation (Zobel 1961; van Buijtenen 1962). In 1963, Mitchell showed that a 0.02 change in specific gravity will affect the modulus of rupture by about 1000 lbs/in^2. An increase of 0.02 in the specific gravity means a change of 100 lbs in the dry weight of a cord of pulpwood, or approximately 50 lbs of pulp.

Similarly, for oak in Europe, wood density is positively correlated to mechanical strength and shrinkage (Polge and Keller 1970). For spruce (*Picea abies*), Lewark (1981b) states "... specific gravity and knottiness are the determining factors in selection for strength. The present front position of specific gravity should be maintained... ." Bending strength tests on boards from four exotic pine species in Queensland showed the importance of specific gravity. For Douglas-fir (*Pseudotsuga menziesii*), Wilson and Ifju (1964) state "specific gravity did prove to be the most important single factor influencing tensile strength." Similarly, Bunn (1981) said "basic density is probably the single most important intrinsic wood property for most products, particularly if we are contemplating adopting short rotations."

Therefore, specific gravity largely determines the value and utility of wood and overshadows the importance of other wood properties. It is used to determine the tonnage of dry fiber produced per unit area, or per unit volume as was done for loblolly pine by Zobel et al. (1969b) where conversions and dry wood weight tables are presented. It is so important that in most forestry programs dealing with conifers which have pulp and paper as the final product, specific gravity is usually the only wood characteristic that is genetically manipulated.

One example of the effect of specific gravity is indicated in Table 1.2, which gives some values for 15-year-old and 21- to 25-year-old trees with extremes of specific gravity.

In addition to conventional forest products, specific gravity plays a major role in the use of wood as an energy source. Higher specific gravity wood contains more substance per unit volume and thus more BTU per unit volume. In addition, specific gravity and moisture content are usually negatively related within a species, so the higher the specific gravity the lower the moisture content (Zobel et al. 1968). This relationship of moisture content to wood substance is of key importance in energy production because net energy is largely a function of the amount of moisture that must be removed from the wood before it is converted into usable energy. The moisture content is shown according to the age of stand in Fig. 1.7. The older stands have a high wood density and a low moisture content.

Table 1.2. Wood qualities of high and low specific gravity slash pine trees of different ages[a]

	15-Year Age Class Specific gravity			21- to 25-Year Age Class Specific gravity		
	High	Average	Low	High	Average	Low
Moisture content[b] (%)	85.0	99.7	117.0	77.5	91.3	89.5
Unextracted specific gravity	0.565	0.497	0.455	0.565	0.536	0.500
Extracted specific gravity	0.550	0.485	0.435	0.545	0.510	0.475
Percent extractives[b]	2.5	2.8	4.7	4.1	3.5	4.0
Tracheid length (mm)	4.10	3.93	4.10	4.65	3.77	4.00
Yield, % dry basis	45.20	46.27	45.27	45.04	46.31	46.09
Yield, % wet basis	24.29	22.52	20.61	26.30	24.04	23.74

[a] Results are only general because of the very large sampling errors due to tree to tree differences within a class where as few as two to four trees were sampled. Values are simple averages for cross-sectional samples taken at the 5- and 15-ft levels above the stump. (From a study by Brunswick Pulp and Paper Company in Georgia).
[b] Based on dry weight of wood.

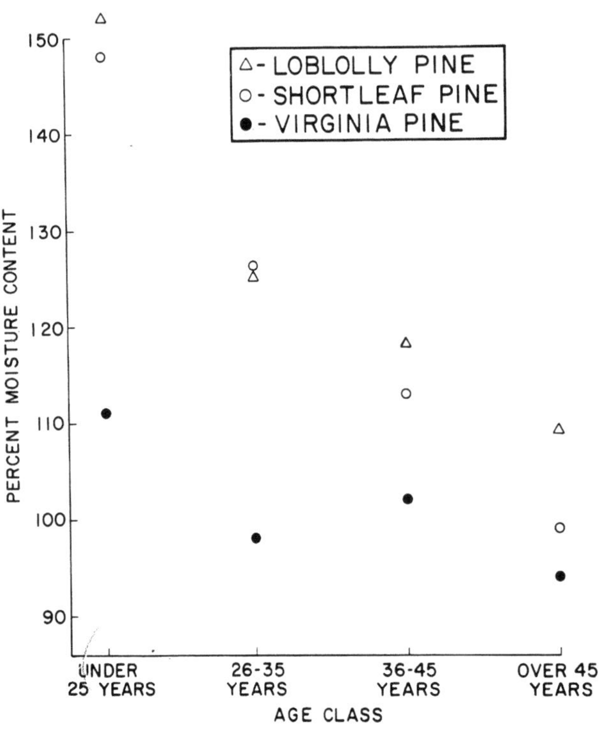

Fig. 1.7. Within a tree and within a species there is usually a negative correlation between specific gravity and moisture content. Shown is the moisture content for wood from different age stands. The older stands have a high wood density. Moisture content is calculated on the basis of dry wood

Trees of the genus *Leucaena* (Anon 1978; Brewbaker and Hutton 1979) are considered to be ideal for energy production because of their dense wood with a low moisture content. These trees have received a great deal of publicity as being a major potential energy source in many parts of the world. This moisture relationship is shown for four species of hardwoods used for energy in the southern

Table 1.3. The relationship between wood density and moisture content shows extreme values among hardwood species from southern USA[a]

Species	Specific gravity	Moisture content (%)[b]
Sweetgum	0.46	121
Sycamore	0.46	110
Ash	0.55	51
Oak	0.59	78

[a] Whole merchantable tree values are shown.
[b] Based upon dry weight.

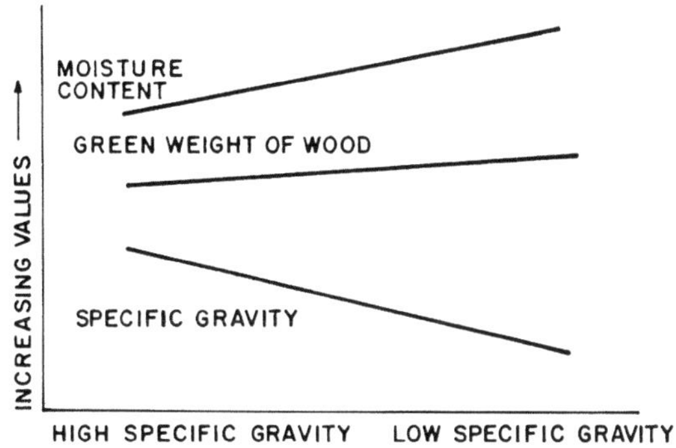

Fig. 1.8. A schematic representation of the relation between moisture content, specific gravity, and green weight of wood. The high specific gravity wood yields more usable fiber per unit green weight of wood than the low density wood

USA (Table 1.3). Note that *Fraxinus* (ash) and *Quercus* (oak) have a high wood density and a low moisture content compared to *Liquidambar* (sweetgum) and *Platanus* (sycamore). The first two species are much favored for energy production.

Wood specific gravity is an important component of overall biomass productivity of a forest. Biomass assessments based only on volume (cord, cunit, or cubic meter) are of limited value because from volume information there is no adequate indication as to the actual amount of wood substance available per unit volume. This is especially true when biomass assessments include bark (and sometimes even branches and leaves) as is frequently done in the tropical countries. These parts of the tree often have quite marked differences in specific gravity compared to the tree bole, and their volume can be considerable. For small eucalypts, bark will make up about 10% to 12% of the total volume, but in young pine 30% to 40% bark volume is common. The specific gravity and percentage volume of each component (bole wood, bark, branches, leaves) must be known before biomass can be accurately determined.

Species differences in specific gravity of hardwoods can be great. To take an extreme in commercial forest trees, balsa (*Ochroma* spp.), one of the lightest of all woods, has a specific gravity about 0.12 (density of 7.5 lbs/ft^3) while *Krugiodendron ferreum*, the heaviest wood in the United States, has a specific gravity of 1.3 (81.1 lbs/ft^3) (Brown and Panshin 1940).

Measuring wood by green weight is also a poor and inaccurate way of determining biomass. In loblolly pine, for example, low density wood has about the same green weight per unit volume as high density wood (Fig. 1.8). Yet the high

Table 1.4. The relationship between dry and green weight of wood of differing specific gravities for *P. taeda*

Specific gravity	Weight in lbs/ft³	
	Dry weight	Green weight
0.385	24.0	60.1
0.396	24.7	59.8
0.409	25.5	59.3
0.438	27.4	57.0
0.458	28.6	54.6

density wood will produce from 10% to 15% more usable fibers from the same green weight of wood (Zobel et al. 1969b; Britt 1970). Dry and green weights in pounds per cubic foot for several specific gravities of loblolly pine are shown in Table 1.4.

1.5 Cell Length

The length of the cells sometimes has a marked effect on product quality and the use of wood. Cell length (tracheid length in conifers, fiber and vessel element length in hardwoods) is generally the cell dimension next in importance to wall thickness in determining the final product value. It is important to remember that most personnel in production operations use the term fiber length to refer to both tracheids and fibers (Megraw 1985).

Often the importance of cell length has been overemphasized in the literature. A common belief is that short cells are responsible for deficiencies in certain paper properties when, in fact, the main problem relates to deficient cell wall thickness expressed through specific gravity (Barefoot et al. 1970). However, cell length does become important in the juvenile wood of some conifers and in the short fibers and vessels in some hardwood species. When using hardwoods, it is often necessary to add about 15% to 20% long fibers (tracheids) to manufacture papers that have a satisfactory tear strength.

Cell length varies greatly both within and among trees. The characteristic is strongly genetically controlled (Wheeler et al. 1965). It can also be altered by changing growth patterns silviculturally (Bannan 1967). There are two reasons why work aimed at changing cell length within species has not received much support despite the large percentage variability and strong genetic control:

1. The fibers of hardwoods are short. Even though a large percentage improvement is possible, the increased lengths in the improved wood will make little difference in the quality of paper produced. For example, if an intensive improvement program in maple (*Acer*) were undertaken, perhaps as much as 0.10 mm increase in fiber length could be obtained. The percentage improvement would be large, but the fibers of normal maple are so short (about 1.0 mm) that increasing them by 0.10 mm would have very little positive effect on the properties of the paper produced.

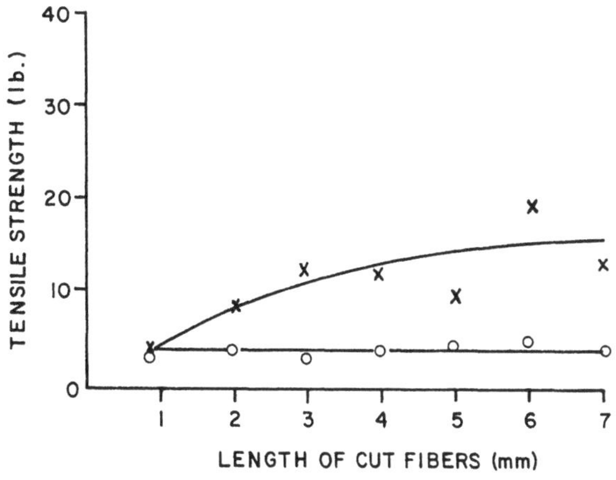

Fig. 1.9. Short cells can have an effect on product quality, but beyond a certain length (about 3 mm) an increase in length has little effect as illustrated for fiber tensile strength

2. There is a threshold value of cell length where increasingly longer cells have little effect on the quality of the product; this threshold value is about 2 mm for making kraft paper with an acceptable tear strength. Also, longer cells do not significantly increase tensile strength (Dadswell 1960; Fig. 1.9). During manufacturing, long cells break more than the shorter ones. In pines, for example, it is possible to increase tracheid length as much as 0.5 mm through selection and breeding. However, if the average is already about 3.0 mm, as is common in mature conifer wood, an increase to 3.5 mm will have only a minor effect on the quality of the final product.

It is when cells are near the "threshold length" that cell length variation can be of primary importance with regards to acceptable product quality. In this situation, methods to increase fiber length in certain longer fibered hardwoods and in the short tracheids of juvenile conifer wood become of real value. Improved technology has overcome cell length limitations to some degree. For example, some *Eucalyptus* species are now being used to make good grades of papers formerly considered possible only from "long-fibered" conifer woods (Higgins and Puri 1976). Such technological adjustments to make better use of short fibers will become increasingly important as more juvenile conifer wood becomes available or as short-fibered hardwoods constitute an increasingly large proportion of the furnish to pulp mills. The change toward shorter fibers in the wood available to the mills is already a fact, and usable paper is now being made from mixed tropical hardwoods with fibers that are quite short (Gomez and Mondragon 1974).

Cell length can also be important in affecting the quality of solid wood products. Long cells with steep fibril angles (see Sect. 1.6) make stronger and more stable boards; this is one of the reasons that juvenile wood of loblolly pine is weak and somewhat unstable (Schmidt and Smith 1961; Pearson and Gilmore 1980). The actual fiber length to strength relationship is not fully understood and questions are raised as to how much of the low strength and instability of juvenile wood of conifers results from short tracheids, and how much is caused by thin cell walls and flat fibril angles.

1.6 Other Wood Characteristics

Although all characteristics of the cell have some effect on the quality of a manufactured product, most are minor in importance compared to specific gravity. Many cell dimensions are in some way related to specific gravity so a clear-cut determination of the effects of individual cell characteristics is most difficult. Yet tissue properties and cell dimensions are variable and important, especially in the hardwoods. Their importance and extent were studied for red and white oak by Maeglin and Quirk (1984) who found that the proportion of fibers (by weight) varied from 74% for fast-grown trees to 38% for slow-grown individuals, while ray parenchyma constituted 24% and 54%, respectively.

Thickness of the cell wall is of prime importance in all wood qualities, but since this property is so strongly related to wood specific gravity, its effects are usually largely explained by the influence of that characteristic. Wood with thick cell walls tends to produce paper with a poor printing surface and poor mullen (burst) strength. The thick-walled cells act like small cellulose rods, do not bend easily, and do not collapse upon pulping, which inhibits chemical bonding. Thus,

Fig. 1.10. The wall thickness and coarseness of the cells largely determine the utility of the fibers during pulping. Shown are flattened fibers of *Pinus caribaea* conforming to the ribboning effect (Courtesy of Beatriz Vera Pozzi Redko, Companhia Florestal Monte Dourado, Brazil)

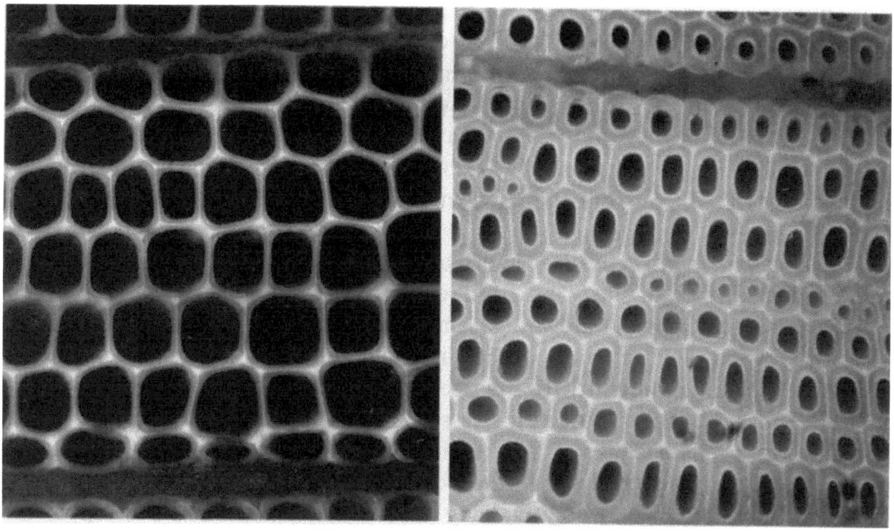

Fig. 1.11. Cell width can be of importance in determining product quality. Great differences occur in the same tree as shown by a section through the earlywood and latewood portions of the cell. Note the flattening in the thick-walled latewood cells (*right*) close to the end of the years growth (Zobel and Talbert 1984)

the surface of paper made from thick-walled cells is not smooth and ink tends to run rather than to produce a sharply defined printed letter or picture. The thinner-walled cells collapse upon pulping, bond well together chemically, and produce a smooth paper surface (Fig. 1.10). Wall thickness also has a major effect on the bending, tear, and tensile strengths of paper.

Therefore, although wall thickness varies greatly within and between species and within a tree and has considerable importance for most paper and solid wood qualities, most efforts to change or improve it will automatically occur when the desired specific gravity is obtained. Few studies directed solely to variations in wall thickness have been made; most relate to changing the proportion of thick-walled cells, through the control of latewood percentage or the amount of thin-walled juvenile wood by changing rotation ages.

The *width* of the cell can have a considerable effect on product quality. Woods like those of the spruces are considered good because of their relatively long and narrow cells, while the pines of the southern USA and the tropical pines have less desirable, coarser, and wider cells. Despite the expressed desire to have long, thin cells (Artuz-Siegel et al. 1968), cell width has been little emphasized within species, although choices among species are often influenced by the width and lumen size of the cells. Usually the tangential width of the cell is quite constant throughout a growth ring, but the radial width changes greatly during the growing season, being largest in the earlywood and smallest, with sometimes quite flattened cells, in the latewood (Fig. 1.11). In the radial direction within a tree, conifers show a great variation in cell diameter, with the larger cells usually having the thinnest walls.

Cell width plays a major role in the wood quality of hardwoods because of the vessel elements (see Fig. 1.6). Often vessel elements appear to be as wide as they are long and, when in abundance, have a major effect on paper properties. They cause difficulty in some printing systems where the vessels "pick out", as described earlier. Vessels are of value in giving bulk to paper, i.e., the paper is thicker without much added weight. Woods that have many large vessels, such as the oaks (*Quercus*), confront the paper manufacturer with unique problems but also with some advantages.

Ray or parenchyma cells have an effect on the quality of both solid wood and pulp products. Much of the wood grain (or figure), often quite showy as in beech (*Fagus*) and some tropical hardwoods, is related to the size and distribution of the rays in the wood. The ray and parenchyma cells themselves are thin-walled and very short and contribute little to the strength properties of paper, although they provide a smoother sheet. Just as for vessels, there is interest in trees with a lesser amount of ray cells because of their variability and adverse effect upon specific gravity, yield of paper, and strength of boards.

The proportion of the wood made up by various cell types is most important. For example, conifer wood is often 93% tracheids and only 7% parenchyma cells. On the average, the hardwoods have about 50% fibers and fiber-tracheids, 20% parenchyma, and 30% vessel elements. However, this can vary greatly according to species (Rydholm 1965).

Cell dimension ratios of many types have been used to better categorize wood according to its potential utility in various products. The ratios are indicators of printability and of tearing and tensile strength for paper. Species such as the northern spruces (*Picea* spp.) have a good length to width ratio and are usually considered to have ideal tracheids for pulping. Fiber size and morphology ratios will not be discussed here. Some, like the Runkel ratio, have been strongly recommended by some researchers but are not considered of importance by others.

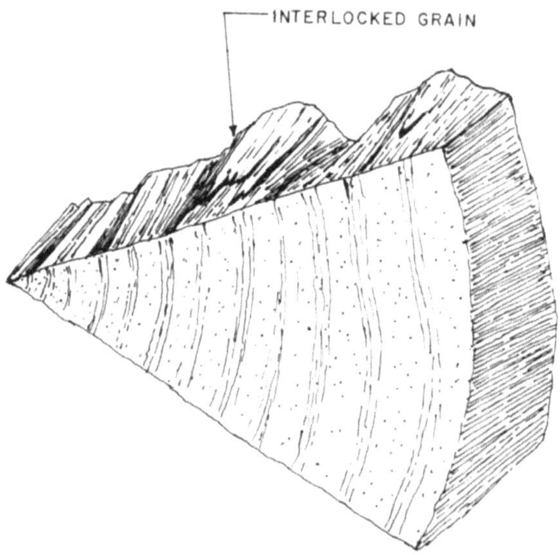

Fig. 1.12. Unusual grain patterns can give wood a desired figure or they can cause major problems as shown for interlocked grain in a hardwood. Proper seasoning and finishing without warping, cracking, or checking is impossible with this type of wood

Grain characteristics can have a major effect upon the value and utility of wood. Especially important for solid wood products is spiral grain and interlocked grain; when this is severe it becomes almost impossible to manufacture solid wood products that do not warp, split, or crack (Fig. 1.12). Since interlocked grain appears to be strongly inherited in *Liquidambar styraciflua* and other hardwoods, Webb (1964) suggests that its effects can be reduced by selection and breeding. For spiral grain, Champion (1929) and Dadswell and Nicholls (1959) used the figure of 7° of spirality as being the upper limit of acceptability for quality solid wood products. Often exotic conifers, especially in certain environments, attain or exceed this limit. Cown et al. (1983) found spiral grain in excess of 5° to be common in *Pinus caribaea* grown in Fiji. In their book on wood properties of *Pinus radiata*, Bamber and Burley (1983) have a chapter on spiral grain and grain orientation, its occurrence and importance. They also report that spiral grain is most severe near the center of the tree. Spiral grain appears to be the most troublesome defect affecting sawn timber. The causes and effects of spiral grain were summarized by Noskowiak in 1963. In 1973, Harris published a detailed study of the causes of spiral grain in radiata pine. Spiral and interlocked grain occurs everywhere but seems to be accentuated under tropical conditions, as is evident in numerous eucalypt species. Spiral grain and wave phenomena in wood formation are discussed in a recent book by Harris (1988).

Collapse, a defect on drying caused by irregular contraction of the wood (as distinct from normal shrinkage), makes the wood take a wavy form, and internal splits are evident. Collapse, which varies greatly from tree to tree, occurs when free water is removed (not bound water, as in normal shrinkage) and can be as much as 28% for tangential, 8% for radial, and 39% for volumetric shrinkage (Sesbou 1981). This cause of variation in wood can be of great importance for solid wood products and has resulted in a lack of usage of certain species.

Fibril angle (microfibril angle) is a most important characteristic, especially for solid wood products where the fibril orientation has a major effect on the stability of the wood upon drying and following manufacture. The cell walls are composed of small structural units called fibrils which consist of small bundles of cellulose molecules. Normally, the fibrils in the S_2 layer are oriented almost parallel to the cell axis. The fibril angle is the deviation from the cell axis and refers to the mean helical angle that the microfibrils of the S_2 layer of the cell wall forms with the longitudinal axis of the cell (Megraw 1985). When large fibril angles occur (flat fibrils) excess longitudinal shrinkage up to 9% occurs. When a board contains wood with both normal and flat fibril angles, the board is unstable and will crack, warp, and check because of the differential shrinkage within the board (Harris and Meylan 1965).

1.7 Cell Chemistry

A great deal is known about cell chemistry and its effect on products but little has been done with respect to its variability and the potential to change it. Much is known about chemical differences among woods of different species, and a great deal has been learned about the chemistry of juvenile wood and reaction wood (Timell 1986). Few studies have been made to relate chemical differences among trees within a species directly to the properties of the final product; those that have been reported have not shown as close a relationship of chemical properties with product quality as is evident for specific gravity (Byrd et al. 1965).

Hildebrandt (1960), nevertheless, emphasizes the importance of chemical composition and its influence on mechanical wood properties. For example, as the amount of lignin increases, the crushing strength and brittleness increases, while the tension strength, resistance to rupture, and shock decreases. Hildebrandt claims that a warm climate favors lignin production and tropical woods are therefore rich in lignin.

Resin production is very variable between and within species. In a study of four pine species in the tropics, Brito et al. (1980) found that *P. kesiya* (Benguet pine) produced more resin than *P. oocarpa* (oocarpa pine), *P. elliottii*, or *P. caribaea* (Caribbean pine) when grown in Brazil. It also had the best resin and turpentine properties, while *P. oocarpa* had the worst rosin. *Pinus elliottii* also had good turpentine.

Wood chemical studies have included cellulose, hemi-celluloses, lignin, extractives, and mineral constituents. The often used term holocellulose refers to the total polysaccharide content in wood. The chemical constituents of wood are defined and listed for a number of species by different methods of extraction by Rydholm (1965). His data show that conifers have more lignin than hardwoods; he reports an average of 27% lignin content for conifers; six northern conifers had 29.4% lignin. Seven major hardwoods averaged 21%, while birch and aspen contained only about 18.5% lignin.

There are numerous chemically related examples of wood variation. For instance, Scurfield et al. (1974) show where and how silica is deposited in wood. The hypothesis is that silica is a by-product that is deposited when cell protoplasts break down, and the silica is related to the formation of phenolic compounds. Yet members of the genera *Acacia* and *Eucalyptus*, which accumulate large amounts of phenols, do not deposit silica in their wood. Nevertheless, some species, especially some tropical hardwoods, produce so much silica that their woods are hardly usable even though the cell morphology is good. Production of silica is a major source of variation in wood and no effective controls are yet known; this is described in later chapters.

Variability resulting from differences in extractives or chemical depositions influence specific gravity and cause great differences in the utility of wood. Frequently, hardwoods such as *Eucalyptus* produce gummy exudates (called gummosis) consisting of phenolic substances and resins. There is so much formed that the utilization of the wood is affected and bleaching and pulp manufacturing become difficult. There are great differences among species. For example, *Eucalyptus viminalis* (manna gum) and *E. globulus* (bluegum) are strongly subjected to gummosis. There are also great variations from tree to tree. The importance of the tree to tree variability in resin formation was emphasized by Plumptre (1983) for *Pinus caribaea.*

Although wood normally contains extractives, the amounts can vary greatly, and frequently resin pockets are formed. These are intercellular cavities containing resin and are frequently associated with checks and splits or knots (Fig. 1.13).

Fig. 1.13. Frequently, resinous deposits are associated with large limbs as is evident for pine. Such deposits lower the quality of solid wood products and create problems in pulping. Wood density is greatly altered

1.8 Factors Controlling Wood Properties

The various factors that cause trees to produce different types of cells are interactive, and rarely is there a simple explanation for variation. Physiologically, the kind of wood produced is controlled by auxins during cambial development. Auxin variation can result from either environmental or genetic differences and their interactions. A discussion of growth patterns and wood development was presented by Larson (1969b) who also discussed the physiology of wood formation and the controls involved. At best, the situation is complex and many of the processes that determine wood development are only incompletely known. Larson emphasizes the control through auxin production by the vegetative organs of the tree. The development of tracheids is determined by the distance from the crown, number of rings (age), and the position of the wood within the growth ring.

Only a few of the possible mechanisms that cause changes in wood will be discussed here, but there will be special emphasis on the causal factors which result in the change from earlywood to latewood cells. One of the earlier studies in this area was by van Buijtenen (1955) who attempted to associate cell characteristics to various combinations of environmental factors; the results are described in Section 1.8.1.

The literature contains many references to the effect of environment on wood properties other than specific gravity. For example, Richardson (1964) states that tracheid length is influenced by temperature, but that increased cell length is not related to increased carbohydrate accumulation. Lumen diameter can also be affected by temperature. He found that cell wall thickness was largely a result of carbohydrate availability in conjunction with hormonal controls. Working with *Larix decidua* (European larch) seedlings, Wodzicki (1964) found a remarkably good correlation between the formation of thick-walled tracheids in a short-day light regimen with separate 12-h low intensity and 12-h high intensity light treatments. Trees subjected to 16 h of light did not produce thick-walled tracheids.

1.8.1 The Change from Earlywood to Latewood

A number of studies have been made as to why a tree changes wood production from earlywood to latewood; when this change occurs it has a major effect on specific gravity. Various factors have been cited as to the cause for the change, but one of the most exhaustive discussions was by Larson (1960). He believes that the regulation of cell diameter is hormonal and presumably mediated by an auxin originating in the bud, growing shoot, or nearby foliage. It is known that high auxin production in the apical meristem will perpetuate the production of large-diameter, thin-walled cells.

Anything that causes a marked reduction in apical activity, resulting in less synthesis of auxin, will reduce the cell diameter; this is normally accompanied by an increased wall thickness and thus a higher specific gravity. Larson (1969b) also credits nutrition, day length, and moisture availability with influencing the transition from earlywood to latewood. Large earlywood like cells are produced during active shoot elongation. Larson states that the increase in latewood cell wall thickness is due to a physiological process separate from that causing a decrease in cell diameter, although the two occur more or less simultaneously in normal latewood development.

Similar results were obtained by Mitchell (1961) who studied the relationship of height growth to initiation of latewood in several coniferous species. He found a close correlation between the cessation of height growth and the formation of latewood in saplings of Douglas-fir (*Pseudotsuga menziesii*) and grand fir (*Abies grandis*), but this relationship was not observed in young trees of western white pine (*Pinus monticola*) or lodgepole pine (*P. contorta*). Also working with Douglas-fir, Kennedy (1961) found that latewood formation was determined by the amount of auxin production, which is related to bud formation and cessation of height growth.

An abrupt cessation in terminal shoot growth leads to an abrupt drop in auxin production and thus initiates latewood formation. It is now generally accepted that earlywood is formed during the period of height growth of the tree and that latewood production occurs only after a resting bud has been formed. One example was given by Savina (1956), who found that latewood growth starts only after height growth is terminated. Numerous authors such as Kennedy (1961) and Kraus and Spurr (1961) point out that the timing of the change from earlywood to latewood varies according to the individual tree. This concept will be expanded in the chapter related to the genetics of wood; trees that switch from earlywood to latewood formation early in the year produce the highest specific gravities.

Another environmental factor that is commonly associated with a change in specific gravity is moisture availability; for example, Gilmore et al. (1966) state that the change from earlywood to latewood can be attributed almost entirely to the availability of soil moisture. They feel that this occurs because a high moisture stress within the plant results in decreased auxin production. The same reason is suggested by Kraus and Spurr (1961), who conclude that the transition in xylem structure related to the two kinds of wood may well be triggered by a reduction in soil moisture.

From the numerous publications with different theories, it seems clear that the stimulus controlling initiation of latewood still remains uncertain.

Kennedy (1961) states that the choice of a single factor as the cause of the change from earlywood to latewood is too great a simplification. It is certain that in most instances numerous factors in combination will stimulate the formation of latewood; these usually result in a change or cessation in height growth. For example, van Buijtenen (1955) studied various factors that could initiate changes in the characteristics of cells. In rapidly growing seedlings he found that an increase in temperature caused a reduction in cell wall thickness, while high soil moisture was associated with large tracheid diameters, whereas moisture did not affect wall thickness. He also found that long photoperiods did not have a consistent effect on tracheid diameter and wall thickness, even though they caused an increase in tree height. Van Buijtenen concluded that an internal change in the tree itself may be responsible for the cessation of height growth and initiation of latewood formation rather than some specific environmental factor.

Schultze-Dewitz (1969) has reported that latewood is formed earlier in trees growing in the lowlands than in the mountains in Germany, a reflection of the length of the growing season. He also claims that latewood cells are formed prematurely in compression wood, as described in Chapter 4.

It is evident that different trees respond differently to the same stimuli, and thus the development of different wood characteristics appears to be under fairly strong genetic control. As previously mentioned, one of the triggers to the changeover is moisture availability. Kraus and Spurr (1961) report that individual trees each have a characteristic transition pattern related to a decline in moisture availability.

Trees grown with excess moisture are subject to similar physiological stresses as trees that develop under drought; the roots are unable to function normally. Silvicultural operations, such as bedding or drainage, create a more favorable moisture regime for the roots, allowing normal growth and thus more normal wood formation. Although wood response to such silvicultural practices is not well documented, it has been our observation that specific gravity is higher when excess moisture is controlled, which allows normal nutrient uptake by the roots and thus an extension of the normal growth pattern.

In his study of Douglas-fir, Kennedy (1961) found that the latewood percentage was strongly correlated with the period of latewood formation but not particularly with the duration of the growing season. An early cessation of earlywood growth followed by a longer period of latewood formation results in higher specific gravities. The physiological mechanism regulating the amount of postcambial radial enlargement of the differentiating tracheids may be the critical one in determining when latewood tracheids will be formed.

Associated with the changeover of cell types is the time of flushing. Late flushing spruce have lower specific gravities than those that flush early; this holds both for populations and individuals within populations (Dietrichson 1964). The major effect is a reduction in the density of the latewood, resulting in more uniform wood within the annual ring. In contrast, in Douglas-fir, Kennedy (1961) found that trees that flushed early started producing latewood early and had high specific gravities. Similarly, for old Norway spruce (*Picea abies*), Thiercelin (1970) showed that late flushing trees had wood with a low specific gravity. This response is different from those trees that start production of latewood at an early date, which results in an increase of specific gravity.

Most reports discuss the importance of factors that affect the proportion of earlywood and latewood, which in turn partially controls specific gravity. All who work with the hard pines have observed the differences in specific gravity between trees that have "brown"-colored latewood and those that have "black" latewood. The color of the latewood is determined by the cell wall thickness, which, along with the percent of latewood, will allow an experienced person to estimate quite accurately the specific gravity of a piece of wood merely from observation.

It is evident from the authors cited and from the numerous other studies which could not be discussed, that the control of the transition from earlywood to latewood formation is complex but that the process can be somewhat altered by forest management activities and genetic manipulation.

1.9 Different Woody Tissues

Wood qualities can vary greatly from location to location within a tree. Two of the major causes for differences will be introduced here but will be covered in detail in later chapters. The first is reaction wood (compression wood in conifers, tension wood in hardwoods) which is formed when the tree is not growing vertically and is sometimes caused by physiological changes resulting from insect or disease attacks, fire, and other factors. The second is related to the age and location of the cambium relative to the tree center and is summarized as the juvenile wood concept. Both of these are mentioned in this introductory chapter only to indicate that they are important additional sources of variability in wood.

1.9.1 Reaction Wood

When a tree is not parallel to the gravity vector, a kind of wood called *reaction* wood is formed (Fig. 1.14). In the conifers, this wood occurs on the underside of the leaning stem, and tends to "push" the tree straight; such wood is called *compression* wood and has been described in great detail by many authors. The latest and most comprehensive of these is the three-volume treatise "Compression Wood in Gymnosperms" by Timell (1986). In the hardwoods, the reaction wood is called *tension* wood and is formed on the upper side of the leaning tree and tends to "pull" the tree straight. A more detailed discussion of the effects of reaction wood on wood variability is given in Chapter 4.

Both compression and tension woods have characteristics that differ considerably from normal wood. For example, Kaeiser and Boyce (1964) found that gelatinous fibers associated with tension wood in *Populus deltoides* (eastern cottonwood) affected specific gravity and influenced the characteristics of the normal fibers associated with them. Reaction wood is undesirable for most final products. It is sufficient to summarize that the best way to reduce the amount and severity of reaction wood, which contributes so much to wood variability, is to

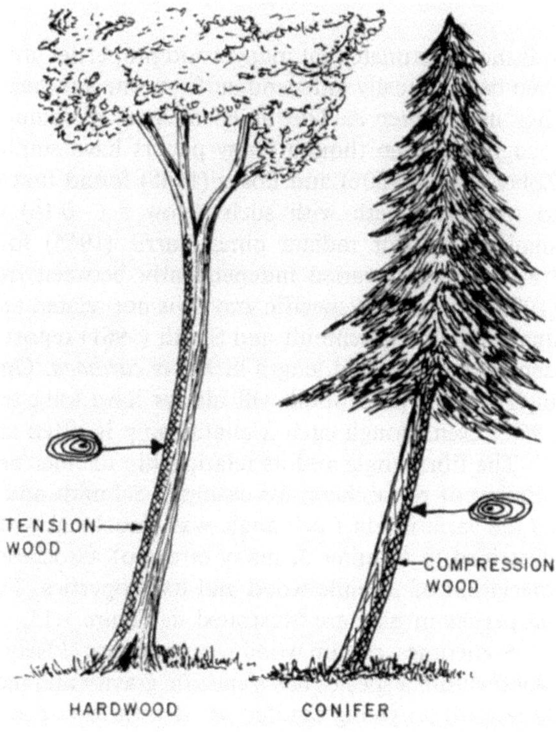

Fig. 1.14. When a tree is not straight it produces reaction wood. In conifers, reaction wood is called compression wood and forms on the under side of the leaning tree. In the hardwoods, reaction wood is formed on the upper side of the tree. These are illustrated schematically

straighten the trees through genetic or silvicultural methods or to use genotypes which produce wood with small amounts of reaction wood.

Reaction wood is also produced in association with knots and limbs. The steeper and the larger the limb, the greater is the relative amount of reaction wood in the tree stem. Thus, another way to reduce reaction wood is to produce or develop trees with small and flat-angled limbs.

1.9.2 Juvenile Wood

The wood first laid down by the cambium near the center of the tree has characteristics that differ from wood formed at a greater number of rings from the pith, or tree center. The first-formed wood is called juvenile wood; it is much more important in the pines than in some of the other conifers, such as the spruces or firs. It generally is not severe in the hardwoods. Juvenile wood of pines can differ so much from mature wood that the tree essentially has a different wood near its center than closer to the bark, or near the top of the tree compared to its base.

The main way to reduce the amount and effect of juvenile wood is to alter the age of harvest, usually referred to as the rotation age. It is possible to breed trees with differing kinds of juvenile wood, but this approach has only been initiated. Juvenile wood will be covered in depth in Chapter 3.

1.10 Relationships Among Wood Properties

It is most fortunate that many wood properties are only slightly correlated or may even be genetically independent from one another. This concept is developed further in Chapter 8. The independence of some wood properties is often not recognized, even though many papers have emphasized this fact. For example, Zobel et al. (1960c) and Posey (1965) found that specific gravity was correlated to tracheid length with such a low r (−0.16) that it was of little practical significance. For radiata pine, Harris (1965) found that specific gravity and tracheid length varied independently between trees. In loblolly pine, Megraw (1985) reports that specific gravity is not related to either tracheid length or fibril angle; similarly, Schmidt and Smith (1961) report no relationship between basic density and tracheid length in *Pinus caribaea*. One will not find, therefore, that high density pine wood will always have long tracheids or vice versa (Brazier 1967), even though such a relationship is often assumed.

The fibril angle and its relationship to fiber properties has been studied by a number of researchers; for example, Schmidt and Smith (1961) found that 76% of the variation in fibril angle was associated with differences in cell length. As described in Chapter 3, many errors of association result from the lack of appreciation of juvenile wood and its properties. The independence of four wood properties in pine are illustrated in Figure 1.15.

Sometimes, certain wood properties are closely related. For example, resin-free wood within a species has a specific gravity and moisture content that are inversely related, so young conifer wood with low specific gravity is heavy because of

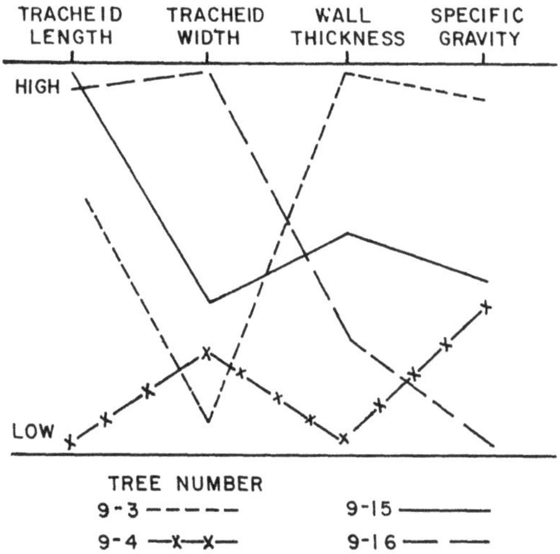

Fig. 1.15. Sometimes wood properties are related but usually they are genetically independent from one another. Illustrated are the relationships among four wood properties for four *Pinus taeda* trees. Note that the wood properties appear to be unrelated

the additional moisture present (see Fig. 1.8). Bunn (1981) emphasized the importance of moisture content when he stated that calculations of yield per green ton for trees of different basic density are not a reliable indication of the yield of dry wood (see Table 1.3). In loblolly pine, one study by the senior author showed that wood with a specific gravity of 0.42 had 129% moisture content, 0.45 had 108%, and 0.47 had 93% when the moisture content was calculated according to the dry weight of wood.

Another pair of strongly related traits are tracheid length and fibril angle of loblolly pine. Megraw (1985) states that fibril angle follows the same, but inverse pattern within a tree as tracheid length. It is to be expected, of course, that specific gravity usually has a close relationship to wall thickness, and sometimes to cell size. For example, Zobel et al. (1960b) found a fairly good correlation between specific gravity and wall thickness in loblolly pine. In addition, they found $r = -0.58$ between specific gravity and tracheid width. In spruce, Schultze-Dewitz (1959) reported that the length of tracheids is inversely proportional to the amount of latewood. In a study on conifers in general, Larson (1973) found radial lumen diameter and tangential wall thickness closely correlated with specific gravity. He reported that radial cell diameter is regulated by growth hormones produced by the actively elongating needles. Wall thickness is correlated with the available photosynthate after the needles have matured.

In hardwoods, the relationships among dimensions of the several different cell types are often very weak and independent from one another as reported for sweetgum (*Liquidambar styraciflua*) by Webb (1964). This, however, is not always true. For example, Saucier and Taras (1966b) found that the high specific gravity trees of red maple (*Acer rubrum*) also had short fibers and Amos et al. (1950) reported that *Eucalyptus* spp. had a negative correlation between cell length and cell width. This means that long cells will have a higher basic density. In poplars the basic density increases with the length of the cells.

1.11 The Literature Related to Wood Variation

In a publication such as this it is possible to cite only a few of the many books and thousands of published articles dealing with wood variation. A few of the more useful general books on wood properties and wood variability are listed in Table 1.1. Since these books are usually available in most libraries, they will not be used as a major reference source; in this book, we will concentrate upon published and unpublished articles and wood studies that are of special value related to wood variability.

As for all scientific subjects, the interest in wood variability has changed over the years. Although a few valuable studies have recently been made, it is particularly noticeable that the bulk of the papers on wood variability were published in the late 1950s to the early 1970s. Much of our knowledge about wood variation is dependent on work done during an approximately 15-year period, with many of the studies having been made nearly 20 years ago.

Only those articles that make a specific contribution to a given subject will be cited. For example, there are many hundreds of references concerning the effect of growth rate on wood specific gravity. The authors have cited some that emphasize negative relationships and some that cover the positive relationships in various species. The mass of literature on growth rate-wood property relationships will then be summarized to the best of the authors' abilities. Few articles are specifically concerned with growth rate and specific gravity; most occur as isolated statements in reports related to other wood characteristics.

Chapter 2

The Effect of Provenance Variation and Exotic Plantations on Wood Properties

2.1 The Effect of Differing Environments on Wood Properties

As a result of evolutionary selection over the eons, variation occurs among segments of a species that grow in different environments within a species range (Callaham 1964; Burley and Wood 1976). These portions of a species with differing characteristics are called geographic sources or provenances[1]. For example, trees from cold climates at higher latitudes or high elevations usually develop into slow growing individuals, with straight stems and small limbs with relatively right-angled branches, making them better adapted to ice and snow. Geographic growth differences are sometimes genetically set; for example, when high-elevation trees are grown at lower elevations they often maintain their slow growth and good form (Langlet 1967).

To fully understand the effect of provenance variation on wood properties, it is essential to have a clear understanding of what is meant by environmental and genetic control of wood. In actuality, all wood properties are determined by an interaction of the genetic potential of the tree with the environment in which the tree grows. When we say that a wood property is under environmental control, we mean that it varies considerably with a change in the environment. When a wood property is under strong genetic control, it may vary without regard to the environment under which the tree is grown or its wood properties may stay constant despite the trees having been grown in differing environments.

The situation is often very complex when assessing wood properties of different geographic sources within a species. Both environmental and genetic control of wood can be important at the same time. For example, wood specific gravity of natural populations of trees of the same species can vary greatly in different regions as a result of strong environmental control. If, in such a situation, seed from a high-density stand is used in plantations in another area, the planted trees will not necessarily produce high-density wood. The resulting stand of trees may have either a high or low specific gravity depending upon the environment in

---

[1] There is some confusion in the literature with respect to the terminology of seed source. Some workers consider geographic source, provenance, and seed source to be synonymous. Others feel that the terms geographic source and provenance are identical and they are defined as "the area from which the original seed collections were made", but these workers prefer to use seed source to describe where the seed was obtained, with no indication of the genetic origin of the seed (Barner 1966; Zobel and Talbert 1984). For example, *Pinus contorta* seed collected from natural stands in Alberta, Canada, and grown in Ireland is obtained from the Alberta geographic source or provenance. When seed is collected from the trees grown as exotics in Ireland and planted elsewhere, it is referred to as the Irish seed source. Provenance refers to the original collection area in Alberta. In this chapter, geographic source and provenance will be used interchangeably to refer to the genetic origin, while seed source will refer to the physical location from which a particular seed lot was obtained.

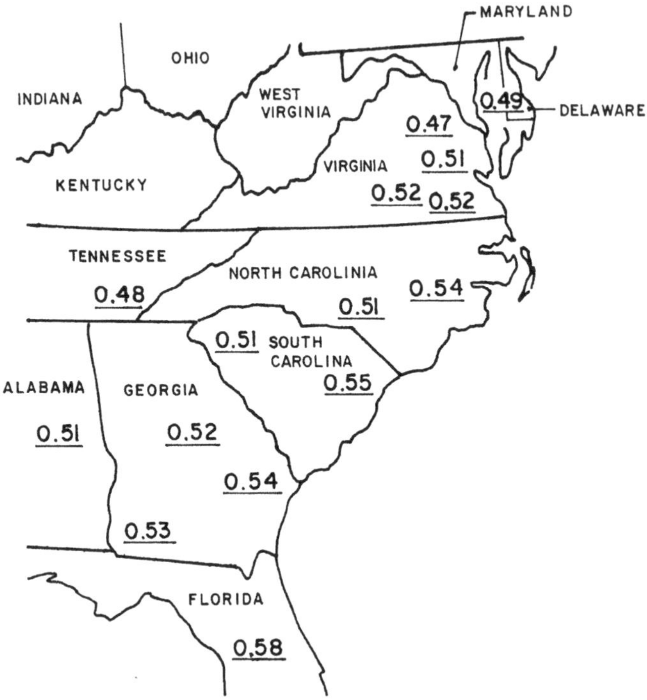

Fig. 2.1. Both environmental and genetic differences cause variation in wood properties within a species. Stand average differences in the coastal and piedmont loblolly pine sources are shown for the southeastern USA. Specific gravity is less in the north and in the piedmont, but tree to tree differences within a stand are the same in both provenances (Zobel and Talbert 1984)

which the plantation is established. Yet, regardless of the environment in which the trees are grown, within each population some individuals will have relatively high while others will have low densities when compared to the average of the stand. This is the result of a strong genetic control of wood specific gravity among individual trees. Thus, the environment exerts strong control over the average wood specific gravity of the population, while genetic control results in tree to tree differences within the same specific environment (Fig. 2.1).

Wood characteristics often vary as greatly within the range of a species as does growth and adaptability. Frequently, trees do not produce the same wood properties when grown in considerably different environments. This response can be illustrated by *Pinus caribaea* var. *hondurensis* which McWilliam (1955) separated as coming from "exposed pine ridges" that produce a high wood density and a high percent of latewood and "broken pine ridges" with lower density wood and lower percent latewood. He recognized these differences as being environmentally induced rather than being due to genetic causes, and ascribed the limited latewood formation directly to unfavorable growing conditions.

The same variability occurs when similar genetic material is planted in different environments. One of a number of reports showing this was by Taylor (1973a) for *Eucalyptus grandis* planted in different areas in South Africa. Differences in specific gravity were considerable from plot to plot using a standard *Eucalyptus grandis* seed source on good and poor sites from each area sampled. Taylor reported that average specific gravities for the ten plantations assessed varied from 0.40 to 0.53. Similar large responses to differing environments have been especially prevalent in *Pinus radiata* in Australia, according to a whole group of studies on this species. For *Pinus radiata,* Cown and McConchie (1982a) concluded that log densities were influenced more by geographic region where the trees were grown than by tree age, despite the large impact that age usually has on wood properties.

In South Africa Boden (1982) found that the wood densities of *Pinus patula, P. elliottii,* and *P. taeda* were all negatively correlated with latitude and altitude where grown, despite the latitude differences being only 1°11'. Altitude also had a great influence on wood density. Results for *Pinus banksiana* by Grigal and Sucoff (1966) showed that the average plot specific gravities varied from 0.373 to 0.435.

Although the wood of a provenance is often different in the new environment, compared to that from the original site, sometimes genetic control is strong for a geographic seed source. For example, Schütt (1962) found that the variations in wood arising from the place of origin for three different geographic sources of lodgepole pine (*Pinus contorta*) seed were greater than those among individual trees within sources. Similar results were obtained by Polge et al. (1972) who summarized as follows for a 6-year-old beech (*Fagus sylvatica*) "for the characteristics which directly influence the technological value of wood, the principal source of variation appears to be the provenance of the plants."

One general statement that can be made with little fear of contradiction is that "anything that affects the growth of a tree can also affect its wood properties." This is true when trees are grown within normal forest environments, but it is much more evident when trees are grown in a different geographic region within their species range or when they are used as exotics. Major care must be taken not to assume that the wood of a species will be the same when it is grown in a plantation in a different environment. One example of such a change with environments was detailed by Brazier (1983) for *Pinus caribaea.* He reported on the wood density of this species when grown in four countries; it varied from 415 kg/m^3 in New Zealand, 550–670 kg/m^3 in Queensland, 650–800 kg/m^3 in Fiji, and 790–900 kg/m^3 in Malaysia. It is difficult to imagine that wood with such grossly differing densities can come from trees of the same species. Huge errors have been made by assuming that trees with given genotypes and planted in different environments will have wood similar to that produced in the environment from which the seed was obtained (Fig. 2.2).

There are two general questions associated with geographic variation and wood. The first concerns the type of wood produced in different environments within the natural range of the species. For example, how does the wood of natural stands vary by latitude, elevation, soil, or rainfall differences within the species range (see Fig. 2.1). These patterns have been most studied, and for a few species some rather positive rules can be formulated.

The second question concerns the kind of wood produced when seed of a given geographic source is used in exotic plantations, sometimes in vastly different

Fig. 2.2. Trees with similar genetic backgrounds often grow differently when planted in contrasting environments. For example, the *Pinus elliottii* shown produced a very high density wood when grown in Zululand, South Africa, compared to slash pine grown in its indigenous range. Differences were so great as to make the Zululand grown trees unsuitable for some products

environments from its indigenous range. The resultant changes in wood that may occur are very little appreciated or understood. Sometimes wood properties obtained when trees are grown in a new or different environment can be predicted. Harris (1965), for example, reported that wood density in radiata pine in New Zealand decreases by about 0.1 g/cm^3 with each increase in degree of latitude and increases by about 0.01 g/cm^3 with each 100 m increase in altitude. Usually, however, changes to be expected cannot be that accurately anticipated.

Extensive experience has shown that *the only real way to determine what the wood will be like when trees of a given provenance are grown in a new environment is to grow the trees there and directly determine the properties of the wood produced.* This was emphasized by Plumptre (1983) who found that differences in wood of *Pinus caribaea* were not seriously affected by the seed provenance but were very sensitive to the particular site on which the provenances were grown. As another example, Gupta and Kukreti (1983) report that the wood of *Acacia mearnsii* from its native range has better properties than the same material when grown in India. In a broad study made in Venezuela, Delgado (1983) forcefully reported "the results obtained – that plantations established solely on the basis of silvicultural characteristics – may end as economic failures. Wood quality of the species is normally taken for granted, and the expectation is that it will be comparable to that of the naturally grown species – this assumption may be ill founded. *Cedrela angustifolia,* for example, showed, an extremely low wood density and severe collapse when grown in a new environment." Another example is represented by stem cracks in Norway spruce where Dietrichson et al. (1985) found that the continental sources of spruce grown on good sites in Norway are most frequently degraded by cracking.

Although not absolute, a general rule is that the average differences in plantation wood grown from different geographic seed sources are not strongly genetically determined but usually are a response to the different environment into which the trees have been moved (Talbert and Jett 1981). Many errors have been, and are still being made, because of the belief that the wood produced by a given geographic source will be the same in the new plantation as it was in the natural range of the trees. For example, loblolly pine (*Pinus taeda*) has a high specific gravity in the southern part of its range and a relatively low specific gravity in the northern part (see Fig. 2.1). Based upon this fact, plantations have been established in the north with seed collected from the high specific gravity southern trees with the anticipation that the wood density of the trees planted in the north would then be high. Tests show that the trees from seed of the high-density southern populations grown in the north produce wood with the same or even lower specific gravity compared to loblolly pine from seed of the local northern stands. Moving the southern seed from high-density stands northward accomplished nothing in the way of increasing wood density.

It is most difficult to separate a discussion about wood as affected by geographic source from one on exotic plantations since geographic source plays such an important part in exotic forestry. Nevertheless, we will discuss these subjects in two separate sections in this chapter to try to clarify some relationships, but it is obvious there is a very large overlap and there can be no clear-cut division between the two. Of course, many of the general concepts related to provenance differences apply equally well to exotics; however, these will not be repeated in detail in both sections. Frequently, wood variation is greater in exotics because of the more extreme environments under which the trees are grown.

2.2 Provenance Variation in Wood from Natural Stands Within the Range of a Species

The many studies of geographic variability in wood within the natural range of a species indicate that in some species the differences can be quite large (see Fig. 2.1). Variation exists in several wood characteristics, although most information available relates to specific gravity and tracheid length. Environmental causes for these differences are similar to factors that affect growth rate, such as elevation, latitude, soils, rainfall, and others. In addition, bud flushing and growth cessation characteristics are specific growth characteristics that have been reported to be closely related to provenance and wood qualities (Dietrichson 1963; Mergen et al. 1964).

2.2.1 Examples of Provenance Variation in Wood Properties in Natural Stands

Studies of the variation of wood within natural stands in the range of various species are too numerous to cover in detail. Therefore, a few examples are given below for various groups of species to illustrate the kind and amount of wood differences that exist in different geographic areas within a species.

1. *Pseudotsuga menziesii, Picea* spp., and *Abies* spp. The western North American species grow under conditions of great diversity of site, elevation, and latitude. Such conditions are ideal for the development of provenance differences, and this has been found to be true for a number of growth and form characteristics. Wood properties can also vary greatly by provenance but the patterns of variation are often not clear. In fact, Wahlgren (1965) states: "Douglas-fir does not show the clear-cut patterns of geographic variation found in the Southern pines." He reports a slight trend of increased specific gravity from north to south along the coast west of the mountains but a reverse trend east of the mountains. The trends are said to have little practical importance. It is difficult to find distinct patterns of wood variation related to the environment in any of the western conifers, with the possible exception of elevation.

Based upon a broad division within the species, strength properties of boards of Douglas-fir (*Pseudotsuga menziesii*) analyzed by Noskowiak and Snodgrass (1961) were assessed for both the coast and inland provenances. The commercial range of the species was stratified into 19 more or less homogeneous regions. Specific gravity and major timber strength properties (elasticity, crushing, bending, shearing) were statistically significantly larger for the coastal than for the inland provenance of Douglas-fir. The southern inland region contained the weakest wood, and the authors concluded that the coast and inland regions represented different populations within the species relative to wood properties. In another rangewide study of Douglas-fir, Drow (1957) found that a separation of wood into northern and southern Rocky Mountain areas is needed, while the more western provenances of Douglas-fir from the Cascade Range and the Sierra Nevada had similar specific gravities.

Elevational differences in wood properties are of special interest for Douglas-fir because of the topographic extremes within which the species grows. They can be large; for example, Lassen and Okkonen (1969) found that wood from trees growing at low elevations and with dry summers had an average specific gravity of 0.55, while trees grown at high elevations with wet summers averaged only 0.44.

A study was made by Hakkila (1969) of differences in the wood of pine (*Pinus sylvestris*) and spruce (*Picea abies*) among stands in the north and south of Finland; tests were made from 85 localities, each of which was represented by 150 trees. Geographic patterns from north to south were found to be constant and predictable, although no east to west trends were evident. Similarly, Nylinder and Hägglund (1954) found that cellulose yields of Norway spruce were different from different locations, with lower yields as latitude and elevation became greater. In Norway, Klem (1965a) surprisingly found no differences in *Picea abies* within natural stands within the country, including stands growing at different elevations. In contrast, in Poland, lowland Norway spruce had a higher specific gravity than upland spruce (Anonymous 1968). Holubcik (1969) also found trees growing at the upper tree line in eastern Europe to vary considerably from the average. In Poland, the geographic source variation was reported to be higher than the tree to tree difference within source (Anonymous 1968). This is unusual since the tree to tree variation within provenances is usually found to be higher than average differences among provenances. Trees growing on the best sites in Austria have relatively thin-walled cells but a high proportion of latewood.

An example of the effect of latitude and elevation on *Picea abies* was given by Rydholm (1965) where he generalized that wood density of Scandinavian spruce decreases somewhat (but erratically) with latitude and altitude, while fiber length and chemical composition is not altered by the differing conditions in which the trees are growing.

In white spruce (*Picea glauca*), Holst (1960) found large differences in specific gravity among sources in Canada, the highest being 0.47 and the lowest 0.31. However, tree to tree variability within source was almost as great as the major geographic differences. In Alberta, Taylor et al. (1982) reported large specific gravity differences among white spruce stands, as well as from tree to tree. Larger tracheids were present in the northern stands. For the same species, also in Alberta, large stand differences were found by Wang and Micko (1984). In Newfoundland, Hall (1984) did not find any geographic trends in wood density for black spruce (*P. mariana*). *Abies amabalis* (Pacific silver or amabalis fir), growing in western Canada (Kennedy and Swann 1968), had a wood specific gravity which increased from north to south.

2. *Pinus sylvestris.* In a study of Scots pine (*P. sylvestris*) from South Tyrol, southeast Finland, Lapland, and Eberswalde, Schultze-Dewitz (1968) found that tracheids from Lapland were only 2.8 mm long, compared to Eberswalde where they were 5.5 mm long. Similarly, cell wall thickness in Lapland was only half that of the trees from Eberswalde. It is not usual to have provenance differences within species as extreme as this example for Scots pine. Hildebrandt (1960) reports that both pine and larch at high elevations, near the timberline, produced wood with exceptionally thin-walled cells. He considers this to be a deficiency symptom, i.e., a result of a short growing season, little summer rain, and cool springs and summers. He goes on to generalize that the relationship of thin cell walls with higher elevations or higher latitudes is common to forest trees.

3. *Pinus nigra.* A study on black pine by Tsoumis and Panagiotidis (1980) assessed latitudinal differences in wood on 331 trees, 55 or more years of age. Latewood percentage and wood density tended to be higher in the southern locations, while tracheid length was larger in the northern region, although some of the differences were not statistically significant.

4. *The southern pines.* A great deal of work has been done on the geographic variation of wood of the group of several species referred to as the southern pines of the southeastern United States. Their wood usually follows a rather predictable pattern (Zobel et al. 1972; Fig. 2.1). The inland and northern sources have lower specific gravities and frequently shorter tracheids than the southern and coastal sources (Jackson and Strickland 1962). This pattern has been reported in many studies. One of the earliest by Zobel et al. (1960b) showed the average specific gravity of the most southern source of loblolly pine in Florida to be 0.58 and that

of the northern source in Maryland to be 0.47. The inland piedmont sources usually had specific gravities 0.03 less than the coastal sources from the same latitude. Differences in wood density were so large that industrial timber and land-buying policies were modified to give preference to loblolly pine areas which produced the highest specific gravity wood.

The wood density of the southern pines can be considerably different even when grown at the same latitude because of soil and other environmental differences. This was shown by Megraw (1985) who found differences in the wood density of loblolly pine between North Carolina and Arkansas to be 5 to 8%, that in North Carolina being the highest. In North Carolina, loblolly pine has wood densities of 22.6 lbs/ft^3 when grown in the piedmont, 24.0 lbs/ft^3 in the coastal plain, and 20.9 lbs/ft^3 in the sand hills or fall line which lies between them. In an exhaustive study of wood variation in the southern pines, Mitchell (1963) also reported a strong trend toward increasing specific gravity from northwest to southeast for all the major species.

The magnitude of the differences that can occur within a species is illustrated from work by the authors given in Table 2.1. It is of significance that the wood property differences shown are large enough to be of great practical and economic importance.

Although the north to south and coastal to inland trends are common in the southern pines, they do not always occur. For example, Talbert and Jett (1981) found the usual geographic trends in specific gravity for mature wood of loblolly pine, but the patterns of variation with location were very weak for juvenile wood. When wood samples were taken in the northern range of loblolly pine, Whitesell et al. (1966) located very few wood differences in the four different zones tested. They concluded that the geographic differences were so small that they could not be detected statistically because of the very large tree to tree differences within each zone (or provenance).

The same geographic pattern has been found for slash pine (*P. elliottii*) and loblolly pine, although this species occurs over a much more restricted range than loblolly pine (Zobel et al. 1972). For example, Goddard and Strickland (1962) reported that slash pine specific gravities increased from north to south and from west to east. The specific gravities were 0.55 in Georgia and 0.63 in Florida. Longitude and latitude accounted for 54% of the total variation in wood density; when early and heavy summer rainfall was added, 83% of the variation could be explained.

Latitude per se seems to be of little consequence; it is the environmental variables changing with latitude that control growth which are of importance. For a variety of slash pine (*P. elliottii* var. *densa*), trees growing within 25 miles of the Atlantic or Gulf coasts had a high specific gravity (0.639), while the specific gravity of inland sources was 0.597, showing the usual coast to inland trend (Saucier and Dorman 1969) which is usually related to the amounts of rainfall and distribution patterns.

The pattern of wood variation by geographic area has been found to be generally similar to loblolly and slash pine for the other major southern pines. For example, shortleaf pine (*P. echinata*) (Gilmore 1968) had lower specific gravities from the northern locations than from the southern areas. Gilmore

Table 2.1. The effect of geographic source on wood specific gravity for natural stands of loblolly, slash, and longleaf pines[a]

Loblolly pine in the southeastern coastal plain of the United States (values are listed from south to north)

Source	Specific gravity
Central Florida	0.58
Middle Georgia	0.54
South Carolina	0.54
North Carolina	0.53
Virginia (South)	0.52
Virginia (North)	0.47

Coastal vs piedmont (inland) loblolly pine in the southeastern United States

Source	Specific gravity
Coastal Georgia	0.54
Piedmont Georgia	0.52
Coastal South Carolina	0.54
Piedmont South Carolina	0.51

A transect from coast to mountains at the same latitude

Source	Specific gravity
Coastal loblolly – North Carolina	0.53
Piedmont loblolly – North Carolina	0.51
Low Mountains – Tennessee	0.48

Latitudinal differences in specific gravity for slash and longleaf pines

	Specific gravity	
	Slash pine	Longleaf-pine
North Florida	0.50	0.52
Middle Georgia	0.47	0.49
Upper Georgia	0.45	0.48

[a] Values given are a summary of results from various studies made by the senior author.

(1968) reported the same pattern for eastern white pine (*Pinus strobus*), also native to the southeastern United States. Based upon his findings and related research, he stated that the hypothesis of increasing specific gravity with decreasing latitude was confirmed. For longleaf pine (*P. palustris*), Saucier and Taras (1969) found a similar pattern, with specific gravity increasing from north to south. In contrast, Strickland (1965) reported that longleaf pine in Florida does not vary in the same way as slash pine because his study showed no geographic patterns of wood differences.

No definite geographic pattern was found for the minor southern pine, pitch pine (*P. rigida*) (Ledig et al. 1975), which has a restricted latitudinal but a considerable elevational range. Tree to tree differences were larger than source differences; however, enough source variation was present so that the authors felt it should be considered. As commonly observed, stand to stand variation within a geographic source was very small. Similarly, in Virginia pine (*P. virginiana*) Thor (1964) found no differences in wood properties among sources except for tracheid length for which there were average differences among stands as great as 1 mm.

For the geographically restricted southern pine, sand pine (*P. clausa*), Strickland and Goddard (1965 a) reported definite stand to stand differences, with specific gravities ranging from 0.485 to 0.552. They hypothesize that limited pollen distribution and scattered stands have produced this unusual pattern of large stand variation in wood properties, which is opposite to the normal trend. Additionally, in the two sources of sand pine (Ocala and Choctawhatchee) they found a real difference in wood specific gravity, the former averaging 0.44 at breast height, while the latter was 0.49. Some authors consider these two to be geographic races within sand pine, while others have suggested they are subspecies. A similar stand to stand variation within limited geographic differences was found for pond pine (*P. serotina*) throughout its range in the southeastern United States (McElwee and Zobel 1962). In their exhaustive study of this species Taras and Saucier (1970) did not find any geographic trends in specific gravity.

5. *Cold climate pines.* Generally, considerable geographic source patterns are found for most cold-climate pines with the exception or red pine (*Pinus resinosa*), which shows only a limited geographic variability. Jack pine (*P. banksiana*) has some source variability in Saskatchewan, but most differences in wood specific gravity were among individual trees within locations (Roddy 1983), with specific gravities ranging from 0.36 to 0.47, while source variation was negligible.

6. *Tropical pines.* Despite their wide utility in exotic plantations, not many significant geographic studies have been made of wood in natural stands of the tropical pines. In those done, the major differences seemed to be related more to elevation than to latitude (Fig. 2.3). One example in *Pinus taiwanensis,* in which both specific gravity and tracheid length decreased considerably at higher elevations (Ta-wei et al. 1972).

When wood qualities were assessed for *Pinus caribaea* var. *bahamensis* from different islands, Nikles (1966) reported that most variation in specific gravity was among individual trees, although small differences among islands were evident. For tracheid length, most of the variation was related to groups of similar islands. There are indications of provenance differences in *Pinus caribaea* var. *hondurensis* but these have not been well quantified (Plumptre 1983). Considerable differences were found in specific gravity among sources of *P. tecunumanii,* growing at similar latitudes but at different sites in Guatemala (Equiluz-Piedra and Zobel 1984). However, most of the variation was from tree to tree within sites; *P. tecunumanii* has very large differences in specific gravity from tree to tree.

Fig. 2.3. In the tropical pines, much of the wood variability of the indigenous pines appears to be related to elevation rather than to latitude or other environmental factors. This low elevation source of pine in Guatemala has a considerably higher wood density than when the same species grows at higher elevations

Because of the very wide use of tropical pines as exotics, it is surprising that their wood has not been more widely studied in the indigenous ranges. Most reported studies are on large, mature specimens and their wood qualities are not comparable to wood from young stands. Emphasis has centered on assessing the wood of the tropical pine species when grown as exotics, usually at very young ages.

7. *Temperate hardwoods.* Some studies of geographic variation of wood properties in temperate hardwoods have been made. One on aspen (*Populus tremuloides*) by Einspahr and Benson (1967) showed a south to north trend of decreasing specific gravity but no trend for fiber length. Also for *Populus tremuloides,* Valentine (1962) found that real differences existed among 26 different stands in New York. As is usually found, individual tree-specific gravities ranged from 0.293 to 0.471, while area means varied only from 0.343 to 0.432. The latter can be considered large for stand to stand differences in a diffuse-porous hardwood.

For sweetgum (*Liquidambar styraciflua*), Webb (1964) reported no differences for specific gravity among provenances and sites. Exactly the same lack of geographic variation in specific gravity among provenances of sweetgum were found by Johnson and McElwee (1967). Several other studies on this species have all shown little definite geographic trends in wood properties. Most of the variation consists of large tree to tree differences within sources. There was, however, a geographic trend for fiber length and the interlocked grain index; both were smallest in the north (Webb 1964). With the same species, Hunter and Goggans (1968) found that specific gravity and colored heartwood varied with latitude and physiographic provenance, but the variations were small. Most variation was among trees within plots.

In yellow poplar (*Liriodendron tulipifera*), Kellison (1967) found geographic differences in wood specific gravity and site to site differences within geographic regions. Great tree to tree differences occurred within sites; they were twice as large as the geographic and stand differences. Most of the latter occurred in trees growing on very acid soils, an unusual habitat for this species. Other than the major specific gravity difference related to soils (Taylor 1965), Kellison found that the piedmont provenance had the highest specific gravities, while the coastal plain and mountain provenances were lower in density and equal to each other. Using the same species, Sluder (1970) also found definite differences in wood density between geographic areas.

In a comprehensive study of a number of species in the southeastern United States, Taylor (1977) found no geographic pattern in either specific gravity or fiber length for shagbark hickory (*Carya ovata*), post oak (*Quercus stellata*), blackgum (*Nyssa sylvatica*), or yellow poplar (*Liriodendron tulipifera*). He did find distinct differences among locations for mockernut hickory (*Carya tomentosa*) and sweetgum (*Liquidambar styraciflua*) but no general pattern of geographic variation was present. In a study throughout the range of water tupelo (*Nyssa aquatica*), specific gravity and fiber length showed only small differences among four stands from North Carolina, South Carolina, and Georgia (McElwee and Faircloth 1966). For red alder (*Alnus rubra*), Harrington and De Bell (1980) found no geographic variability in wood properties.

Working with Eastern cottonwood (*Populus deltoides*) in Oklahoma and Arkansas, Posey et al. (1969) reported an interesting pattern over a 600-mile sampling distance. The specific gravity increased from east to west, while fiber length decreased. Unlike the case with many species, they found that variation between geographic sources accounted for a higher percentage of the total variation than the variation among trees within stands.

In general, the temperate hardwoods appear to have a very limited amount of provenance variation in wood properties.

8. *Tropical hardwoods.* In a study of tropical hardwood species, Chudnoff (1976) surveyed the 1756 major species making up about 50% of the volume of the forests in six countries in Central and South America. Although results of such a broad study can only be a generalization, Chudnoff found a high proportion of the wood produced within the 2000 and 4000 mm (80 to 160 in.) rainfall zone to be very dense, while in the 1000 to 2000 mm range (40 to 80 in.) he obtained a good distribution of specific gravity classes. In the very wet areas (4000 to 8000 = 160 to 320 in.) the wood was mostly in the 0.40 to 0.59 specific gravity range. Chudnoff summarizes as follows "...the density distribution patterns found in this study are not the result of site influence on similar sets of species but rather are based on groups of species that have captured advantageous niches in different tropical environments."

Fig. 2.4. Wood often varies within a species depending on the geographic source as it grows in its indigenous range. Most of the eucalypts show little such variability, however, and good stands such as the *E. regnans* shown growing in Tasmania have wood that is generally usable. Most studies on provenance effects on wood of the eucalypts have been done after the species have been used as exotics

Other studies of tropical species relating wood specific gravity to different environments have also been reported. For example, Whitmore (1973) found no differences in balsa wood specific gravity from the different Holdridge life zones commonly used to differentiate environments in the tropics. Working with four species growing in six different life zones in Costa Rica, Howe (1974) reported specific gravity changes according to life zone for *Cordia alliodora,* but little relationship for *Virola koschnyi, Cedrela mexicana*, and *Brosimum* sp. In a study on teak (*Tectona grandis*), Nair and Mukerji (1957) found large differences in specific gravity and strength properties of trees from different areas, whether natural or planted. Boone and Chudnoff (1970) obtained very dense wood (0.9) from mahogany (*Swietenia macrophylla*) from southern Mexico and Nicaragua and a lower density of about 0.7 from Guatemala and Costa Rica.

Studies have been made with a number of other hardwood species. One example is the genus *Terminalia.* In India, *Terminalia paniculata* varied according to locale from 49 to 55 lbs/ft^3 and *Terminalia tomentosa* from 52 to 74 lbs/ft^3 (Karnik et al. 1970). The wood of these two species is dark and heavy. In western Africa, *Terminalia superba* has very variable, light-colored wood (from 25.5 to 45 lbs/ft^3) and has special use in veneers (Britwan 1977).

One surprising fact is the lack of studies on the variability of the wood of the eucalypts when growing on different sites. Studies that have been made show that there is some effect of both elevation and latitude on the wood, but it is minor compared to the conifers and to some of the tropical hardwoods. As a result, the wood within a eucalypt species is generally usable without too much concern for the provenance it represents (Fig. 2.4).

2.2.2 A Summary of Provenance Variability in Wood from Natural Stands

Based on the literature on many species from many areas, a few general concepts appear to apply for geographic variation of wood properties for forest trees, although there are always exceptions:

1. Wood from trees growing at high elevations has lower specific gravity and shorter fibers (tracheids) than trees growing at lower elevations. This rule is more applicable to conifers than it is to the hardwoods.

2. Wood from trees grown at high latitudes has lower specific gravity and shorter fibers (tracheids) than does wood from trees from lower latitutes. This trend is very evident for the southern pines and some European conifers which have been extensively studied; a latitudinal trend is found in the coastal plain of the southern USA but less for inland sources. Latitudinal trends are evident for many different species including *Pinus taeda, P. palustris, P. echinata, P. elliottii, P. sylvestris, Picea abies,* and a number of others. In a summary paper on the southern pines, Mitchell (1964) emphasized the strong tendency for increased wood density from northeast to southwest for all species. Similar trends were found for pine in Finland where an east to west pattern was also evident (Hakkila 1968). When summarizing the western North American species, Mitchell (1963) stated that strong patterns of geographic location were not as evident in the west as in the east.

3. Wood quality is better from higher latitudes compared to the lower latitudes because of the better tree form and less degrade from crooked trees and large limbs with accompanying reaction wood.

4. Wood specific gravity is lower and cell length shorter from inland sources were there is a more continental climate, when compared to coastal sources where more maritime conditions prevail (Zobel et al. 1972). This pattern sometimes does not hold for the hardwoods.

5. Geographic trends in wood quality exist in the temperate hardwoods, but they are much less distinct and often follow no apparent pattern as do most conifers. In a number of hardwood species, little or no difference in wood properties exists among provenances. Results are so varied that no generalized statement can be made for any trends or patterns of temperate hardwoods growing in different geographic locations.

6. Some differences according to provenance are found for the tropical hardwoods, such as for *Swietenia* and *Terminalia.* Trends in wood quality tend to follow rainfall patterns rather than latitude or longitude.

7. In warm climates, where rainfall is abundant during the latter part of the growing season, wood of the conifers tends to have a high specific gravity because

growth continues for a longer period during the period of latewood formation. In climates where moisture is limited during the latter part of the season or where cold wheather causes early growth cessation, the specific gravity tends to be low.

8. Minor site and soil differences within a broad geographic climate have only a small effect on wood properties. However, really extreme environmental differences, such as between deep sands and clays, or between mineral and organic soils, may result in the development of distinct provenances with quite different kinds of wood. In general, however, stand to stand differences within a geographic region are not large. It appears that the broad general environmental conditions of temperature, day length, and rainfall distribution patterns and amounts more strongly control wood qualities than do lesser differences in soil nutrients or soil texture.

9. The tree to tree variability within a source is usually, but not always, considerably greater than the provenance or stand differences within a provenance.

A dangerous assumption too frequently made is that the wood of a species is the same throughout the range of the species regardless of the environment in which the trees have grown. This error has resulted in great economic losses. For example, yield and quality in one part of the species range has been used to predict the properties of wood that a mill will have available in another part of the range. Frequently, after the mill has been built, yield and quality are not as calculated, and problems then occur. A prime example is the previously cited one for loblolly pine in the southeastern United States (Table 2.1). The wood of *Pinus taeda* with a specific gravity of 0.58 in Florida differs greatly from the wood with 0.47 gravity found in the northern part of the species range in Maryland and Virginia. The wood of the Florida source will have a much greater pulp yield per unit volume or per green ton than that from the north; paper properties will differ because of the different thickness of the cell walls and different tracheid lengths. The wood of loblolly pine from the north is much more suitable for newsprint, fine paper, and tissues than wood from the southern source. Loblolly pine from the south produces much stronger and heavier boards than that from the north. Of special interest to those who buy wood by green weight, the lower density wood of loblolly pine from the northern latitudes or the piedmont has a higher moisture content, so green weights per unit volume of wood are about the same as for wood grown in the coastal plain even though the fiber content is lower.

2.3 Wood Variation Related to Species and Provenance in Plantation-Grown Trees

Studies on the importance of the geographic source of seed and species used in plantations upon the wood produced have greatly increased in recent years. Although some results obtained on older trees provide meaningful wood quality assessments, many reported wood studies are from young plantations, so results must be viewed only as general indications of the relationship between seed source, or even species, and wood properties. Few really good overall patterns have developed that will allow a prediction, and the only absolute rule is that "one cannot accurately predict the influence of the source of seed upon wood until the trees are grown in the particular environment of interest."

This statement is also true for differences among species; many examples can be given. One, cited by Skolman (1963), found that *Araucaria heterophylla, Pinus pinaster,* and *P. radiata* produced strong timber at high growth rates in Hawaii, while *P. taeda, P. elliottii,* and *P. patula* also grew well but produced inferior timber.

Another example was given by Kromhout and Toon (1977) for southern Africa where the densities of *Pinus caribaea* and *P. elliottii* are similar. In contrast, at higher elevations in Colombia, Ladrach (1986c) found that *Pinus elliottii* had a ten-year specific gravity of 0.40, but *P. elliottii* var. *densa* averaged only 0.30. The latter species, so named for its dense wood in its indigenous range, had many trees which could not support their own weight and "crawled" along the ground. Ladrach reported that *Pinus oocarpa* had the highest density (0.429) followed by *P. pseudostrobus* (0.411) and *P. maximinoi* (0.407) when grown in the same environment. The following values, taken from Ladrach (1986c) show the average specific gravity of 8-year-old trees.

Species	Juvenile wood	Mature wood	Whole tree weighted
Pinus oocarpa	0.375	0.477	0.429
P. pseudostrobus	0.352	0.444	0.411
P. maximinoi	0.372	0.420	0.407
P. patula	0.317	0.425	0.374
P. kesiya	0.343	0.390	0.370
P. caribaea	0.357	0.375	0.363

The reason for the lack of predictability about a species is that the growth conditions in the new environment can have such a strong control on the type of wood that will be produced that normal species differences may be masked. Differences among plantations can result either from the genetic control by species or by the differing environments in the plantations in which the species are grown. These two factors are very difficult to separate and often they interact with one another.

The effect of differing test environments on wood properties was clearly shown by Bunn (1981) in radiata pine and by de Guth (1970) for *Pinus elliotti* var. *elliottii, P. elliottii* var. *densa, P. caribaea* var. *caribaea, P. caribaea* var. *hondurensis, P. taeda, P. patula, P. douglasiana*, *P. michoacana* growing at six different locations in northeastern Argentina. Both specific gravity and tracheid length show a strong locality effect, significant at the 1% level. Not all species had the highest densities or longest tracheids at each location, however, indicating the interaction of species with site. An interaction can even be larger with provenances than with species.

As mentioned previously, tree to tree differences are usually large, making it difficult to obtain a good estimate of the real average differences between species and provenances. For example, *P. caribaea* seed sources grown in Fiji had relatively small differences in wood densities compared to the variability among trees (Cown et al. 1983). Sunly and Lavers (1961) found that the majority of the wood variability occurred among individual trees at the same site, even though some variation in wood properties was present among different geographic regions in which Sitka spruce (*Picea sitchensis*) was grown. Although exceptions do occur, just as for provenances from natural stands, the usual observation is that *tree to tree variation in wood properties is greater than that resulting from provenance or species differences.* This relationship is observed within many different species grown in many different environments.

The considerable amount of genotype × environment interaction frequently reported for height growth or volume may be minor or even lacking for wood properties. This result was reported for loblolly pine by Barker (1973) and for the "Lost Pines" of Texas (loblolly pine) by van Buijtenen (1978), who found no genotype × environment interaction for specific gravity at the quite different sites in which the tests were made. For *Pinus oocarpa* grown in Brazil and Colombia, Lima (1987) found that 83% of the families showed a stable performance in specific gravity even though the others exhibited a meaningful genotype × environment interaction. In hybrid poplars, Ferrari and Scaramuzzi (1980) only determined a small genotype × environment interaction for wood density and fiber length but no meaningful interaction for other traits.

There is a closer relationship between wood properties and latitude from which seed were obtained than from longitude except when huge environmental differences such as drought or extreme soil differences accompany changes in longitude.

2.3.1 Examples of the Effect of Provenance upon Wood Properties of Trees in Plantations

In some instances, the provenance from which seed are obtained has an effect on the wood of a plantation, while in others the environment in which the plantation has been established exerts such a strong control that provenance differences, if any, tend to be masked. One sometimes hears of the strong influence of provenance, such as for beech (*Fagus sylvatica*) where it was found that for characteristics which directly influence the technological value of wood, the principal source of variation appears to be the provenance of the plants (Koltzenburg 1966). The reasons for wood variability related to provenance are many; for example, Kennedy (1969) found a delayed cessation of cambial activity of *Pinus banksiana* provenances related to the milder climates.

Some of the known effects on wood properties of provenance in plantation-grown trees as well as the environment in which the plantations are grown are cited below to illustrate trends that have been found:

1. In the southeastern United States, the wood properties of the so-called southern pines have been intensively studied with respect to their relationship to provenance. One such report was given by Lantz and Hofmann (1969) for the coastal plain, piedmont, and sand hill geographic sources of loblolly pine where differing wood specific gravities are found in indigenous environments but the trees produce wood with similar properties when grown together. Wood differences in the several tests were mostly related to the site on which the tests were made and not to the provenance of the seed (Fig. 2.5).

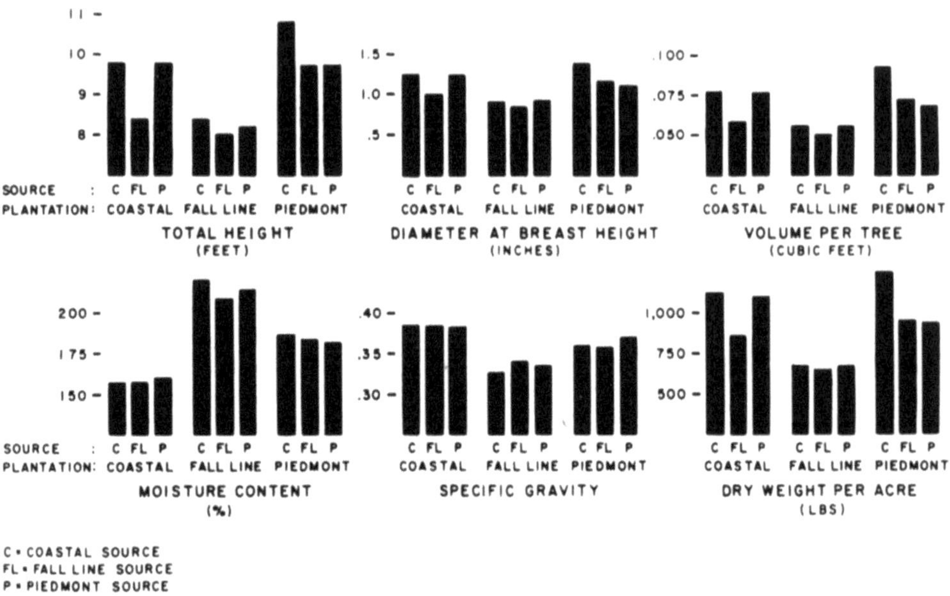

Fig. 2.5. Wood properties are frequently determined more by environment than by provenance source. Shown are the wood densities and wood moisture contents of the coastal, fall line (deep sand), and piedmont sources of loblolly pine when grown on the three different sites in North Carolina. Note the very small effect of provenance but the larger differences caused by the site where the trees were grown

Large areas of slash pine have been planted in the sand hills of the Carolinas, with the expectation that the wood produced would have a high wood density similar to *P. elliottii* in its native range (0.56). However, the slash pine grown in the sand hills north of the species range has a low specific gravity (0.49) similar to and sometimes even less than that of the local loblolly pine; the source of the slash pine seed used to make the plantings seems to make little difference (Zobel et al. 1983).

When loblolly pine was grown north and west of the species range, Gilmore et al. (1966) found that neither percent latewood nor specific gravity was related to source of seed but the wood was different in the different environments where the trees were grown. A comprehensive analysis of regional specific gravity values for loblolly pine in the southeastern United States (Talbert and Jett 1981) showed that there were four definite regions for mature wood specific gravity related to the environment, but essentially none for juvenile wood. As summarized by Zobel et al. (1972), "in general, the environment at the location of tree growth has a greater effect on specific gravity than does the source of seed used."

As is usually the case, there are a few exceptions where the source of seed has an effect on the wood of the plantations. For example, in a study of a number of geographic sources of loblolly pine grown at Athens, Georgia, Strickland (1960) obtained unusually high correlations of specific gravity and latitude of seed provenance, i.e. r = 0.989, and with longitude r = 0.95. In 13 plantations in

Texas, the Arkansas source of *Pinus taeda* always had the highest specific gravity (0.426) and the Southern Louisiana source was least dense (0.401) (Bryan et al. 1984).

In 1967, Saucier and Taras found only one statistically significant difference among nine sources of loblolly pine when grown for 14 years in Alabama; only the northern Maryland source produced high specific gravity wood. However, there was a trend for specific gravity to be higher in trees grown from northern geographic seed sources and to be slightly lower for trees from seed from high rainfall sources. A similar geographic trend was found by Crow (1962) when loblolly sources from within Louisiana were grown together. Seed from the northern part of the state produced trees with wood with higher average specific gravities than seed from the south. This trend is surprising, and diametrically opposite to the trend in natural stands of these species in other regions of the south. This provenance source to wood density relationship in the western part of the species range of *Pinus taeda* has been reported by others. However, when a relationship does exist for loblolly pine between origin of seed and wood specific gravity of trees grown in plantations, the correlations are usually low. The inherent differences in wood related to provenance source are often so low that adaptability, tree growth and form, and disease tolerance should be the decisive factors with regard to which provenance is grown, not the quality of wood that will be produced.

In a study of another southern pine, Saucier and Taras (1966a) found that there were no differences in wood properties among five sources of longleaf pine (*P. palustris*) when grown in Virginia. A sixth source had significantly lower specific gravity but it was susceptible to frost, and the authors stated that the wood difference may have resulted from the influence of cold damage.

In Virginia pine (*P. virginiana*) Rink and Thor (1973) found that different seed sources produced wood with varying specific gravities, but wood property variations among plantations were large despite seed source. They concluded that specific gravity of Virginia pine is strongly affected by environment rather than by seed source.

Species with relatively narrow natural ranges usually do not show much geographic variation in wood properties. This pattern was found for 22-year-old slash pine by Derr and Enghardt (1960) where no variability in either specific gravity or tracheid length was evident among the five scattered geographic sources of slash pine tested.

Although sometimes given the ranking of variety rather than provenance, there were differences in specific gravity between plantations of typical slash pine (*P. elliottii* var. *elliottii*) and South Florida slash pine (*P. elliottii* var. *densa*) (White and Saucier 1966). Despite the name of var. *densa*, at 11 years of age, the typical slash pine grew faster and had a higher specific gravity (0.57) than the South Florida slash (0.53) when grown in adjacent plantations. The reason for the higher density of var. *elliottii* was its higher latewood content.

Another pattern appeared to be present for shortleaf pine (*P. echinata*) when grown in Oklahoma, where Posey et al. (1970) reported the wood specific gravity of trees grown in plantations was related to provenance at the 5% level with one source being the highest in both test plantations. Specific gravity and tracheid length differences were greater among provenances from north to south than for those from east to west (Posey et al. 1970).

For an entirely different wood property, Barrett and Bengston (1964) found that oleoresin yields were essentially similar when five separate sources of slash pine were grown together.

2. Some species appear to have wood that responds to provenance in a way contrasting to the southern pines. For example, Echols and Conkle (1970) report that seed from lower elevation ponderosa pine (*Pinus ponderosa*) produced higher gravity wood when planted at high elevations than did the local seed source and vice versa. The best total tonnage yields were obtained at middle elevations when middle elevation seed sources were used, since growth was rapid on trees which also had relatively high specific gravities. The lower elevation sources produced poor wood when planted at high elevations because of compression wood formed from the leaning trees and deformation caused by wind and snow. In contrast to specific gravity, Echols (1973) found essentially no relationship between elevation of seed parents and tracheid length of the progenies of ponderosa pine. There was a relationship of tracheid length with test sites, whereby trees on the sites that produced faster growth had shorter tracheids.

3. In Sweden, Remröd (1976) reported that seed from the northern provenances of Scots pine (*Pinus sylvestris*) produced trees with somewhat higher density wood, but the differences were of such a small magnitude that he concluded that regional differences in seed source will probably be of little importance for wood of Scots pine. For *Pinus contorta* in Bavaria, differences in cellulose and lignin were related to seed provenance (Schütt 1958). In one of the oldest provenance tests, established in Sweden in 1904, Petrini (1959) obtained the best quality timber from inland Scots pine sources in middle Sweden. One source from southeastern Sweden was also satifactory. The German and southern-Swedish sources had poor stem form and large branches, reducing wood quality.

4. When 29 sources of jack pine (*P. banksiana*) were examined in several test locations, Kennedy (1971) noticed a number of phenological and xylem characteristics that had a significant relationship with provenance, as expresed in terms of growing degree days. However, the relationship with specific gravity was small; latewood percentage was not related to seed provenance and it showed considerable genotype×environment interaction. In 5-year-old trees from widely differing sources of jack pine, King (1967) reported very large differences in specific gravity and tracheid length among sources, with the more northern sources producing the highest specific gravity and the shortests tracheids. In his extensive study on jack pine, Schoenike (1962) found more than 30% difference related to provenance in all wood traits studied, one of which included latewood percent. The slow-growing sources from northern Canada and Nova Scotia had a much higher specific gravity than the fast-growing sources from Minnesota, Wisconsin, and Michigan when grown together in the same environment. This is similar to the report by Schütt (1962) for *P. contorta.*

In red pine (*Pinus resinosa*), which sometimes grows with jack pine, no difference appeared in wood properties from the trees grown in Wisconsin when compared to natural stands in Maine (Baker and Shottafer 1967). Another example was given by Maeglin (1966) who found that red pine grown in plantations

Fig. 2.6. Some species have few provenance differences and very little wood variation regardless of the seed source. Red pine (*P. resinosa*) shown in the photograph growing in Canada is usually reported to have essentially no wood variation related to seed source; however, there is one exception reported by Rees and Brown (1954)

in Wisconsin had wood similar to the stand in Maine from which seed had been obtained. This species is well recognized as being quite uniform in all characteristics, including wood. However, in one red pine study, statistically significant differences were reported in wood specific gravity related to provenance of seed even though the magnitude of the differences was only 0.02 among provenance sources (Fig. 2.6).

5. Working with 25 provenances of *Abies grandis* grown in Europe, Polge (1968) reported strong latewood percentage correlations with provenance in young, 3-year-old seedlings. He found a few French seed sources and coastal American provenances that had a good combination of fast growth and high wood density.

6. In Norway spruce (*Picea abies*), generally no statistically meaningful relationship has been found between specific gravity and source of seed, although occasionally differences are evident. In an analysis of the 1938 spruce (*Picea abies*) plots in Sweden and Norway, Dietrichson (1963) reported a provenance conditioned growth termination; this is related to initiation of latewood formation and possible later damage by frost. This finding was emphasized again by Dietrichson in 1964 when he showed that greatly delayed latewood formation in the western central European provenances (when grown in Norway) resulted in incomplete lignification of the outer latewood zone in some years. The delayed latewood for-

mation was connected with low wood density. When Norway spruce was grown in the northern USA, Stairs (1969) obtained only small differences in specific gravity related to seed source but he felt that these may be large enough to be of importance in large-scale forest plantings.

Only a few studies have been made with white spruce (*Picea glauca*) wood related to provenance. In Canada in 1960, Holst found some differences in wood related to seed source but the differences were small. Similarly with Sitka spruce (*Picea sitchensis*), Deleporte (1983) observed about the same variability in wood properties at an interprovenance and intraprovenance level when the trees were grown in France.

7. The reported results can sometimes be confusing. For example, in an assessment of 46-year-old Douglas-fir from 13 different sources, McKimmy (1965) obtained a variation in specific gravity and latewood percent that was many times greater from plantation to plantation than among seed sources, indicating the strong influence of the environment. However, also for Douglas-fir, McKimmy reported in 1959 that geographic location and site affected specific gravity to only a minor extent. An assessment of 50-year-old families of Douglas-fir for wood variability in different plantations by McKimmy and Nichols (1971) showed that the environmental control of specific gravity was quite strong even though family control was considerable; environmental control was less for tracheid length. The test areas were in three rather contrasting environments, and the results indicated the importance of the environment where trees are grown. However, in contrast, provenance of Douglas-fir and not growth rate influenced wood specific gravity according to Haigh (1961).

8. In eastern Canada, no differences in specific gravity were evident among seed lots of a number of eastern white pine (*P. strobus*) sources tested (Lee 1974).

9. Several clones of members of the genus *Populus* were assessed for the genetic and environmental influences on growth stresses which result in distortion during conversion (Waugh 1972). Environments where the trees are grown were very important and there were definite interactions with environments, i.e., the same genotypes did not react similarly when grown in different environments, despite rather strong clonal effects. A similar observation was made for hybrid poplars by Son and Chung (1972) in Korea where wood variation between locations of plantations was considerable despite the fact that it was also very high among individual trees. Also for *Populus*, Scaramuzzi (1960) obtained opposite and remarkable differences in wood between provenances leading to the conclusion that at least part of the differences were genetically determined. In cottonwood (*Populus deltoides*), no differences in specific gravity were related to the location of the test (Farmer and Wilcox 1966b). Thus, this small sampling of reports indicates that all types of wood property reactions related to provenance and environment have been reported for the poplars.

10. In the southeastern United States, Johnson and McElwee (1967) did not find any provenance caused patterns in specific gravity or fiber length in plantations of sweetgum (*Liquidambar styraciflua*). Similarly, no relationship was present between geographic source and specific gravity in sycamore (*Platanus occidentalis*) plantings in Mississippi (Land and Lee 1981; Land et al. 1983). Nebgen and Lowe (1982) reported for sycamore that family variation was more important

than regional differences; family differences accounted for between 79 and 100% of the total genetic variation in branch angle and specific gravity.

11. Little information is available concerning wood of species and provenances of tropical hardwoods when grown in plantations, except for *Gmelina* and *Eucalyptus.* In these two genera, wood generally varies little with provenance source of seed, but this is not always so. For example, de Souza et al. (1986) found different values for different provenances of *Eucalyptus saligna* and *E. grandis* in Brazil.

Delgado reported in 1983 on tests of six species of hardwoods in Venezuela, four of which were indigenous. He states "wood quality of the species is normally taken for granted and the expectation is that it will be comparable to that of the naturally grown species." A study of the wood of *Bombacopsis quinata, Cendrela angustifolia, Cordia apurensis* and *Tabebuia rosea* (the indigenous species), and *Gmelina arborea* and *Tectona grandis* (the exotic species) show that this assumption is not correct. From the plantation, *Cedrela angustifolia* showed collapse and extremely low wood density and thus would not be suitable for the high quality products expected from *Cedrela* species. Delgado emphasizes that the suitability of a species for plantations must be established on wood qualities as well as on silvicultural response or the plantations may be failures. We also add to this warning the necessity of making sure that the trees are of such an age that a suitable assessment of their wood properties can be made.

In summary, the situation concerning wood properties of trees from plantations as related to the provenance from which seeds are obtained is very complex and confusing and there are few evident or clear-cut trends or relationships. In some instances the seed source has an important effect on the wood of trees grown in plantations, while in others any effect is overshadowed by the influence of the environment in which the plantation is established; in both instances the large tree to tree differences within source and interaction with the environment make a simple conclusion impossible. The first reaction to the many contradictions and inconclusive findings is that the attempt to relate geographic seed source to wood that will be produced in plantations is hopeless. Even though it is not possible to list absolute rules or guidelines, a few useful trends do appear from the literature reviews and experiences of the authors:

1. The more severe the environment, the greater the control it has on wood properties relative to inherited differences expressed through seed of different provenances.

2. In some species, wood properties of provenances appear to be quite strongly genetically fixed, while in others, provenance differences appear to be primarily a response to the local environment. Strong to modest provenance control appears in *Pinus contorta, P. banksiana, Abies* spp., perhaps *Picea* spp., and in members of the genus *Populus.* The wood of other species, like the major southern pines, the tropical pines, Douglas-fir, and some temperate hardwoods appears more easily influenced by the environment where the plantation is established, and the provenance from which seed are obtained has relatively little importance.

3. The complications of species differences and especially genotype × environment interaction of individuals, families, and provenances make it impossible to predict the response of wood to origin of seed with any degree of accuracy.

Although most authors recommend choosing provenances with the desired wood properties, only a few have shown that such action is positively beneficial to any great extent. The only positive statement that can be made is that THE ONLY SURE WAY TO TELL WHAT KIND OF WOOD TREES FROM SEED FROM A GIVEN PROVENANCE WILL PRODUCE IN A DIFFERENT ENVIRONMENT IS TO GROW THEM THERE.

2.4 Wood Properties of Trees Grown as Exotics

2.4.1 General Considerations

The use of exotics is becoming widespread in forestry. Resultant benefits are many, but there are also some major problems, one of which relates to the wood produced by a species when grown in an exotic environment. Because of the sometimes large differences between the environment where the species originated and the new one where the exotic is grown, the wood of an exotic sometimes greatly differs from that which is expected. Even though it can be generally stated that choosing the correct provenance can enhance the desired wood properties in a plantation (Wilcox 1977), this does not always happen in the more extreme environments frequently found where exotics are grown. In his summary paper on the control of wood properties in exotic species, Ladrach (1986b) states that "wood quality varies with species, provenances, tree age, site and plantation altitude for many conifers and hardwoods."

Most of the interrelationships discussed in the previous section on the effect of geographic seed source on wood will hold for exotic tree species, so they need not be repeated in detail here. It is the objective of this section to discuss and document the extent and importance of changes in wood properties that occur in exotic forestry.

Although this is not the place to analyze why exotic species are frequently used rather than the indigenous species, it is significant that one of the major reasons for growing exotics is to obtain a specially desired type of wood not available from the indigenous species where the exotic is to be grown (Zobel et al. 1987). The most common desire is to produce conifer wood where none exists, both for its fiber and solid wood qualities. Thus, pines are often grown to supply long and thick-walled tracheids which give good tear strength to paper and strength to boards, or members of the genus *Cupressus* are used to produce high quality wood (Fig. 2.7).

A major complicating factor in assessing wood of exotics is that in the trees grown, it is not possible to separate the wood variability resulting from the provenance of origin from that caused by the new exotic environment. Provenance of the exotic species used is of major importance for growth and survival when exotic forestry is practiced; as discussed in the previous section, the kind of wood that will be produced by the provenance is sometimes unpredictable and occasionally it will be so different from that expected that it is useless. As was so strongly emphasized earlier in this chapter for geographic source differences, the only sure way to know what type of wood will be produced by an exotic is to grow the desired provenance in the exotic environment.

Fig. 2.7. Exotics are often grown to supply wood not otherwise available, both for pulp and paper and for solid wood products. This is illustrated by the *Cupressus* growing near Bogota, Colombia. Members of this genus are widely grown as exotics in South America and Africa because of the high quality wood produced

The problem is so severe that Hughes (1968), Palmer and Tabb (1968), Zobel et al. (1983), and many others have strongly stressed the absolute necessity of an early evaluation of the suitability of wood grown in exotic environments for anticipated end uses before large-scale planting is done. This key concept has frequently been ignored, especially when conifer plantations have been established. A number of organizations are currently facing crises because their exotic plantations are ready for harvesting or thinning, but the markets available cannot accept the kind of wood that has been produced.

2.4.2 Changes of Wood Properties in Exotics

Although the wood of the exotics may be different from that in the indigenous habitat, it is not necessarily inferior. For example, young wood from tropical pine plantations is not the best for kraft paper products or solid wood products requiring strength, but it can be valuable for other uses. This can be illustrated by a number of studies of the wood of the tropical pines and their utility, e.g., by Palmer and associates at the Tropical Products Institute (Tropical Development and Research Institute) in London. Many references have been cited by Palmer (1977) who summarized the wood properties of 12 species of forest trees grown as exotics in the tropics.

It is evident that wood differences occur not only because of the influence of provenance or the exotic environment, but also because the exotic trees are usually fast growing and thus are harvested at a young age, which results in trees having a high proportion of juvenile wood. A major cause of undesirable product quality of exotics can be directly related to juvenile wood combined with reaction wood commonly associated with exotics because of poor tree form from poorly adapted seed sources (see Chap. 4).

Many reasons can be given for a change in wood properties with a change in environment; these are discussed in Chapters 6 and 7 and will not be covered further here; however, one example will suffice to show the trend. In some tropical climates, species such as *Pinus caribaea* will respond to ideal growth conditions by continuing height growth essentially throughout the year. Since latewood is usually not produced in a tree during its period of height growth, the pines in these ideal environments have only a small amount of latewood and thus produce extremely low specific gravity wood not desirable for many products.

Both conifers and hardwoods are used in exotic plantings, and changes in their wood usually do not occur in the same way. Therefore, conifers and hardwoods are discussed separately in the following sections.

2.4.3 Wood Properties of Exotic Conifers – General

In the future, much of the timber available to the forest industry will be from fast-growing exotic conifers. The only positive statement that can be made about the wood of these trees is that it contains a higher proportion of juvenile wood because it is harvested at a younger age and thus has low specific gravity as well as a somewhat reduced tracheid length compared to wood produced in the natural range of the species. This is especially true for species such as the hard pines and Douglas-fir in which juvenile wood differs most from mature wood (Zobel 1981).

Not all species, or provenances within species, respond in the same way to differing sites when grown as exotics. The authors have observed that wood of *Pinus caribaea* and *P. kesiya* seems to be easily affected, sometimes radically changed, when grown in different environments. *Pinus caribaea* can vary from very light and soft with essentially no latewood when grown in certain tropical areas where growth conditions are ideal throughout the year to wood whose rings are mostly latewood with a high specific gravity in other tropical climates (Plumptre 1977)

Fig. 2.8. The *Pinus caribaea* stand shown growing in South Africa has undesirably low wood density. The climate is such that the trees grow in height all year, drastically reducing the amount of latewood produced, which results in a low specific gravity (Zobel and Talbert 1984)

(Fig. 2.8). The authors have observed that other species, such as *Pinus taeda* and *P. oocarpa* tend to produce wood that is more consistent regardless of the environment where it is grown.

2.4.3.1 The Tropical Pines – Pinus caribaea Example

There is debate about the definition of a tropical pine. We do not wish to participate in this controversy, but for simplicity and understanding, we will refer to tropical pines as those growing in areas where freezing weather does not occur. There is an intergradation, of course, into subtropical pines which will sometimes grow under either tropical or temperate conditions. A specific or absolute definition is not necessary and will have little bearing on the factors that cause wood property variation.

Wood of some species, especially *Pinus caribaea*, is so sensitive to the environments where the tree can be grown, that although provenance effects are very important for growth and survival, they have less effect on wood properties than the new environment (Cown 1981). The variability in the wood of *P. caribaea* and its sensitivity to site were emphasized by de Villiers (1974) and Harris (1977a, b). It was found by Cown et al. (1983), for example, that in East Africa, the elevation where the trees were grown was of great importance; trees grown below 300 m had a basic density of 515 kg/m^3 compared to 445 kg/m^3 at higher elevations.

Most information on the wood of tropical or subtropical pines is available for the species *Pinus caribaea, P. kesiya, P. oocarpa,* and *P. radiata.* Other subtropical or temperate species such as *Pinus taeda* and *P. elliottii* have been grown extensively in tropical and subtropical areas, and their wood has been extensively studied.

By far the greatest amount of information on a tropical species grown as an exotic refers to the wood of *P. caribaea*, especially in reports by Palmer and associates. There are too many to cite individually so they have been summarized in Table 2.2.

Based upon these studies, it is evident that even though the wood of *P. caribaea* varies greatly depending upon where it is grown, the wood is usually useful and the wood from older plantations frequently makes paper with qualities similar to that from the southern pines of the USA. Other intensive studies on the wood of *P. caribaea* were carried out by Ibraham (1972) and Falkenhagen (1979). Plumptre (1983) emphasizes that this species grows in a very wide range of sites and under a variety of conditions. Its wood varies greatly because it is genetically very variable and is very responsive to differences in site, climate, and silviculture. He states that numerous types of wood can be obtained from *P. caribaea* if the right site, provenance, and silviculture are followed; this includes correct thinning and pruning. Failure to take wood quality into account when growing *P. caribaea* can result in a product which has much less value than its potential. This is especially true if the products produced are to compete on the world market. It is sufficient to say that any generalizations about the wood of *P. caribaea* cannot be made without a definition of the age and conditions under which the trees have been grown as well as the provenance that has been used.

The potential variability of the wood of species grown as exotics was described in a study on *Pinus caribaea* planted in New Zealand, Queensland, Fiji, and western Malaysia by Harris (1977b). He found a great diversity in the kind of wood produced; the basic density varied from 400 to 700 kg/m^3. He noticed a trend for increasing gravity with decreasing latitude, but great variations related to the local environment also occurred within regions. Harris stated that "few species have wood that varies as widely as *Pinus caribaea* when grown in different environments. Furthermore, this species has a great amount of tree to tree variation, as much or more than that present in other tropical pines." Similar results relative to the effect of environment on specific gravity was found for *Pinus caribaea* by Plumptre (1977) when it was grown on 18 different sites in East Africa. Wood from the site with the greatest density was 1.5 times denser than from the site with the lowest density.

Table 2.2. A summary of some wood properties of *Pinus caribaea* as an exotic analyzed by Palmer and associates (Published by the Tropical Products Institute, London, England: now Tropical Development and Research Institute[a])

Country or region where trees were grown	Report number and pages	Date	Results
Malaya	Res Pam 55, 24 pp.	1966	Sulfate pulping gave low yields and pulps were inferior to commercial pulps. Tree age was critical
Sabah	L 12, 23 pp.	1967	Wood from 5-year-old trees is better than wood from some other areas where *P. caribaea* is grown. It yields reasonably good pulp but it is somewhat weaker than that from the species in the southeastern USA
Fiji	L 14, 27 pp.	1968	Great variation exists from tree to tree. Best pulps come from high density trees which are nearly as good as southern pine pulps from the USA. The average pulp would be difficult to export
Trinidad	L 15, 49 pp.	1969	*P. caribaea* grown in Trinidad is unlikely to yield an unbleached kraft pulp with first-class strength properties, but it is satisfactory for an integrated paper mill
Sabah	L 25, 37 pp.	1972a	Nine-year-old wood was more suitable for bleached pulp than 5-year-old wood. The pulp is similar, but slightly weaker than that from southern pines and much weaker than Scandinavian pulp
Fiji	L 27, 60 pp.	1972b	Wood from merchantable (10-year-old) trees had pulping properties similar to that from the southern pines
Sabah	L 28, 13 pp.	1973a	Wood of 9-year-old trees was suitable for dissolving pulps
Jamaica	L 30, 24 pp.	1973b	Tear and yield of paper increased when wood from higher density trees was used
Fiji	L 46, 16 pp.	1977	Fast-grown trees showed no significant differences in wood density, chemical composition, pulp yield, ease of digestion, or characteristics of the unbleached pulp compared to slow-growing trees
Sri Lanka	L 57, 21 pp.	1982	Pulps were similar to those from Douglas-fir and the sourthern pines
Tanzania	L 66, 32 pp.	1984	The resin content was low; pulp yields were lower than with *P. patula* but about the same as with *P. elliottii*. The pulp was weaker than that from southern pines of the USA

[a] Not listed in the references.

The large variability among individual trees was stressed by a number of foresters such as Burley and Palmer (1979) who reported mean densities from 441 to 734 kg/m^3, 0.9–4.2% variation in alcohol-benzene extractives, and 3.6–6.9% variability in total extractives. Such variation plus strong inheritance enables the breeder to rapidly develop *P. caribaea* with wood having the desired properties. Palmer and Tabb (1971) emphasize the need for an increase in specific gravity in *P. caribaea* and that high density wood generally yields pulps superior in all strength properties investigated. The effects of the large amount of juvenile wood present can be reduced, as has been possible for loblolly pine (Zobel et al. 1978).

Although provenance had a major effect on growth, there was essentially no effect of provenance on wood density or tracheid length in *Pinus caribaea* (Palmer and Tabb 1971). For *Pinus taeda, P. elliottii,* and *P. caribaea* Birt and Harris (1970) reported only very small wood differences related to provenance from which seeds were collected. There is much discussion about the rate of growth affecting wood quality of *Pinus caribaea,* but Palmer and Gibbs (1977) found no difference between fast and slow grown trees in Fiji. They recommended growing the trees as rapidly as possible. The major cause for defect is the large proportion of juvenile wood because the trees are of a merchantable size at young ages.

It is evident that the wood of the exotic *Pinus caribaea* has properties sometimes drastically different from wood of this species when grown in its indigenous habitat. One major source of difference is the fast growth of *Pinus caribaea*, resulting in harvesting of the exotic trees at a young age. Thinnings frequently consist entirely of juvenile wood.

The intrinsic wood properties of the three varieties of *Pinus caribaea* (*hondurensis, bahamensis,* and *caribaea*) show some differences but are similar enough so that their wood can be considered together. The main differences in wood among these varieties results from differences in tree form; *P. caribaea hondurensis* has the poorest form with a generally crooked bole and heavy, sometimes steep-angled limbs. As a result this variety has the most compression wood and knot wood. Some foresters prefer the *bahamensis* and *caribaea* varieties because of their more homogeneous wood, resulting from better tree form, even though the variety *hondurensis* usually has a considerably higher growth rate.

The wood of young *Pinus caribaea* may have limited utility both for quality solid wood products as well as for fiber products, and there is real concern about its use, as reported by Brown in 1971 for Jamaica. He found that it is useful for some papers and quite desirable for others such as tissues, groundwood, high quality writing papers, and thermomechanical pulps. The pulp made from young trees is usually inferior to the southern pines and Canadian and European conifers, but wood from older trees with less juvenile wood is quite good. New and improved manufacturing techniques will be most helpful in utilizing the wood of *Pinus caribaea* more efficiently when grown as an exotic. A further problem with *Pinus caribaea* is its propensity to produce spiral-grain wood. Although this defect occurs in some trees, no clear pattern with source, age, or position within the stem could be established in Fiji.

In summary, the wood of young *Pinus caribaea* is currently considered to be marginal in quality but it is useful, and as better formed trees with better wood density are developed and grown to longer rotation ages, wood quality will cease to be a serious problem. This species is certain to become a major contributor to the world conifer wood supply in the future (Zobel 1984). There are exceptions, however, where the wood will be difficult to use, such as the extremely high or extremely low specific gravity woods of some trees produced in certain tropical climates.

Different species do not always respond identically to the same environment. For example, Falkenhagen (1979) found that *Pinus elliottii* had higher densities and shorter tracheids than *P. caribaea* when grown in South Africa even though their woods do not differ significantly in their indigenous ranges. There was little difference in wood properties (specific gravity and tracheid length) among provenances of *Pinus caribaea.* Wood of *Pinus elliottii* can have a high density in an environment where very low wood density is produced in *P. caribaea.*

Work on wood of the tropical and subtropical exotic conifers is certainly not restricted to *Pinus caribaea.* For example, Palmer and associates have studied *Pinus oocarpa, P. merkusii, P. kesiya, P. patula, P. taeda,* and *P. elliottii.* They compared woods among species and found that *Pinus elliottii, P. oocarpa,* and sometimes *P. patula* produced wood with the greatest utility. In his studies on *Pinus elliottii,* Allen (1977) found specific gravity to be directly correlated with growth rate as measured by height growth. He emphasizes that the wood of this species is too dense when grown in Queensland, Australia, and work is necessary to reduce the specific gravity.

For *Pinus taeda* grown in Brazil, Barrichelo et al. (1977) found no difference in wood density related to provenance, while van der Slooten et al. (1976) showed differences among areas, apparently related to the rainfall distribution pattern in 7- to 14-year-old plantations in southern Brazil. In New Zealand, Shelbourne (1971) detected very little difference among provenances of *P. taeda* even though the provenance pattern of growth rates was normal. He emphasized that provenance selection for wood properties would produce very little gain.

Pinus oocarpa (and its close relative, *P. tecunumanii*) is rapidly becoming one of the most favored tropical pines. Its wood appears to be the best of the tropical pines for general usage, especially for solid wood products. When grown in Argentina, de Guth (1973) found the wood quite usable with a specific gravity of 0.39, slightly lower than for *Pinus taeda* and *P. elliottii* but greater than *P. caribaea.* Similarly in Brazil, Foelkel et al. (1975) found that the wood of slash pine was satisfactory, and that wood density, tracheid length, and pulp yields increased with age.

There are many reports on the wood properties and variability of the wood of *P. oocarpa.* All are favorable for its use for both fiber and solid wood products. The wood is not too dissimilar from that of *Pinus taeda* and it tends to be similar even when the trees are grown in somewhat different environments.

Another species widely used and well known is *Pinus patula.* Its wood is also useful but differs depending on where it is grown. For example, in Colombia, Ladrach (1984) found it to be the lightest wood formed at 3000-m elevation and the densest wood at 1800 m. There was less difference between juvenile and mature woods at the higher elevation. Ladrach concluded that the specific gravity and tracheid length of *Pinus patula* were related to tree age, tree size, and plantation elevation.

Table 2.3. Wood property variation with elevation and age for *Pinus patula* grown in Colombia[a]

Age (years)	Elevation (m)					
	Specific gravity			Tracheid length (mm)		
	1800	2200	3000	1800	2200	3000
7	0.33	0.32	0.30	3.88	3.80	3.22
10	0.37	0.36	0.33	4.16	4.06	3.89
15	0.42	0.41	0.39	4.68	4.57	4.35
20	0.45	0.44	0.42	5.24	5.11	4.86
25	0.47	0.46	0.44	5.78	5.64	5.37
30	0.48	0.47	0.44	6.31	6.16	5.85

[a] Adapted from Ladrach (1986b). Values shown are whole tree values.

One exotic species that is becoming of increasing importance in the tropics is *Pinus kesiya.* It is adaptable and grows well on poor, heavy clay sites. Its wood has not been well studied but it seems quite usable. In northern Rhodesia (Zimbabwe) *P. kesiya* from the Philippine source was found to have better wood than *P. patula* and was as useful as *Pseudotsuga menziesii.* The Philippine and Assam sources had better wood than the other sources; the Philippine was best all-around in growth, form, and wood properties (Shelbourne 1963).

The effect of the area where trees are grown was shown for *Pinus taeda, P. elliottii,* and *P. caribaea* in a 14-year-old provenance test in New Zealand. None of the species produced wood nearly as dense as found in the indigenous ranges of the species. *Pinus taeda* formed the densest wood, while *P. caribaea* produced practically no latewood (Birt and Harris 1970). When 18 different clones of radiata pine were tested on four different sites in New Zealand, site effects were found to be large on the genetically similar trees. Density was very little affected by nongenetic differences in ring width in radiata pine (Burdon and Harris 1973).

In a paper summarizing factors that change the wood of tropical exotic forest trees, Ladrach (1986b) reported major differences in wood properties related to where the trees were grown. The effect of age and elevation on wood properties of *Pinus patula* grown in Colombia is shown in Table 2.3. However, Ladrach found no differences in specific gravity with *Cupressus lusitanica* at 1730, 1850, and 3000 m elevation. A pattern related to an elevation similar to that of the pines was found for some hardwoods. One example was *Leucaena leucocephala,* where 12 provenances had a specific gravity of 0.561 near sea level, but only 0.461 at 1000 m elevation. At the same location, 13 provenances of *Gmelina arborea* had a specific gravity of 0.399 at sea level and 0.468 at 1000 m. These samples were provided to illustrate that changes in environment, in this instance elevation, can cause great variation in wood properties, but the patterns cannot always be predicted.

Not many studies have been made on geographic variability in wood properties of *Pinus radiata,* although there are numerous reports of large stand to stand differences related to differing environments. In one case, however, Harris (1967) reports that tracheids are longer at low latitudes. When radiata pine was grown from two sources in two different locations and from seed selected from local plantations (Burdon et al. 1972), only slight differences in specific gravity were related to source and only a slight relationship was found to the areas where testing was done. The authors concluded that wood density appeared to be highly heritable. In Greece, Matziris (1979) found for radiata pine that there was no relationship between seed source and wood density. Trees growing in the warmer climates in Greece produced wood of higher density.

2.4.3.2 Conifers from Temperate Regions

When conifers from temperate climates are grown in exotic environments, their wood is sometimes similar to that of indigeneous species. For example, in a study of the exotic *Pinus contorta* by Persson (1982), the wood was similar to the native Scots pine and Norway spruce and the sulfate pulp yields and qualities were also similar. Persson suggests the strong potential use of lodgepole pine in thermomechanical pulping. He emphasizes that the sawn timber quality of lodgepole pine grown as an exotic is poor compared to Scots pine. When grown in France, lodgepole pine from seed obtained from high elevations in Oregon was best with regards to both growth rate and wood quality (Polge 1963).

When several sources of lodgepole pine (*Pinus contorta*) were planted in 16 different sites in New Zealand, some relationship of wood specific gravity to seed source was observed; one source consistently produced higher specific gravity wood. However, the general results were that trees grown on sites with low mean annual temperatures had low wood specific gravities. The reason given was that the latewood cells in the cooler climates had a low density (Harris 1973a). When grown in France, lodgepole pine seed from higher altitudes produced the best growth and the best wood density (Polge 1963). Also working with *Pinus contorta* Schütt (1962) made the interesting observation that differences in both tracheid length and specific gravity were related to provenance. However, he still feels that no reliable criteria can be used to choose a desired place of seed origin.

In a study of a number of provenances of 17-year-old Scots pine grown in New Hampshire, Echols (1958) reported that both tracheid length and specific gravity were highly correlated with latitude of seed origin. Dorn (1969) found specific gravity differences among provenances of Scots pine when grown in northeastern North America, but results were confounded with the rate of growth of the provenances. Echols (1958) related the wood qualities of *Pinus sylvestris* to growth rate; the differences in growth that result from exotic environments should therefore affect wood density. Faster growth produced lower density wood (see Chap. 5).

Working with maritime pine (*Pinus pinaster*) in Australia, Nicholls (1967c) found that wood from trees from the seed of one Corsican provenance was generally less dense that that of the other three provenances tested. He summarizes: "wood quality can be improved by the choice of parents with desirable wood characters, but expected gains may not be realized if proper attention is not also given to growing conditons... ."

Norway spruce (*Picea abies*) is commonly used as an exotic. Worrall (1975) reported that there were only minor differences in wood related to provenance when spruce was grown in northeastern North America. On the basis of the large within-provenance variation and fairly high repeatability values, he felt that selection of individuals might be more profitable than selection of provenances for this exotic species. No differences in wood properties among different sources of Norway spruce when grown in the northeastern USA were observed (Stairs 1969). Knudsen (1958) also did not notice any difference in technological properties of Norway spruce from seed of eight widely separated provenances when grown in the United States. Nanson et al. (1975), however, report small to large differences in specific gravity and tracheid length with respect to provenance of seed used based upon young Norway spruce trees from the 22 provenances tested in France. The densest provenances were those whose seed came from the high altitudes. Japanese larch (*Larix leptolepis*) is also commonly used as an exotic. When grown in Michigan, Lee (1975) found no differences in the wood related to seed source. He reported that Japanese larch produced a usable wood similar to that of *Pinus taeda*. A number of recent mill studies have produced similar results for some exotic larches.

2.4.3.3 Summary – Wood Properties of Exotic Conifers

It is difficult to summarize the information on the wood of conifers when grown as exotics. In general, most pulps produced from conifers grown in the tropics are inferior to pulps produced from temperate species, and they will be difficult to sell on the world market (Palmer and Tabb 1968). However, products such as writing and tissue paper, as well as newsprint, can be efficiently made from the wood of young plantations of tropical exotic pines and are quite desirable. Paper made from young exotic conifers often does not meet the standards for tear strength necessary for acceptance on the world market (Dadswell and Wardrop 1959; Barefoot et al. 1970).

The strength and finishing properties of solid wood products made from juvenile wood, which occurs in large amounts when some exotics are harvested, are inadequate (Pearson and Gilmore 1980). In addition to their sometimes inherently poor wood qualities, the exotic conifers frequently are crooked and limby, producing a high proportion of knots resulting in compression and knot woods, both of which are undesirable. The importance of provenance seems to vary from study to study, such as for *P. contorta*, but it seems clear that the environment where the exotic is grown usually has a marked effect on the kind of wood produced, which often supersedes differences caused by the provenance of seed used.

Generally, differences of exotics from their indigenous ranges are greater in the tropical pines than for those in temperate regions. The only certain method of predicting the wood qualities to be produced by an exotic is to follow the recommendations for provenances, i.e., grow the tree in the exotic environment and then assess its wood qualities. This was stressed by Plumptre (1983) for *Pinus caribaea* where he found a very large site×provenance interaction for wood properties. This is especially true when different provenances of an exotic species are involved. For example, Wright et al. (1970) point out the problem posed with growing Douglas-fir in Argentina. This exotic species of unknown origin grows very well in the mountain forest regions of the country. Although tests in Michigan and Nebraska show an immense amount of provenance variation within the species, Wright et al. state that the United States data should not be used to forecast which sources grow best south of the equator; only tests in the Andean foothills will identify the most valuable provenances. This conclusion is even more important when wood properties are involved.

2.4.4 Wood Properties of Exotic Hardwoods

With the exception of *Eucalyptus*, work on wood properties of hardwoods grown as exotics has been relatively limited compared to that devoted to the conifers. The eucalypts have been grown under many different conditions, some in environments very unlike those in the indigenous range of the species. The attitude has developed that hardwoods, in general, have wood that is quite stable wherever the species may be grown, as usually occurs in *Eucalyptus*. This may be a false assumption. Although the large fluctuations found in species such as *Pinus caribaea* do not normally occur in hardwoods, wood variation within exotic hardwoods can be significant.

Wood properties of hardwoods usually have the same overall pattern as conifers, varying greatly from tree to tree within a species and sometimes also considerably with the geographic range where the species originated or where it was later grown as an exotic. Because juvenile wood development is only moderate in most hardwoods, the variation in wood with the age of the tree is usually small, making it much simpler to categorize wood properties regardless of age. Similarly, the differences in wood properties from the base to the top of the tree are less, resulting in more wood uniformity within a given forest stand.

Hardwoods grown as exotics have considerable utility for energy and chemicals, a trend that greatly increased during the petroleum crisis (Goldstein 1979, 1980a). How important this usage will become in the future depends on the energy supply and economy, but, in the long run, the use of wood energy and wood chemicals will undoubtedly increase a great deal; when this happens, fast-growing exotic hardwoods will be a major source of supply, and wood properties suited for these uses will be needed.

Another major usage of exotic hardwoods is for firewood, especially in the developing countries. It is reported that about half the wood consumed in the world is used as firewood, and that especially in the developing countries firewood is becoming increasingly important and scarce (Zobel 1984). As a result, not only refuse wood is used, but good trees that could be employed for other products are harvested for firewood. Efforts are expanding to grow firewood rather than merely to harvest natural forests, and fast-growing exotic hardwoods are being primarily considered for that purpose. High density, low moisture content wood is desired, and species or individuals with this type of wood will be preferred.

Genera such as *Leucaena*, which contains species with super-fast growth and high density wood, are good energy suppliers; they are increasingly being used for firewood (Anonymous 1978; Brewbaker and Hutton 1979). This genus, often referred to as the wonder tree, grows as an exotic in many environments, but its reported outstanding growth occurs only on the best sites. Other species, especially in *Eucalyptus*, are being widely used as exotics for firewood, and this usage will rapidly develop in the near future. Wood properties most suited for energy production are known but development of this type of wood has only begun.

For convenience of the reader, hardwoods used as exotics have been combined into a general heterogeneous group including all but the eucalypts; the latter are treated separately.

2.4.4.1 Exotic Hardwoods Other than Eucalyptus

The wood of the hardwoods varies greatly from the very light and soft wood of balsa to the heavy tropical woods, some with specific gravities greater than 1. However, the species most used in exotic plantations have wood in the medium specific gravity range.

One of the hardwoods that is planted widely as an exotic because of its rapid growth and excellent wood is *Gmelina arborea*. Wood of *Gmelina* has been widely studied by industry, which finds it to be suitable for both pulp and solid wood products. However, few wood quality data on this species have been published except for those by Palmer (1973), Palmer and Gibbs (1974), Ladrach (1986c), and a few others. Use of *Gmelina arborea* as an exotic is greatly increasing, and certainly more studies on its wood variation will be made. It is a species that appears to have a relatively uniform wood wherever it is grown as an exotic, as reported by Palmer (1973). He pulped *Gmelina* grown in seven different countries and found little difference in wood properties; however, further testing is needed to determine accurately the wood variation within this species. Present indications are that a fast growth rate does not have any adverse effect on the wood properties of *Gmelina* (Palmer and Gibbs 1974). This species needs improvement in stem and limb form, which in turn will reduce the amount of knot wood and tension wood. In *Gmelina* from the Soloman Islands, Palmer et al. (1984) found the wood to be suitable for pulp export. Its qualities were as good or better than USA southern hardwoods. This agrees with many unpublished studies.

Mahogany (*Swietenia*) and teak (*Tectona*) have been grown quite widely in exotic plantations in several countries; however, only a few publications regarding wood quality variation are available. Chudnoff and Geary (1973) reported little relationship between the wood of *Swietenia macrophylla* grown in its natural range and that of progenies from the same trees in the exotic plantings. In four test environments, the specific gravity of the progeny varied only from 0.50 to 0.54 from parents which had a spread in specific gravity from 0.48 to 0.84, indicating the effect of the climate where the exotics were grown.

Teak (*Tectona grandis*) seed from Burma, grown in Trinidad, produced trees with lower specific gravity than the indigenous teak. The Trinidad teak averaged five rings per inch, while the Burma teak grew much more slowly at 11 rings per inch (Scott and MacGregor 1952). The authors suggested that growth rate differences probably accounted for the specific gravity differential rather than the source of seed. The origin of teak seed has an apparent effect on wood properties; for example, Kedharnath et al. (1963) found that one of the four different seed origins of teak tested had a quite different fiber length from the other three.

Other tropical hardwoods have been planted as exotics and their wood has been tested. For example, Palmer et al. (1983) tested the wood of four low-density species (*Albizia falcateria, Anthocephalus cadamba, Schizolobium parakybum,* and *Terminalia brassii*) and three medium density species (*Gmelina arborea, Maesopsis eminii,* and *Terminalia superba*) when grown in plantations in Fiji. All, except *Terminalia superba,* were pulped satisfactorily by the sulfate process, yielding 47–50% of a bleachable grade pulp. *Terminalia superba* had a high lignin content of 31.8% and could be pulped to only 41% yields even when high levels of chemicals were used.

One group of trees planted extensively as exotics are the many species and hybrids within the genus *Populus.* The members of this genus have been intensively managed for years and a considerable amount of research has been done with respect to wood variability. Clone to clone differences are sometimes considerable, but there are few reports about differences in wood between trees of a clone when grown in indigenous and exotic environments. Because specific gravity is low, the general opinion is that only small variations will occur. However, Scaramuzzi (1960) reports "remarkable" differences among provenances, which he feels might be at least related to genetic factors. Other species, such as the alders (*Alnus*), have also been grown as exotics. They appear to produce wood in the new, exotic environment, that is very similar to that in their indigenous range (Harrington and de Bell 1980).

2.4.4.2 Wood of Eucalyptus When Grown as an Exotic

A number of the approximately 500 species of *Eucalyptus* (Higgins 1985) are widely used as exotics throughout the tropical and subtropical areas of the world. A discussion of the wood property variation among species is not pertinent to this chapter but suffice it to say that the differences range from low specific gravities to very high. The wood property variations even within a species may be huge. For example, Dillner et al. (1970) found that the density of *Eucalyptus globulus* grown in Portugal extended over a wider range than that of pine, spruce, or birch. Although this may be an extreme statement, it appears that nearly all desired wood properties can be found among the different species of the eucalypts except fibers as long as those of the conifers.

Many provenance studies have been made on the eucalypts, but there are remarkably few that included wood properties in their assessment. When wood properties were considered, they usually were related to the pulping qualities of the species. One exhaustive report related to the wood of *Eucalyptus* as an exotic was made by Barrichelo and Brito (1976) who analyzed the pulping characteristics of the wood of a number of species of *Eucalyptus* grown in Brazil.

In South Africa, Taylor (1973 a, b) reported on the wood of *Eucalyptus grandis* grown in a range of environments. His results confirmed that the eucalypts grown as exotics show no major wood differences when grown in differing environments, a pattern quite the opposite from many conifers. The same result was obtained for the effect of provenance on wood, as reported by Chudnoff (1961) for *Eucalyptus camaldulensis* in which the wood was similar between the local and Australian sources.

Although altitudinal provenances produce trees with differing growth characteristics, Doran (1974) found no difference in wood of *Eucalyptus regnans* grown from seed from different elevations; there were very large differences among open-pollinated families. Even though there were considerable differences in growth conditions and growth rate, the mechanical and physical properties of *Eucalyptus camadulensis* grown in Israel were almost identical with those of the Australian timber (Chudnoff 1961). In general, *wood qualities related to seed source and provenance are usually not of major consideration in the eucalypts*; the major variation is among individual trees within a stand.

Despite the reported uniformity, *Eucalyptus* sometimes produces wood different in the exotic environment than in the indigenous range of the species. For example, Ferreira et al. (1979) found that seed from the Coff's Harbour provenances of *Eucalyptus saligna* and *E. grandis* produced plantation trees with lower wood specific gravities than did those from the Rio Claro seed source when grown in Brazil. In Hawaii, Skolman (1972) found large differences between plots of *Eucalyptus robusta.* He considered the causal factors to be site differences as well as within- and between-tree variation. Another example of provenance having an important effect on eucalypt wood is provided by the New South Wales sources of *Eucalyptus nitens*, which had a lower specific gravity than the Victorian sources; the latter are preferred in Zimbabwe for utility hardwoods (Mullin et al. 1980).

The wood of the eucalypts, when grown in short rotations as an exotic, is not desired for solid wood products because of its propensity for collapse with shrinkage upon drying. Collapse results from irregular contraction of the wood, which has internal splits that are invisible externally; the sawn board has a somewhat wavy form. In their study of 24 provenances of *Eucalyptus camaldulensis* in two tests in Italy, Sesbou and Nepveu (1978) found that both shrinkage and collapse, as well as specific gravity, varied with provenance. They noted, for example, that provenances from the Murray River Basin had a low density and high shrinkage. The faster growing provenances had the highest shrinkage. In the same study, Sesbou (1981) reported that when *Eucalyptus camaldulensis* was grown in Sicily, shrinkage in volume varied from 22 to 60% by provenance and from 17 to 50% in Calabria, while density varied from 451 to 584 kg/m^3 in Sicily to 440 to 592 kg/m^3 in Calabria. Density varied much less in Morocco, namely from 465 to 537 kg/m^3.

In *Eucalyptus pilularis* grown in São Paulo, Brazil, differences in wood specific gravity varied among sources from 0.50 to 0.55 after 5 years (Pastor 1977). These provenance-related differences are large enough to be meaningful.

Sometimes there are recognized differences within a species that are not considered to be provenances. An example is the smooth and rough-barked types of *Eucalyptus urophylla* when grown in South Africa (Darrow and Roeder 1983). The smooth-barked types had higher basic density than the rough-barked ones, making them superior wood producers. However, the rough-barked types often grew more rapidly, such as in the Transvaal, which enabled them to be the best wood mass producers despite a lower density.

In summary, it appears that the changes in wood of *Eucalyptus* grown from differing provenances within a species are usually small, although there are some examples of provenance-related differences. Certainly sufficient differences have ben reported to justify a recommendation to test the wood of different sources before accepting the idea that provenance has no meaningful effect on the wood properties of *Eucalyptus* trees grown in exotic environments. When the same source is grown in different environments it must also be tested. This was found for *Eucalyptus deglupta* at Jari in Brazil, where a dark stain occurs in the interior of the tree up to 30% of its height, which will probably restrict its utilization and reduce yields.

Chapter 3

Variation Within and Among Trees

3.1 Introduction

Perhaps the subject most widely studied in forestry with respect to wood is the variability within and among trees. Both are large; this provides an advantage to the forester for developing better wood, especially since inheritance patterns of wood properties are usually strong, and silvicultural practices can influence wood. However, it is a distinct disadvantage to the user who encounters problems in developing efficient methods for the manufacture of variable woods. Often the user of wood must "overbuild" in the sense that the construction must be designed for the weaker, not the stronger wood members, so that much potential strength is "wasted".

In making pulp, the method of cooking the wood cannot be designed to handle efficiently all kinds of wood that are found in the furnish supplied to the digester, so some of the wood ends up being "overcooked" and some "undercooked", both of which lead to poor yields and often impaired quality. A major objective in forestry is to reduce wood variability by one of any number of methods, such as reducing the magnitude of the transition from earlywood to latewood (Megraw 1985) or by reducing tree to tree differences (Zobel et al. 1983).

Most trees have a pattern in wood qualities from the center of the tree outward, from the base to the top of the tree, within an annual ring, and sometimes even on different sides of the tree in relation to the sun and temperature (Dadswell 1960; Polge and Illy 1967; Sluder 1972). The within-tree patterns of variation have been intensively studied for many species. There is no norm; each species in each environment can develop its own pattern.

The kind of wood formed is influenced by the stage of development of the tree when the wood is laid down and the proximity to the foliage and growing buds (Larson 1969b). For convenience, Dadswell (1958) divided tree growth into the youthful or immature period, the adult or mature period, and the senescent or overmature period. Most emphasis in this chapter will pertain to the first two periods since they are the stages that will be found in a well-managed forest. Special emphasis will be placed on the immature period because wood produced during this stage of development is becoming much more prominent in world forestry as harvest ages are reduced and better forest management, including thinning, is practiced (Zobel 1984). As Bendtsen (1978) states, "differences between wood properties of trees from natural and "manmade" forests are associated with the short-rotation and resulting high proportion of juvenile wood..."

The question is immediately raised as to what can be done to modify the large amount of variability present in wood either among or within trees of a species. The within-tree variation is the result of a fixed developmental pattern which is difficult to change to make wood more uniform. However, the within-ring differences can be modified somewhat by controlling the growth. Even the grosser within-tree changes are subject to some modification, as outlined later in this chapter, but this is difficult, and changes are not large. One obvious way of obtaining more uniformity among trees for a given industrial operation is to restrict the wood source to a given species from a given environment, as was disussed in Chapter 2, but this is rarely either feasible or successful.

Another most important and effective way of decreasing variation in the wood supply is to use wood of approximately the same age, and much of this chapter deals with this aspect. The effects of age occur as the result of the proportion of juvenile to mature wood. Another effective method is to develop trees with more uniform wood through breeding and silvicultural manipulation, including fertilization. In its extreme form, as described by Zobel et al. (1983) for *Eucalyptus*, the wood produced from a given stand of trees can be essentially similar, at least from tree to tree if not within the tree, by use of methods of vegetative propagation.

As for all subjects related to wood, there is abundant literature involving within-tree and among-tree variability. There are many thousands of good references; only a few can be cited here. To help the reader, these will be summarized in tables. References relevant to between-tree and within-tree variation have been cited covering a broad time period, with many species in many parts of the world, to show the contrasting patterns that exist. From these, and from our tables and experiences discussed, the reader can draw conclusions about the trends of variation within and among trees.

Nearly every published study of wood makes some reference to within-tree or among-tree variability, although frequently nothing in the title of the paper would indicate that it contains information on wood variation patterns. Some of the very best information on variability in wood is contained in these apparent casual references, many of which we have cited.

Much of the tree to tree variability within stands of the same age on similar sites will be discussed in Chapter 8, which deals with the genetic control of wood. In this chapter, we will refer only to enough studies on tree to tree differences so that the magnitude and variation patterns are indicated and appreciated. Especially the within-tree variability will be covered in detail because it is generally not well understood. It is necessary to have knowledge of within-tree differences as a basis for understanding later chapters. The tree to tree differences are better known but often not appreciated, or are even ignored, resulting in some major errors in the use of wood.

3.2 Tree to Tree Variability

Generally, the tree to tree differences in wood properties within a species or within a provenance are large. For example, in *Pinus caribaea*, Lantican and Hughes (1973) found that all wood properties studied differed markedly from tree to tree. This is the usual pattern. Such variation occurs within both conifers and hardwoods and is of a magnitude that is oftern greater than that between species. For example, Harris (1961) stated that differences in specific gravity among radiata

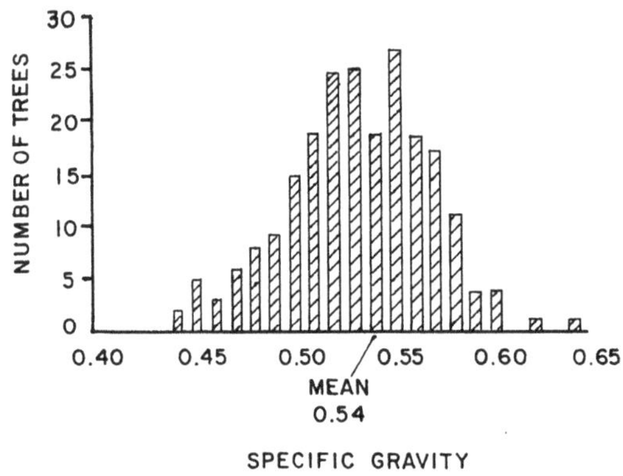

Fig. 3.1. Within a stand of trees of the same age growing on similar sites, there is a large tree to tree variation in wood-specific gravity. Shown are specific gravities for a number of loblolly pine trees of the same age growing on the same site. Normally such a stand will contain trees that vary in specific gravity as much as 0.20 (Zobel and Talbert 1984)

pine trees of the same age grown at any one site can amount to 60%. For *Eucalyptus camadulensis*, Chudnoff (1961) reported the usual finding that there is more variability among trees on the same site than the average differences between sites. To complicate this is the size of variation within a tree. As stated by Tajima (1967), "the variation within a tree is often greater than that between trees of the same species grown under the same conditions."

The variation pattern from tree to tree within a species is relatively constant from provenance to provenance. For example, the range of specific gravity in mature loblolly pine is always about 0.20 if 50 trees of the same age are randomly sampled on the same site. Thus, if the average is 0.55, one can be relatively certain that trees will be found within a stand with specific gravities as low as 0.45 and as high as 0.65. This is illustrated graphically in Fig. 3.1 for a large number of loblolly pines of similar age grown in the coastal plain of North Carolina.

Tree to tree variation is so large that it commonly masks other causes for wood variability (McKimmy 1959). It results from environmental influences and genetic differences; these are covered in Chapters 6, 7, and 8.

Table 3.1 illustrates the amount of tree to tree variation, including a number of species growing in different environments, in plantations, and in natural stands.

In summarizing Table 3.1, the following is evident:

1. Tree to tree variation is so large that it is of major importance for all wood products. It is usually necessary to sample at least 30 trees for each assessment in order to obtain a valid estimate of average wood properties for a group of trees.

2. The magnitude of the individual tree variation makes if difficult to assess accurately site, environmental, or silvicultural effects on wood properties.

3. The amount of tree to tree variation differs considerably among species. While no species has unitorm wood, some are much more uniform than others.

4. Differences between related species such as the southern pines are small compared to individual tree differences (Zobel et al. 1972).

5. Much of the tree to tree variation is genetically controlled.

6. Earlywood and latewood differ greatly and each can vary from tree to tree. The kind and amount of latewood is most important in influencing product quality, but differences in earlywood can also have a major influence.

Table 3.1. Examples of the magnitude of tree to tree variation in wood within stands of species growing in differing environments

Species	Reference	Comments
Conifers		
Pinus patula *P. radiata*	Scott and Stephens 1947	Large individual tree differences exist, from 27 to 35 lbs/ft^3 for whole stems of *P. patula* and 28 to 40 lbs for *P. radiata*. Difficulties in correlating wood properties with environmental factors result from the large tree to tree differences within stands
Pseudotsuga menziesii	McKimmy 1959	Variation of specific gravity among trees was so large as to offset the effects of factors such as site, crown class, and geographic location
Picea rubens	Holst 1960	There was as much as 0.155 variation in specific gravity within open-pollinated progeny from one tree; the total range among all trees was 0.163
Abies balsamea *Tsuga heterophylla*	Hawthorn 1961	Balsam fir has a greater variation in specific gravity among individual trees than hemlock
Pseudotsuga menziesii	Littleford 1961	Trees of the same size and age from the same site had large differences in bending strength (a range of 1200 p.s.i.). At high rates of growth one tree is more competent than another in producing high strength per unit weight
Picea sitchensis	Sunly and Lavers 1961	Large variations occur among trees at the same site
Pseudotsuga menziesii *Tsuga heterophylla*	Wellwood and Smith 1962	Almost all of the slight variation in cellulose content can be ascribed to tree to tree differences
Pinus radiata	Warren and Mirams 1963	Differences of up to 60% were found for wood density among trees in a plantation at the same site, varying from 21.9 to 35.9 lbs/ft^3
Pinus clausa	Strickland and Goddard 1965a	The range of specific gravities of individual sand pine was from 0.36 to 0.61
Pseudotsuga menziesii	Wahlgren 1965	The western softwoods show about the same range in specific gravity variation among trees as the southern pines
Pinus banksiana	Grigal and Sucoff 1966	The variation among trees within a plot was high; from 0.28 to 0.52 in specific gravity. Variation was small from plot to plot
Picea abies	Kennedy 1966	The range of variation within clones is dwarfed by the differences among clones
Pinus caribaea	Brown 1971	Highly significant differences were found among the trees within the Honduras provenance when grown in Jamaica, showing potential for selection of individuals for breeding high density offspring
Pinus rigida	Ledig et al. 1975	Differences among pitch pine trees within stands accounted for 50–60% of the variation in all wood properties; of this, 28–71% of the total variation of specific gravity and tracheid length was among trees within stands
Larix leptolepis	Isebrands and Hunt 1975	There were large differences among trees; several had high specific gravities
Pinus radiata	Cown and McConchie 1982a	Wood density among trees ranged from 348 to 469 kg/m^3
Pinus patula	Palmer et al. 1982b	The density of *P. patula* wood produced in Kenya varied from 325 to 430 kg/m^3
Picea glauca	Taylor et al. 1982	There were large specific gravity differences among trees within stands
Pinus caribaea	Cown et al. 1983	Tree to tree variability was appreciable and unrelated to tree diameter
Pinus caribaea	Plumptre 1983	Variability among trees in a given stand was great

Table 3.1 (continued)

Species	Reference	Comments
Pinus banksiana	Roddy 1983	The range of tree specific gravities was 0.360 to 0.467 (average 0.405) for 95 trees in northern Saskatchewan
Pinus taeda	Megraw 1985	Much of the very large tree to tree variation is genetically caused
Pinus kesiya	Ladrach 1986a	Variation of specific gravity at 4 years was 0.32 to 0.39, at 8 years, 0.35 to 0.54 and at 14 years, 0.34 to 0.57
Hardwoods		
Populus spp.	Göhre 1960	Variations among trees were so great that it was not possible to establish wood property correlations with site
Eucalyptus camaldulensis	Chudnoff 1961	There is more variability among trees at the same site than between sites
Quercus falcata	Hamilton 1961	There is a great amount of variation in wood properties among individual trees
Liriodendron tulipifera	Thorbjornsen 1961	There are highly significant differences in specific gravity among individual trees in an area as small as 0.2 acre; individual tree variation contributes about 70% of the total wood variability. The large tree to tree differences in fiber length account for 49% of the variation at 11 years and 90% at 31 years
Populus tremuloides	Valentine 1962	Individual trees varied in specific gravity from 0.293 to 0.471. Most variation in specific gravity is among individual trees within areas
Liquidambar styraciflua	Johnson and McElwee 1967	Contrary to expectations, there were only small tree to tree differences in specific gravity of juvenile wood. Tree to tree differences in fiber lengths were considerable
Liriodendron tulipifera	Kellison 1967	Among-tree differences in specific gravity were large, accounting for 49% of the variability compared to 21% for stand differences
Eucalyptus alba *E. saligna*	Ferreira 1968	In both species, the whole tree average basic density variation is very high; for *E. saligna* specific gravity varied from 0.433 to 0.634
Liriodendron tulipifera	Taylor 1968	Specific gravity values are significantly different among trees
Eucalyptus regnans *E. deglupta* *E. calophylla*	Rudman et al. 1969	Differences among trees are large, sometimes more than 30% in specific gravity
Platanus occidentalis	Taylor 1969a	Wood properties are quite different from tree to tree. This includes specific gravity, cell diameter, fiber length, and proportion of ray tissue
Carya pecan	Taylor 1969b	There are differences in specific gravity among trees; fiber length differences among trees are also significant. This is also true for the proportions of vessels, fibers, and parenchyma tissue. High specific gravity trees are high at every sampling point
Liriodendron tulipifera	Sluder 1970	Differences among trees within plots accounted for 52% of the total specific gravity variation
Celtis occidentalis	Taylor 1971	Specific gravity varies greatly from tree to tree, as does the proportion of vessels, fibers, and ray parenchyma
Eucalyptus grandis	Hans et al. 1972	There was a marked variation among trees for both wood density and fiber length
Eucalyptus robusta	Skolman 1972	Specific gravity from tree to tree varied from 0.331 to 0.869
Populus spp.	Son and Chung 1972	Variation in specific gravity among individuals was high

Table 3.1 (continued)

Species	Reference	Comments
Eucalyptus grandis	Taylor 1973b	Specific gravity varies significantly from tree to tree. Trees with a high specific gravity have high specific gravity at each sampling point
Eucalyptus regnans	Doran 1974	Variation in wood density from tree to tree was very large
Eucalyptus saligna *E. grandis*	Ferreira et al. 1979	Specific gravity variation among trees is high and is more important than variations among plantations
Platanus occidentalis	Land and Lee 1981	Individual tree specific gravity values ranged from 0.39 to 0.54; 80% of the variation was among trees within stands
Eucalyptus spp. *Acacia mearnsii*	Palmer et al. 1982b	The eucalypts varied in density from 470 to 570 kg/m^3. *A. mearnsii* had a density of 590 kg/m^3
Platanus occidentalis	Land et al. 1983	Genetic variation among trees within stands was highly significant for both specific gravity and moisture content
Populus tremuloides	Yanchuk et al. 1983a	For only 12 stems, wood density varied from 0.402 – 0.438 g/cc

7. Use of vegetative propagation can produce trees with similar wood within clones, while clonal differences can be very large.

8. The large amount of variability among trees and strong genetic control make breeding for improvement of wood properties very successful.

9. Tree to tree variability in juvenile wood is usually less than in mature wood.

In summary, it is sufficient to say that tree to tree variation is so large for all wood properties and species that it makes studies of wood difficult and utilization inefficient. Foresters must recognize the variation, learn to live with it, and reduce it whenever possible, either genetically or by use of methods such as fertilization or vegetative propagation. If tree to tree variation is not recognized and accounted for huge errors will be made in wood studies or in predicting strength and quality of products in wood utilization.

3.3 Variation of Wood Properties Within Trees

There are several patterns of variability within trees that are of importance. The first is the within-ring differences, the second the changes from the center of the tree to the outside, and the third the differences associated with different heights in the tree. Within-tree variability can be very large. As stated by Larson (1967b), "more variability in wood characteristics exists within a single tree than among trees growing on the same site or between trees growing on different sites. These patterns must be understood; they are always present and are hard to eliminate." Numerous publications describe the various within-tree variation patterns; the recent work of Megraw (1985) for loblolly pine and that of Koch (1985) for hardwoods clearly illustrate some of the many within-tree patterns.

3.3.1 Wood Differences Within Annual Rings

Wood variation within an annual ring is generally known as earlywood or latewood. According to Megraw (1985), "the greatest variability in specific gravity occurs within each annual ring." For example, Yazawa and Fukazawa (1959) stress that the specific gravities of earlywood and latewood are very different. They report a ratio of 2.8 to 1.0 in Sugi (*Cryptomeria japonica*) but this ratio varies greatly by species. Not only does specific gravity vary but, according to Andrews (1986), greater differences in chemical composition were found between earlywood and latewood within an annual ring than between sapwood and heartwood of the same Douglas-fir tree. It can be the greatest source of variability in wood. Earlywood, which is laid down early in the year and latewood, which is formed later in the year, can be very different (Fig. 3.2). Earlywood is generally formed during the period of active leader growth; during and after the bud resting stage, latewood is formed (Kennedy 1971). As described by Larson (1973), an intermediate, or transition, wood is formed in some species which can be categorized as neither earlywood nor latewood. It occurs between the other two types of wood and is responsible for much of the variation in wood properties in juvenile wood.

The specific gravity can vary for both the earlywood and latewood themselves, and this variation contributes greatly to the overall tree specific gravity, independent of the proportional amounts of earlywood and latewood (Pillow 1952). For example, the authors have observed in *Pinus taeda* that trees or rings with dark latewood (i.e., thick walls and small lumens) have a higher density than those with

Fig. 3.2. Differences in cells between earlywood and latewood are shown in the photomicrograph of a *Pinus taeda* grown in Brazil. Note the near lack of cell lumens in the latewood (*below*) and the large, thin-walled cells in the earlywood (*above*) (Courtesy of Companhia Florestal, Monte Dourado, Brazil)

lighter brown latewood. A visual estimate of the specific gravity of a tree can be quite reliably obtained from the color and relative proportion of the latewood. The contribution of earlywood and latewood specific gravities to overall wood specific gravities was worked out mathematically by Warren (1979). The major contributor (34%) is the proportion of latewood weighted by the square of the difference between latewood and earlywood specific gravity. But even this is not simple because, as Megraw (1985) expresses it, "latewood specific gravity (for loblolly pine) increases rapidly with ring number from pith. . .until, by five rings or so from the pith, values are almost up to their characteristically high level". After that the proportion of earlywood to latewood more clearly determines overall ring specific gravity.

One of many estimates of the specific gravity of earlywood and latewood is that of Tajima (1967) who states, "the density of the latewood is 2 to 2.5 times heavier than that of the earlywood." Douglas-fir has earlywood and a specific gravity of about 0.20 and latewood about 0.62, with an abrupt transition between the two (Wilson and Ifju 1965). In a comparison of Douglas-fir and southern pines, Lutz (1964) showed the specific gravity of earlywood and latewood pine to be 0.29 and 0.65 and Douglas-fir 0.28 and 0.69, respectively. He points out that the very low specific gravity of the earlywood of these "dense" woods is less than that of *Tilia* (basswood) (0.32), which is one of the lightest, commonly used woods; and the latewood is denser than *Cornus* (dogwood) (0.64), one of the densest North American woods. Ponderosa pine ranged from 0.20 in earlywood to over 0.85 in latewood, but the proportion of earlywood was relatively large (Echols 1971).

The effect on product quality is considerable; Biblis (1969) found that specific stress and specific stiffness of latewood are more than 50 and 60% higher than corresponding values for earlywood. Gladstone et al. (1970) reported that latewood of *Pinus taeda* had higher holocellulose, alpha-cellulose, and glucan contents than earlywood, and latewood gave 2–7% higher pulp yields from a given weight of dry wood than earlywood. According to Brazier (1983), especially in the conifers, the soft and hard earlywood and latewood zones give the wood a coarse texture. This leads to deflection of nails, splitting, and distortion on drying. As often emphasized, soft earlywood and hard latewood make wood finishing difficult.

In the soft pines, the differences between earlywood and latewood are much less, and there is a long, gradual transition from one to the other that makes it difficult to differentiate accurately between the two zones. The within-ring variation of coniferous genera other than *Pinus, Larix,* and *Pseudotsuga* is frequently similar to that of the soft pines so that the two zones are more similar than in hard pines and Douglas-fir, but they still differ greatly. The specific gravities of earlywood and latewood in *Thuja plicata* are 0.12 to 0.6, in *Abies amabilis* 0.25 to 0.6, in *Picea sitchensis* 0.2 to 0.5, and in *Tsuga heterophylla* 0.2 to 0.5 (Wilson 1964). In *Cryptomeria japonica,* a relatively light wood, the ratio of latewood to earlywood specific gravity was 2.8 to 1. The controlling factor for wood properties among trees was the varying latewood percentage which greatly affects wood quality (Yazawa and Fukazawa 1959).

It has been found through experience that generally the properties of the latewood have the major influence on the quality of finished wood products. The

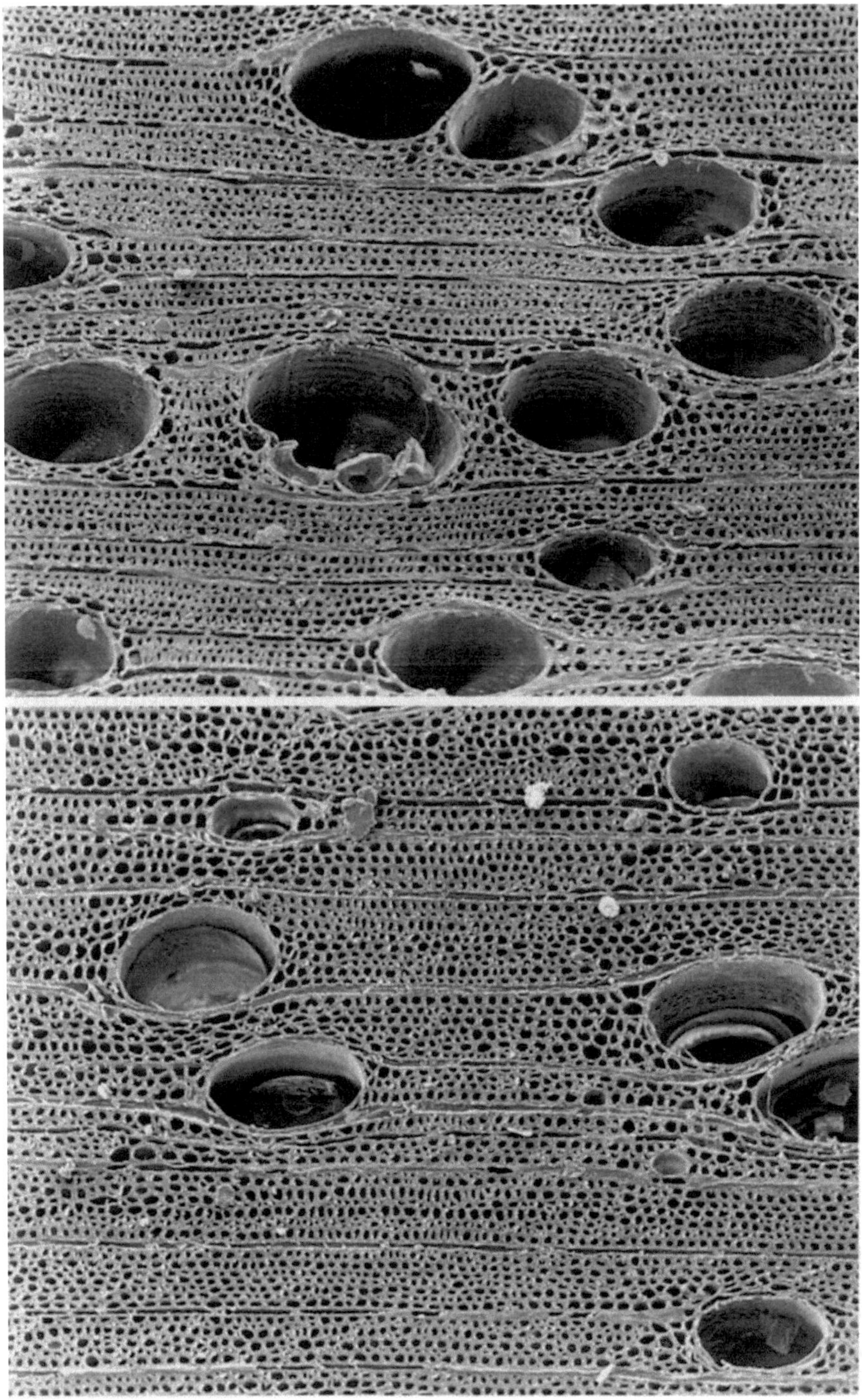

Fig. 3.3. The pattern and size of the vessels within an annual ring have a major effect on the properties and usefulness of hardwoods. Shown is the vessel distribution and size in *Eucalyptus deglupta* (*above*) and *E. urophylla* (*below*), both grown in the Amazon at Jari in Brazil (Courtesy of Companhia Florestal, Monte Dourado, Brazil)

earlywood usually varies little from ring to ring; for example, Tajima (1967) found for *Pinus densiflora* that the density of earlywood was nearly the same from pith to bark. The amount of earlywood can be important; as Paul and Smith (1950) and Wilson (1964) state for the southern pines "...variation of 10% in earlywood content within a species...gave greater differences in pulp quality than were exhibited among species."

The hardwoods can also have great differences in cell type and structure within an annual ring. Especially the ring-porous hardwoods have a section of large-sized vessels formed in the early part of the growing season, followed by many smaller, thick-walled fibers (see Fig. 1.2). In contrast, the diffuse-porous hardwoods have much less within-ring variability and the distribution of cell types and the characteristics of the types tend to be quite uniform throughout the ring. This, however, is not always true as shown by Dodd (1985) for *Acer pseudoplatanus*, where, although vessel size was nearly constant, it declined sharply toward the end of the annual ring.

Distribution of cell types within the ring has a great effect upon wood properties, especially in the tropical hardwoods (Fig. 3.3). No generalized statement can be made for within-ring patterns of variation for hardwoods as a group except that they are complex.

The within-ring variation is determined by the growth pattern within the plant and the forester can only attempt to modify it somewhat. For example, one method of modifying the within-ring variability (Polge 1969b; Isebrands and Hunt 1975) is through fertilization. The authors cited found the modifications to be quite large for spruce and pine with the earlywood of fertilized trees having thicker walls and the latewood thinner walls, resulting in more uniform properties within a tree, the within-ring variation having been reduced (see Chap. 7).

3.3.2 Wood Variability at Different Locations Within a Tree

Wood varies from tree to tree but also within a tree from the center outward and from the base of the tree to its top. The magnitude of within-tree differences can be very large and are important. For example, in his comments on the quality of the wood of tropical pines grown in South Africa, de Villiers (1974) states that "problems experienced with the utilization of the timbers of these species...are due, in the first place to the presence of a radial density gradient in the stem."

Nearly all discussions relevant to within-tree variation center on the patterns found in the hard pines and species such as *Pseudotsuga menziesii* where the patterns can be very strong and thus very important for all aspects of wood studies or wood utilization. However, this emphasis on the hard pines masks the importance of within-tree patterns in other genera. It is generally accepted that the ring-porous and diffuse-porous hardwoods have different patterns, and that these can differ considerably from the conifers. It is not always so clearly understood that genera such as *Abies, Picea, Cupressus*, and several others exhibit patterns different from those of the pines. As just one of numerous examples (some of which are listed in the following Tables) Ladrach (1986c) reported the usual pattern in Colombia for *Pinus patula, P. oocarpa*, and *P. kesiya*. He then states, "*Cupressus lusitanica...* has a different physiology of wood formation than the pines. Wood density does not vary within the tree but tracheid length does." He found longer tracheids near the bark and near the top of the tree, which also occurs in the pines.

Because the within-tree patterns partly depend upon the characteristics of the juvenile wood and its location, juvenile wood will be discussed in some detail in the following sections.

3.4 The Juvenile Wood Concept

Perhaps the greatest overall cause of wood variation among conifers is the presence of juvenile wood and its relative proportions to mature wood. Sometimes juvenile wood is referred to as corewood and the mature wood as outerwood (Fig. 3.4). Although juvenile wood occurs in both conifers and hardwoods, it is usually much less evident or important in the hardwoods. It is a great cause of concern, and numerous articles have been written about it, such as that by Zobel (1984) which dealt with juvenile wood of tropical pine plantations or that by Senft et al. (1985), which emphasizes the poor quality of the wood produced. Senft's article leaves the impression that fast growth per se causes formation of weak wood in fast-growing plantations, while in fact it is the high proportion of juvenile

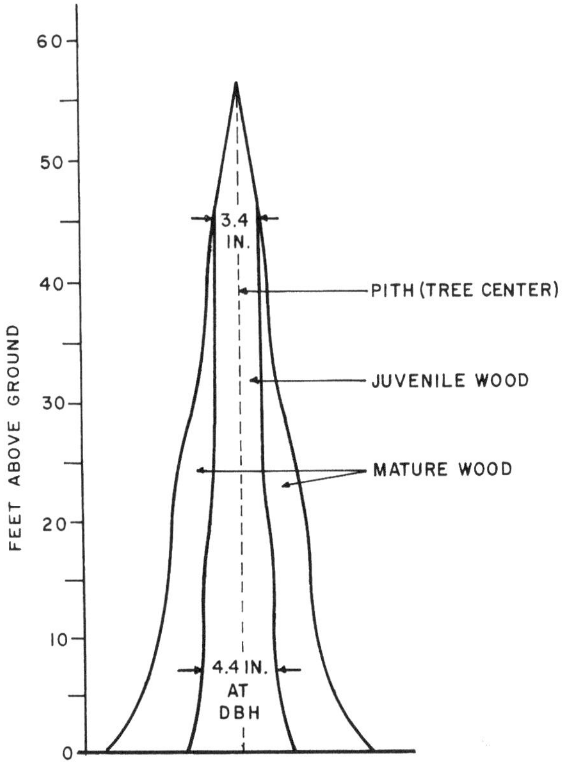

Fig. 3.4. Juvenile wood occurs near the tree center, or pith, and roughly forms a cylinder up the tree. Shown is a sketch of the juvenile zone of a number of 17-year-old southern pines, indicating the high proportion of juvenile wood toward the top of the tree and a lesser amount in the basal part (After Zobel and Talbert 1984)

wood in young trees that are merchantable early that results in the problems in product quality. Various aspects of the patterns of wood variation within the tree have been well discussed by Megraw (1985). Bamber and Burley (1983) put the problem well in perspective when they state, "what was not generally appreciated was that these factors (fast growth and low wood density) are not necessarily causally related, and that if wood of the same age within a species is compared, density differences largely disappear. The reason why wide rings appear to be related to low density is because wide rings generally occur in the corewood and it is the corewood, or juvenile wood, which is low in density."

Juvenile wood has been extensively studied and there is abundant literature available concerning the cause of juvenile wood (Paul 1960; Zobel and Kellison 1972; Zobel and Talbert 1984). It is produced near the center of the tree and is related to the number of rings from the pith. It is controlled by auxin production in the tree crown (Larson 1962) and results from close proximity to the foliage (Larson 1969b). Some refer to it as crown wood. The most accepted concept is that it is directly related to the age of the cambium. As stated by Pearson et al. (1980), "this reinforces the point...that it is age (growth ring from the pith) which is crucial rather than rate of growth."

There are numerous hypotheses about the cause of juvenile wood. In addition to those above, Richardson (1959) feels that changes in net assimilation rate with tree age account for the differences in wood density between juvenile and mature woods. Plumptre (1983) summarizes for *Pinus caribaea*, "the pattern of juvenile wood formation and transition to mature wood varies with the genetic make-up of the tree, the site, the climate and the silviculture practiced". This was recorded for Japanese larch (*Larix leptolepis*) by Tsutsumi et al. (1983) who state that site class plays a significant role in determining the width of juvenile wood as does stand density. They reported that juvenile wood made up from 20% of the volume of the butt log of a tree on a low quality site to 40% on a high quality site.

The age of the tree is usually not considered to be of importance in determining the properties of juvenile wood, since the characteristics of juvenile and mature woods are generally related to the number of rings separating the cambium from the tree center, or closeness of the crown, regardless at what height in the tree. It is difficult for many practitioners to separate absolute tree age from the juvenile wood concept; many retain the impression that young trees produce juvenile wood (a correct assumption) but conversely feel that older trees produce only mature wood (an incorrect assumption) (Fig. 3.5). The effect of the decreased proportion of juvenile wood in a tree, with tree age, is described at the end of this chapter.

3.4.1 Occurrence of Juvenile Wood

In one sense, use of a term such as juvenile wood is misleading. Wood characteristics within the juvenile zone are not uniform but rapidly changing throughout the zone from the pith outward (Fig. 3.5). Following this rapid change, there is a leveling off, creating a transition zone leading into the mature wood where changes in wood properties are much less rapid. As pointed out by Zobel (1980b), "there

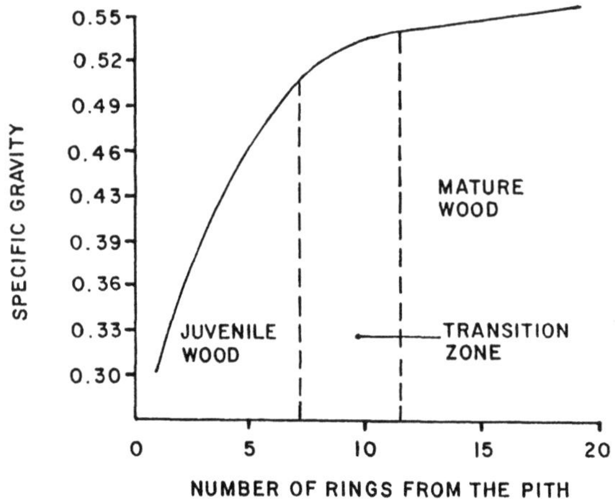

Fig. 3.5. A schematic drawing depicting the change in specific gravity from the center of the tree and outward in *Pinus taeda*. The area of rapid change is juvenile wood, while the area with minor change is mature wood. The undefined zone in between is often referred to as the transition zone

is no absolute shift from juvenile to mature wood in one year and the change from juvenile wood to 'typical' mature wood usually occurs over several years." Nearly all wood properties, both physical and chemical, are very variable within the juvenile zone and much more constant within the mature zone.

As a tree ages, the cambial cells further from the center of the stem produce more mature cells. Although he did not refer to it as juvenile wood per se, Dadswell described the situation well in his 1958 publication on the wood structure variations occurring during tree growth. Dadswell emphasized that most wood anatomists prefer to work with mature wood in preference to that from the central region of the stem. He mentions that the cell changes occur in a distinct pattern; referring to cell length, Dadswell states that it increased from the pith outward for a number of years (juvenile wood) and then reaches a more or less constant value (mature wood).

The central zone of juvenile wood occurs in both fast- and slow-grown trees. In conifers the magnitude of the change in cell length from juvenile to mature wood may be three or four times, but in hardwoods it is much less. In a generalized statement regarding hardwoods in the temperate zone, Knigge and Koltzenburg (1965) state that "...the sizes of all types of cells formed in the youth of the tree, i.e. near the pith, are smaller than those in later years. This clearly regular phenomenon...led to the use of the term 'juvenile wood'..." As cell length increases, filbril angle in the call wall decreases. Also, near the center of the tree, high growth stresses occur. These result in compression failures which can be very numerous and make the juvenile wood weak. Such failures are more common in hardwoods than in pines.

The period of juvenile wood production varies among wood characteristics within the same tree. For example, Loo et al. (1985) report that there was no correlation between the age of transition from juvenile to mature wood for specific gravity and tracheid length. Juvenile wood formation with respect to specific gravity occurs for 5–6 years, as in *Pinus elliottii*, *P. caribaea*, and *P. radiata* to about 10 years for *P. taeda* and 20 years or more for *P. ponderosa*. Dadswell (1958) found that the number of years taken to reach a constant cell length varies from species to species and is somewhat related to the life span of the tree. He states that cell length increases the first 12–15 years in *Pinus radiata,* and in *Eucalyptus gigantea* cell length may increase for 20 years. Dadswell generalizes that the cell length of all species usually increases for about 20 growth rings at all heights and at different growth rates. Conversely, he notes that the change in wood density occurs much earlier and varies greatly among species. In loblolly pine, Loo et al. (1985) report genetic variation in the time of transition from juvenile to mature wood. They found that frequently fast growth was negatively related to early age of transition. The same trend for a slight extension of the juvenile period related to faster growth has been frequently noted by the authors. Thus, silvicultural practices such as good site preparation, competition control, fertilization, and wider spacing all tend to extend the period in which the tree produces juvenile wood.

Regardless of the tree age, a juvenile zone occurs at the top of the tree as well as at its base which is formed within a given number of rings from the tree center or pith. Thus, the juvenile zone somewhat resembles a cylinder at the center of the tree (Fig. 3.4); the top logs consist primarily of juvenile wood, while butt logs from the same tree contain a preponderance of mature wood.

Table 3.2. Wood specific gravity above a 4-in. merchantable top for 20-year-old loblolly and slash pines, compared to wood from the merchantable bole of the same tree[a]

Species	Specific gravity	Density lbs/ft^3
Loblolly pine		
Tree tops	0.36	22.5
Tree boles	0.42	26.2
Slash pine		
Tree tops	0.40	25.0
Tree boles	0.46	28.7

[a] After Zobel (1974). Values shown are a general average and vary by location of the plantation.

Table 3.3. Percent moisture content of juvenile and mature woods of a number of 30-year-old loblolly pines by 5-ft bolts from base to top[a]

5-ft Bolt (No.)	Juvenile wood Moisture content (%)	Mature wood Moisture content (%)
Basal	110	74
Second	122	85
Third	133	97
Fourth	137	102
Fifth	139	110
Sixth	145	117
Seventh	151	
Eight	153	
Ninth	163	

[a] After Zobel (1984). Percent moisture is based on the dry weight of wood.

Fig. 3.6. In the past, most wood used by the forest industry was obtained from large, mature trees, such as from the logs indicated. Here, juvenile wood is of minor concern, making up less than 10% of the volume of the log. When logs from young trees are used, a high proportion of the wood will be juvenile wood

In his discussion on loblolly pine, Megraw (1985) states that the concept of juvenile wood consisting of a cylindrical core from the bottom to the top of the tree is a useful concept, but that it is not strictly biologically correct because juvenile wood changes somewhat up the tree. For example, tracheid lengths of juvenile wood are greater at the top compared to the base of the tree, an observation made earlier by Zobel (1975a). The magnitude of the difference for specific gravity and moisture content with height in loblolly pine is indicated in Tables 3.2 and 3.3 Differences in specific gravity and moisture content of the basal bolts and the top of the tree are largely due to the relative amounts of juvenile wood. For the southern pines, Zobel (1975a) reported that a 10-year-old tree is essentially all juvenile wood as is the top of a 40-year-old tree, while the butt section of the older tree contains over 80% mature wood (Fig. 3.6).

Although juvenile wood is formed at the top of the trees of all ages, the proportion of juvenile wood in the entire tree, both by volume and by weight, decreases considerably with increasing age of the tree. We find that as much as 85% of the merchantable volume of 15-year-old loblolly pines can be juvenile wood, while in 30-year-old trees it can make up as much as 30% of the volume (see Table 3.15). However, the values are usually considerably less than that (Table 3.18). It is obvious that no generalized figures can be quoted, because the amount of juvenile wood depends upon the growth rate, the site, and the kind of stand (natural or plantation). Generally, plantations will have a higher percentage of juvenile wood, but this is not always true.

The important point is that when young trees or topwood are harvested, the proportion of juvenile wood and accompanying compression wood is much greater than from the lower bole of mature trees. When a large percent of the furnish available to a mill consists of juvenile wood, a marked effect on yield and quality of paper is experienced. The effect of overall tree age on specific gravity and percentage of juvenile wood is covered in Section 3.7.

There have been suggestions, and a few attempts, to alter the percentage of juvenile wood. There has not been much success but it has been possible to develop juvenile wood with quite different properties through selection and breeding and to some extent, through silvicultural manipulation (Zobel 1970).

3.4.2 Juvenile Wood and Growth Rate

One common misconception is that the characteristics of juvenile wood are the result of a high growth rate (wide annual rings) near the tree center. The subject of the effect of growth rate on wood properties is discussed in Chapter 5. It is, however, essential to understand here that it is not fast growth (wide rings, that causes juvenile wood (Pearson et al. 1980). As emphasized by McMillin (1969), juvenile wood from a slow-grown stem has the same low density as that from a fast-grown one. Lack of appreciation of this fact has led to grave errors, many foresters believing that if one grows trees rapidly, the specific gravity will be low. Although this is the situation in some conifers, in most hard pines and diffuse-porous hardwoods there is little correlation between ring width and specific gravity.

The misconception arose when the wide-ringed, low specific gravity wood at the center of the tree (which is juvenile wood) was compared with the narrow-ringed, high specific gravity mature wood nearer the bark. The incorrect conclusion was then made that the wide rings produced low and the narrow rings high specific gravity wood and thus, fast growth resulted in low specific gravity. As emphasized in Chapter 5, comparisons of the effect of growth rate on wood properties can only be made on rings of the same age.

As pointed out by Schultze-Dewitz (1959), it is essential to differentiate between wide-ringed juvenile wood with short tracheids and narrow-ringed mature wood with long tracheids in the same stem. This point was forcefully made by Hiley (1955) who stressed the misconception about wood properties that results when broad rings near the center of the tree are compared with narrow rings further out; this is especially serious when the difference in wood is ascribed to growth rate rather than to juvenile wood. Wood in the juvenile core is almost always different, and sometimes dramatically different, from mature wood, irrespective of the growth rate.

3.4.3 Importance of Juvenile Wood

The importance of having information about the extent, kind, and quality of juvenile wood cannot be overemphasized. As mentioned by Zobel (1984) and Zobel

Fig. 3.7. When growing pines in the tropics, thinning is almost always an essential silvicultural practice. Since most thinnings are done at a young age, most of the wood obtained is juvenile wood. A surplus of juvenile conifer wood is creating major problems in the practice of thinning. Thinnings now being done, such as on the 9-year-old slash pine plantation shown in southern Brazil, will essentially be all juvenile wood

and Kellison (1984), one of the major problems of forestry on a worldwide basis is the excess inventory of juvenile wood of pines. As early as 1964, Polge pointed out the importance of solving the problems related to juvenile wood in conifers. This is especially true in the tropics and subtropics, where the trees grow to merchantable size so rapidly that all thinnings and much of the wood obtained at final harvest are juvenile (Zobel 1984; Fig. 3.7).

The prevalance and importance of juvenile wood was emphasized by Smith (1962) whose opinion is that one cannot produce a large volume of high density wood at an early age; he states that an attempt to do so is quite unrealistic. Improvements in forest management and genetics in both the temperate and the tropical zones result in the production of trees of a harvestable size at younger ages (Zobel et al. 1978). On this point, Bendtsen (1978) states: "...from a utilization standpoint, the principal difference between wood from the man-made and the earlier generation forests is that accelerated growth leads to earlier harvest and a greater proportion of juvenile wood."

Some organizations produce pines of merchantable size after 8 years, and the resulting products are inferior (see Fig. 3.7). Even 12-year-old pine has proven to be marginal for some products (Kirk et al. 1972), although the wood from older trees, even from the tropics and subtropics, is quite satisfactory. The proportion

Table 3.4. Juvenile and mature wood specific gravities from seven species of southern pines

Species	Specific gravity		
	Number of trees	Juvenile wood	Mature wood
Pinus taeda	1298	0.45	0.54
P. virginiana	243	0.46	0.52
P. palustris	176	0.52	0.57
P. serotina	82	0.47	0.54
P. elliottii	199	0.46	0.55
P. echinata	118	0.46	0.53
P. strobus	75	0.31	0.38

of young trees, tops, and peeler cores, will continue to increase until juvenile wood becomes a major source of timber (Zobel 1981). In Australia, Cown and McConchie (1982a) state that as far as wood density is concerned, the proportion of corewood to outerwood is likely to be by far the most important factor governing the value of radiata pine stems. Expressing the concern about excess juvenile wood in plantations, Hillis (1980) stated that the wood quality must be raised to meet the end user's requirements.

It is important to recognize that juvenile wood is a different wood, not necessarily a poor wood (Zobel 1984). It is excellent, for example, for thermomechanical pulp (TMP) or chemical-thermomechanical pulp (CTMP). It is useful for newsprint, some tissues, high quality printing and writing papers, among others. The amount of juvenile wood becoming available makes it mandatory to learn to utilize it efficiently. Usually this type of wood will affect the pulp quality from a mill when it makes up more than 20% of the furnish. There is more difference between the juvenile and mature woods within a pine tree, for example, than there is between the mature wood of many pine species.

Species differences in specific gravity for juvenile and mature woods can be considerable; these are shown for seven southern pines in Table 3.4. Note the different magnitude of juvenile-mature values for species such as *Pinus taeda* compared to *P. virginiana.*

The importance of juvenile wood to pulp and paper properties has been determined in a mill-scale study by Semke and Corbi (1974); some of their results are shown in Table 3.5. The sawmill chips were all mature wood, the roundwood is the whole chipped tree which contained some juvenile, but mostly mature wood. Topwood was obtained when the butt log, which was mostly mature wood, was removed for sawtimber; therefore, the portion of the log left contained a substantial portion of juvenile wood. The 15-year-old trees have a large proportion of juvenile wood and 10-year-old trees are essentially all juvenile. Twelve-year-old pine trees yielded 10% less pulp than 30-year-old trees (Kirk et al. 1972).

Many studies have been made, such as those by Zobel (1979a) and Zobel and Kellison (1984), comparing the wood of pines produced in the tropical and subtropical regions with that of conifers in the more temperate zones. A common comment relates to the poorer quality and yield from the tropical and subtropical areas. Because the trees grow rapidly, they become of merchantable size when

Table 3.5. Wood and paper properties from different sources of loblolly pine wood[a]

Source of fiber	Average cell dimensions			Average sheet properties		
	Length (mm)	Width (μ)	Wall thick-ness (μ)	Sheet density (g/ml)	Tear g·cm/5 cm	Tensile strength (lbs/in)
Sawmill chips	2.80	46.7	5.3	0.536	58.2	21.7
Conventional roundwood	2.48	44.1	4.8	0.562	57.2	24.9
Topwood	2.43	42.1	4.4	0.572	48.7	26.9
15-year-old trees	1.86	42.1	4.0	0.590	45.3	26.5
10-year-old trees	1.62	40.9	3.4	0.613	46.5	26.5

[a] After Semke and Corbi (1974). The fiber sources are shown in order of increasing amount of juvenile wood present.

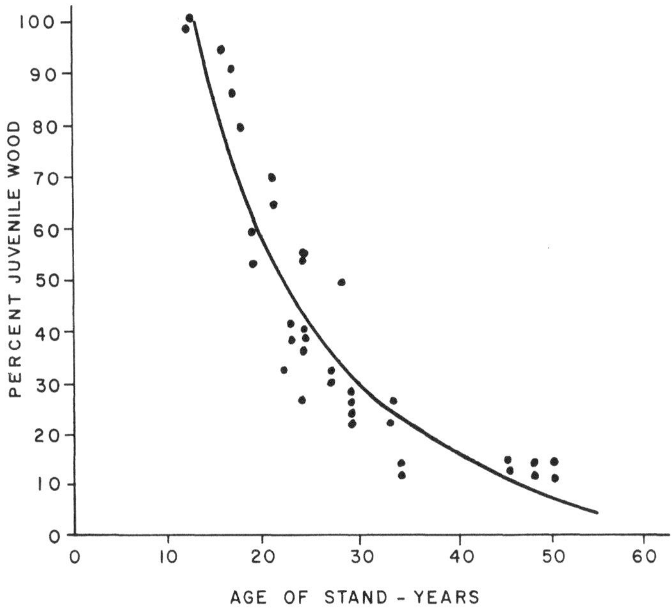

Fig. 3.8. When trees grow rapidly so that they are of harvestable size when young, the percentage of juvenile wood increases dramatically, as shown in the graph. Young thinnings are all juvenile wood

young and have a high proportion of juvenile wood; this is illustrated in Fig. 3.8. The harvested wood gives low yields of pulp and a low tear strength of the paper. Solid wood products are weak and have difficulties inherent with wide rings.

All trees show a juvenile phase in cell dimensions and chemical constituents as well as specific gravity; this is illustrated for tracheid length in two species in Table 3.6. However, the specific gravity of juvenile wood has such a major effect on yields, strength of paper, and utility of solid wood products (Barefoot et al. 1970; Pearson and Gilmore 1980) that it is emphasized over other cell characteristics.

Table 3.6. A comparison of the tracheid lengths of juvenile and mature woods for two pine species

	Tracheid length (mm)	
	Juvenile wood	Mature wood
Loblolly pine	2.56	3.41
Slash pine	2.60	3.50

3.4.4 The Properties of Juvenile Wood

Juvenile wood has been described by many writers. It is categorized as having a "lifeless" appearence due to low light reflectivity and a flat fibril orientation, with a "cheesy" consistency; it contains a very low proportion of latewood (Zobel 1984). Juvenile wood of loblolly pine averages 2.5 to 6.2 lbs/ft^3 less than mature wood (Zobel et al. 1972). The cell walls of juvenile wood of hardwoods tend to be thin, and according to Knigge and Koltzenburg (1965), all types of cells formed near the pith are smaller than those produced in later years.

As described earlier, there is no sharp line of demarcation between juvenile and mature woods, and one gradually grades into the other (Figs. 3.4 and 3.5). The top of an old tree consists essentially of juvenile wood, with characteristics generally similar to juvenile wood formed in earlier years near the base of the same tree, except that sometimes the cells from the top are longer. All the wood formed above a 4-in. (10.2 cm) merchantable top is juvenile wood.

Reaction wood (compression wood in the conifers and tension wood in the hardwoods) is strongly associated with juvenile wood. It is sometimes difficult to separate reaction and juvenile wood, since their properties can be quite similar. When dealing with juvenile wood, it is essential to recognize that reaction wood plays a considerable part in determining its yield and properties. Somewhat related to this is brittleheart or brashness of juvenile wood. According to Plumptre (1983), the juvenile core of *Pinus caribaea* often becomes brash, caused by slip planes in the tracheids and the stresses caused by a layer of dense wood outside a less dense layer.

The properties of juvenile and mature woods are greatly affected by wood extractives. Extractive contents are shown for four species in Table 3.7, where the specific gravity of juvenile wood is more affected by extraction than the mature wood. The trees for which data are given in the Table are only 22 years of age, thus heartwood formation had just started.

Juvenile wood in the hard pines is of greatest importance and has many unique characteristics. These have been fully described and illustrated for the hard pines by Zobel (1980b), who also covered some of the effects of juvenile wood on the final product. Some of the characteristics of juvenile wood are summarized below:

1. Low specific gravity, with thin-walled cells resulting in weak solid wood products and paper with low tear but a high bursting strength (Barefoot et al. 1970; Foelkel et al. 1976; Isebrands et al. 1982). In the juvenile wood region true latewood does not occur (Larson 1960). Dark bands of cells may be produced an-

Table 3.7. Specific gravity and extractives content of juvenile and mature woods of four pine species (trees were 22-years-old, all grown in the same environment)[a]

Pine species	Specific gravity						Percent extractives[b]		
	Juvenile		Mature		Total tree		Juvenile	Mature	Weighted
	Unext.[c]	Ext.[c]	Unext.	Ext.	Unext.	Ext.			
Loblolly	0.43	0.41	0.55	0.54	0.54	0.53	3.1	2.7	2.8
Longleaf	0.50	0.44	0.54	0.53	0.54	0.52	13.6	2.4	3.5
Slash	0.46	0.44	0.54	0.53	0.53	0.52	5.0	1.8	2.1
Pond	0.42	0.40	0.48	0.47	0.48	0.46	6.7	2.6	2.9

[a] Adapted from Cole et al. 1965.
[b] Based on dry weight of wood.
[c] Unex. = unextracted, Ext. = extracted with alcohol-benzene.

nually in the juvenile zone that are referred to as latewood, yet these do not meet Mork's (1928) definition of latewood (twice the cell wall thickness is greater than cell lumen width). The lowest density juvenile wood may be only two-thirds the density of mature wood of *Pinus caribaea* (Plumptre 1983).

2. A high percentage of compression wood in juvenile wood results in short tracheids that break easily during manufacture and are difficult to bleach because of their high lignin content. Plumptre (1983) states that juvenile wood "is a serious defect. It has spiral grain, distorts on drying and has more compression wood."

3. Since juvenile wood has shorter and thinner walled tracheids and fibers, they consequently have larger microfibril angles (Wheeler 1987). Longitudinal shrinkage is negligible when fibril angles are less than 25°–30°, but increases dramatically with increasing angles. A large or flat fibril angle, resulting in a high longitudinal shrinkage, makes solid wood products so unstable that there has been discrimination against southern pine lumber (Pearson and Gilmore 1980) as well as against wood of some tropical pines. The flat fibril angle results in up to 9% more longitudinal shrinkage than is found for mature wood (Meylan 1968). Therefore, if a board is cut such that a part of it contains juvenile wood with its larger fibril angle and the other part mature wood with a smaller fibril angle, one portion shrinks more than the other resulting in warp, check, cracks, and other defects (McMillin 1973). The common method of gang sawing or sawing through the tree center, often applied to conifers, results in low quality lumber products because juvenile and mature woods occur in the same board.

4. Tracheids of juvenile pine wood are short, often being less than 2 mm long, and they are sometimes shorter than the fibers of some hardwoods. For example, MacDonald and Franklin (1969) report that the cell length of most conifers is less than 1 mm in the rings adjacent to the pith; many hardwoods have fibers longer than that. In this instance, juvenile wood does not fulfill the needs of those who use conifers as a source of long fibers. The short length and thin walls of tracheids in juvenile wood are not generally recognized and there has been surprise when conifers, specifically grown for their long tracheids, do not produce the needed

tear strength when wood from young pine thinnings or topwood is pulped. Tracheid lengths of mature conifer woods are satisfactory for all paper and wood strength properties, but those of juvenile wood are often so short that considerable benefit would result from an increase in cell length. Tracheid length comparisons for juvenile and mature woods are summarized for loblolly and slash pines in Table 3.6.

5. Juvenile wood of pine usually is chemically different from mature wood, having a greater percentage of lignin and xylan and a lower content of cellulose and galactoglucomannan. When pulped by the sulfate process, chemical consumption per ton of juvenile pine is slightly greater and manufacturing costs per unit finished product can be increased by 10% (Zobel and Kellison 1972). However, the chemical differences between juvenile and mature woods appear to be of less importance in affecting product quality than the anatomical differences (Byrd et al. 1965). In old trees that contain heartwood, the amounts of resinous extractives and polyphenols that have been deposited in the juvenile zone can be very large. When such deposition occurs, yield, quality of the final product, and manufacturing efficiency may be grossly affected.

6. Moisture content varies greatly from juvenile to mature wood and with height in the tree (Fig. 3.9). This is of key importance in the drying process as well as when timber is bought or sold by green weight or transported by weight. The differences in moisture between juvenile and mature woods and trends are shown in Table 3.3 for loblolly pine. Moisture content as reported is based upon dry wood weight; thus, the 133% moisture of the juvenile wood of the third bolt means that for every pound of dry wood the log contains 1.33 lbs of water. The mature wood of the third bolt (M.C. = 97%) shows that for every pound of wood there will be 0.97 lbs of water. Moisture content of most pines increases considerably with increased height in the tree because of the greater proportion of juvenile wood (Fig. 3.9).

7. Pulp yields per unit volume of juvenile wood are 5–15% lower than those of mature wood, with the paper produced from juvenile wood having qualities quite different from that made from mature wood (Barefoot et al. 1965; Zobel et al. 1973). Juvenile wood produces paper having lower tear and better burst as well as having low cellulose and high hemicellulose yields per unit dry weight of wood (Zobel et al. 1966; Gladstone et al. 1970). Holocellulose yields per unit volume

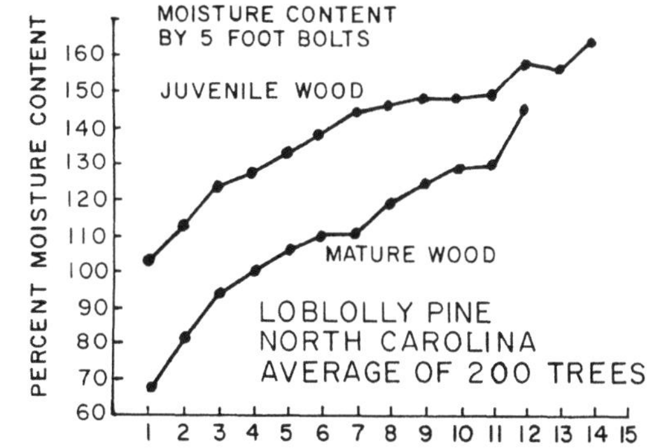

Fig. 3.9. Shown is the change in moisture content in mature wood, as well as in juvenile wood, from the base to the top of the tree. Note how much higher the moisture content of the juvenile wood is than that of the mature wood at each of the 5-ft bolts. When the trees become older, and heartwood formation occurs, this trend may no longer occur

of dry wood are about 3% less, and alpha-cellulose about 7% less from juvenile wood than from mature wood. Put in another way, McKee (1984) states that a normal 4-in. diameter pine has a fiber content approximately 13.5% less than that of a 10 in. diameter pine when the wood is measured by green weight. The lower yield of juvenile wood increases the economic penalty associated with young wood relative to mature wood by 25%. He recommends methods of control over the supply of juvenile wood.

Juvenile wood of conifers other than the pines varies considerably by species as will be described later in this chapter.

In most hardwoods, the juvenile wood has only slightly lower specific gravities or slightly shorter fibers than the mature wood. Hardwoods accordingly have a much more homogeneous wood from the tree center to the outside of the tree, or from the base to the top of the tree, than the conifers (Taylor 1973a; Jett and Zobel 1975; Zerges and Newman 1980). The presence of juvenile wood in hardwoods has therefore been largely ignored or considered to be only of minor importance with respect to its effects on the product produced. Yields are slightly affected, but pulp or paper quality is rarely influenced enough for age of harvest to have an important effect, as is discussed in Section 3.7.

3.4.5 Can the Amount of Juvenile Wood Be Changed?

The production of a juvenile-type wood results from normal physiological growth processes (Larson 1962). There is not too much that the forester can do to avoid it. The amount can be reduced somewhat by changing growth patterns as outlined by Larson (1969b). He feels that nearly all activities that obtain the objective of faster growth result in more juvenile wood. For example, he reports that for red pine, fertilization with nitrogen enlarges the core of juvenile wood. Many foresters agree that fertilization, particularly with nitrogen, prolongs the time that juvenile wood is produced. Improved forest management, such as better site preparation and competition control, results in the trees becoming of a merchantable size at a younger age and thus, the percentage of juvenile wood is increased, although the number of rings of juvenile wood are not greatly altered.

The potential for planting trees closely to suppress growth during the period of juvenile wood production in the butt log of a tree has often been suggested (Zobel 1970). Initial stocking was found to have an effect on 35-year-old loblolly pine, thinned twice, where the 7×7 ft spacing had 31% juvenile wood, the 9×9 ft spacing 37% juvenile wood, and the 12×12 ft spacing 51% juvenile wood (Martin 1984). Martin states that "this analysis supports the feasibility of lowering the final harvest of juvenile wood by tight spacing and intermediate thinning." The size of the juvenile core can be reduced in this way, but at a high cost in total volume production and it generally is economically and biologically impractical. As soon as the tree is released and begins to grow rapidly, the juvenile zone being produced at that time becomes large again. Most of those who have tried dense planting to suppress the size of the juvenile core have abandoned the concept because of the high cost of extra planting, the costs of early thinning, and the extension of the rotation age before harvest is feasible.

It is possible to change the specific gravity of juvenile wood somewhat by breeding. For example, Zobel et al. (1978) were able to improve the wood density of 10-year-old loblolly pine by using seed from only high specific gravity parents. A similar suggestion was made by Ledig et al. (1975). Whether this can also be done for tracheid length is not known. Such methods should be used for certain products for which juvenile wood is unsuitable in order to reduce any adverse results from the greatly increasing amounts of juvenile wood currently being produced. Some, such as Burdon and Harris (1973), feel that the density gradient from pith to bark can be considerably reduced by clonal selection, using those clones with a low gradient. As reported by Ladrach (1986c), the age of transition from juvenile to mature wood varies greatly by individual trees, and this should be taken into account when making selections in a tree improvement program.

3.4.6 Utilization of Juvenile Wood

In naturally grown temperate forests, juvenile wood normally makes up only a small proportion of the total volume of the wood furnish. In plantations, the amount of juvenile wood is greater, especially when young thinnings are used, when basal logs are used for solid wood products, leaving the topwood for chips, or when the plantations are harvested at an early age. In tropical plantations, trees are often harvested before any significant amount of mature wood has been formed and certainly thinnings from tropical plantations consist predominantly of juvenile wood. The high proportion of juvenile wood and its effects on world timber production and quality were emphasized by Zobel and Kellison (1984). Ever increasing quantities of juvenile wood will have a major impact on product quality and yield but its adverse effect, especially on pulp, is being ameliorated by the use of TMP, CTMP, and pressure groundwood.

Juvenile wood of conifers is not suitable for certain paper products but is suited for others. It provides a paper that is weak for products requiring high tearing strength but produces papers that bond well and have good burst, tensile strength, and fold (Kellison and Hitchings 1984). Some tissues made from juvenile wood are excellent, and as a result of the large surface caused by fiber collapse and close chemical bonding, printability and bendability of paper made from juvenile wood are satisfactory (Fig. 3.10). Tear depends on cell length, cell wall thickness, and strength similarly to opacity, which is related to cell stiffness; both attain higher values in paper from mature rather than from juvenile wood (Zobel and Blair 1976). It was concluded from this study that pulp of juvenile wood will be considerably different from pulp of mature wood, but whether the juvenile wood pulp is better or worse depends upon the final product. In any case, a cost penalty is imposed because of yield differences and greater harvesting costs from handling smaller logs which contain predominantly juvenile wood. Differences were quantified by Kellison and Hitchings (1985) who stated that in sulfate pulping it would cost over $ 500000 per year more if juvenile instead of mature wood were used in a mill producing 1500 tons per day. The tear factor of bleached juvenile loblolly pine pulp was 23% inferior to bleached pulp made from mature wood.

Fig. 3.10. Juvenile wood, with its thin walls and large cells, tend to "ribbon" well and produce well-bonded paper with a smooth surface and good bursting strength. The ribboning (flattening) is shown for young loblolly pine which was grown in Chile (Courtesy of Companhia Florestal, Monte Dourado, Brazil)

For mechanical pulp, juvenile wood is often preferred. In radiata pine, Kibblewhite (1984) found that juvenile wood tracheids were shorter, thinner-walled, and had surfaces more damaged than tracheids from slabs (mature wood). Pulps from juvenile wood have excellent optical properties because of the good air to tracheid interfaces.

Results from a study by Barefoot and co-workers (1970) indicated that topwood, whole trees of young age, and juvenile wood from the center of older trees produced very similar pulps. An excellent summary of the qualities of fibers and paper from different types of wood, including topwood and 10- and 15-year old lobolly pine, has been reported by Semke and Corbi (1974). They demonstrated definite differences in tracheid qualities and paper properties (see Table 3.5) in pulps made from the different types of wood.

Several other studies have demonstrated the effect of juvenile wood on pulp and paper properties. One on loblolly pine by Kirk et al. (1972) will be used as an example (Table 3.8). The differences that occur when 12-year-old trees of three specific gravity classes were pulped and compared with 30-year-old trees were quite evident. Note especially that if mill production using the 30-year-old wood is rated as 100, normal juvenile wood (specific gravity of 0.42) will yield 91% as much in the mill, while the lowest gravity juvenile wood (0.37), common in many pines, produces only 76% of mill capacity. Note also that tear is lower

Table 3.8. Pulp characteristics and yields from normal mill chips and juvenile wood of three specific gravity categories[a]

	Normal mill chips[b] Average density	Juvenile wood		
		Low density	Average density	High density
Specific gravity	0.44	0.37	0.42	0.48
Kappa number	27.5	26.5	28.3	28.4
Pulp yield, % dry wood	47.5	44.2	45.7	47.1
Pulp yield, % wet wood	21.7	18.3	20.9	22.2
Lb pulp/ft^3 green wood	13.0	10.0	11.9	14.0
Mill production capacity, percent of normal	100	76	91	107
Physical tests at 750 Schopper-Riegler freeness				
Tensile strength, kg/in	15.01	16.88	16.80	15.87
Burst factor (mullen)	60	73	69	70
Tear factor	120	88	96	109
Fold endurance, M.I.T.	1540	2390	2150	1970
Opacity (%)	67.3	62.0	62.0	64.2

[a] Adapted from Kirk et al. (1972).
[b] Average for 30-year-old trees.

from the juvenile wood, while fold is much better than from the 30-year-old wood.

In his summary, Hitchings (1984) noted for *Pinus taeda* that, on an oven dry weight basis, use of juvenile wood reduces pulp yield from 3 to 5% compared to mature wood. On a volume basis yield differences can be as much as 10—15%.

Boards made from juvenile wood of pine may be quite unsatisfactory, as mentioned earlier. Since they are weak and unstable with excessive warping and checking, juvenile wood is often avoided for solid wood products. This was discussed by Senft et al. (1983) who stated that "green lumber containing juvenile wood tends to warp excessively in a finished structure, creating wall and floor problems." This concept was developed further by Lamb and Wengert (1987) and Senft (1987) who showed that boards made from juvenile wood are very prone to crook, bow, and twist. In hardwoods, it is hypothesized that crook and bow during drying is caused more by longitudinal growth stresses than by juvenile wood alone. Reid (1963) also stated that the core of immature wood has a very low value for sawn timber because of excess warping during seasoning. Such instability has resulted in discrimination against juvenile wood (Pearson and Gilmore 1980). New methods of high temperature kiln drying have helped to produce a more useful product.

Working with Douglas-fir in France, Nepveu and Ngoc (1984) emphasized that the smoothness of planed boards depends upon wood density, and higher density juvenile wood produced a better surface. The authors estimated that an increase of 65 kg/m^3 in density can improve smoothness by an average of 40%. Juvenile wood can be used for several types of particle board to make a light weight product. Alternatively, it can be pressed with heat to make a heavier product (Zobel 1984).

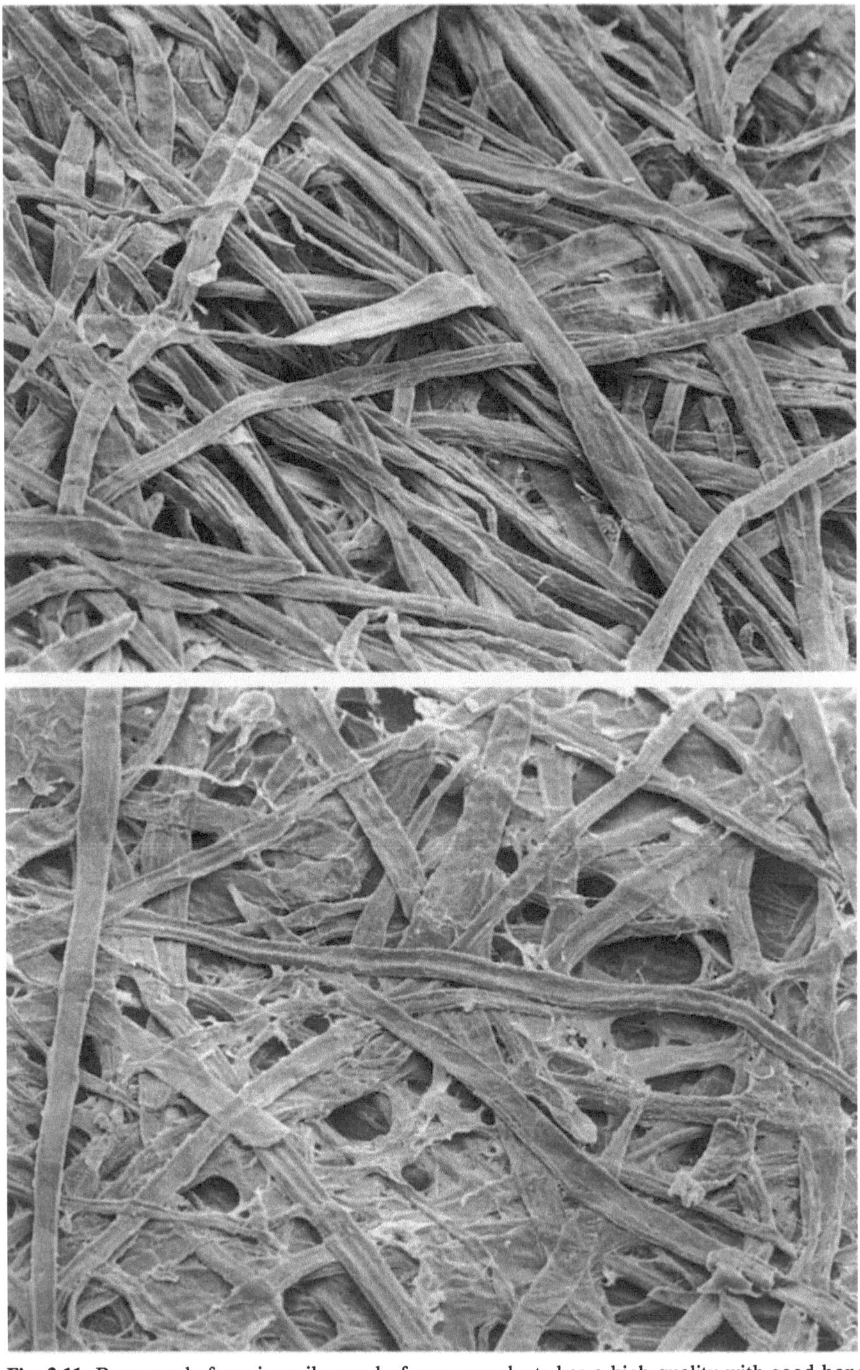

Fig. 3.11. Paper made from juvenile wood of some eucalypts has a high quality with good bonding and good surface characteristics. Shown (*above*) is the configuration of juvenile *Eucalyptus* fibers without beating. *Below* is a sheet after 30-min beating. Note the bonding and surface differences (Courtesy of Companhia Florestal Monte Dourado, Brazil)

An interesting concept for the use of juvenile pine wood is to employ it as a partial substitute for hardwoods. For example, Amidon (1984) feels that the shorter and more collapsible juvenile pine fibers may reduce the need for hardwood in papers where surface smoothness is important. He states that the use of juvenile pine in place of hardwood is quite feasible. A similar conclusion was made by Semke and Corbi (1974) who reported that either hardwood or juvenile pine fibers could be used for printing quality papers except in certain light-weight grades.

When grown in plantations, short-rotation hardwoods, such as the eucalypts or *Gmelina,* are usually harvested by clear-cutting at an age of 5–8 years. These young trees contain mostly juvenile wood, but it does not have a marked effect on their products because the juvenile wood is usually very similar to the mature wood. Young hardwoods can yield high-quality pulp, as reported for *Eucalyptus* in Brazil by Campinhos (1980) and Zobel et al. (1983; Fig. 3.11). In a similar work by Steinbeck et al. (1972), young sycamore (*Plantanus occidentalis*) produced a useful product when pulped by the sulfite process; the major problem was that the tear strength was low.

Similar to the conifers, the fibril angle of juvenile hardwood is flat and thus the stability of solid wood products from the juvenile wood of hardwoods is low and they usually produce boards with many defects and degrade.

A major advantage of juvenile wood in young hardwoods, especially in the eucalypts, is the low percentage of extractives and polyphenols. Most *Eucalyptus* species produce only small amounts of these undesirable extractives before they are 6–8 years old, but older trees usually contain a high proportion of chemical deposits, which makes their juvenile wood undesirable and sometimes difficult to use for pulp or solid-wood products.

3.4.7 Summary of Juvenile Wood

One of the best recent summaries of juvenile wood in conifers is the paper "The Characteristics of Juvenile Wood" by Thomas (1985). His six summary points were:

1. Juvenile wood is formed by a juvenile or immature cambium, whereas mature wood is produced by a mature cambium. Since the tree top always has an immature cambium, juvenile wood is produced in the top regardless of tree age. Both juvenile and mature woods are produced simultaneously: mature wood at the base after cambium maturation and juvenile wood at the top.
2. Juvenile wood occupies the center of the tree stem and varies from 5 to 20 growth rings in size. The variation is primarily species-related with longer-lived trees forming juvenile wood for a longer period of time.
3. Juvenile wood is characterized by shorter cells with larger lumen diameters, thinner cell walls, larger microfibril angles, more compression wood, lower specific gravity, higher lignin content, lower cellulose content, lower strength, and larger longitudinal and lesser transverse dimensional changes than mature wood.
4. Pulp yields from juvenile wood are lower than from normal wood on a dry weight basis and even more on a green weight or volume basis.

5. Paper manufactured from pulp prepared from juvenile wood exhibits higher tensile strength, sheet smoothness, burst strength, fold endurance, and apparent density but lower tear and opacity than paper from mature wood. Paper produced from higher specific gravity juvenile wood approaches mature wood in all of these properties.

6. Because the strength properties of juvenile wood are lower than those of mature wood, caution should be exercised in the utilization of juvenile wood for structural lumber.

Usefulness of juvenile wood was summarized by Zobel and Blair (1976) as follows:

1. Increasing the juvenile fiber content of the pine pulp furnish of a mill improves printing quality and results in better tensile strength, which can be advantageous. However, the lower bulk obtained may be very undesirable.
2. The costs of pulping young wood and topwood will be high; additionally, when wood is purchased by weight, the high moisture content of young wood is very disadvantageous. The cost of pulping will increase as the proportion of young wood is increased.
3. At the same dry wood content, the yield of pulp from 10- to 15-year-old hard pine trees is 5 to 10% lower than from mature tree chips; on a green weight or volume basis, 10–15% less yield from young wood compared to mature wood is not unusual.
4. The tall oil and turpentine yields from southern pines are about three to five times more in conventional roundwood from older trees than from juvenile wood in the Kraft process.
5. Mills with limited digester capacity will encounter difficulties in maintaining production as the proportion of juvenile wood is increased.
6. Juvenile pine can be used as a partial substitute for hardwood, depending on the properties required for the paper product.
7. The adverse effect of low specific gravity juvenile wood can be offset by developing trees with a high specific gravity juvenile wood whose yields are equivalent to those of mature trees.
8. Juvenile hardwood generally has few adverse effects and can be very beneficial for some products.

The best method available to the forest manager to control the amount of juvenile wood in plantations is rotation length. For example, the volume of juvenile wood is about 50% at age 22, 30% at age 32, and 20% at age 40 according to Saucier (1987).

3.5 Variation from the Tree Center to the Bark

The best known and most studied within-tree variability in wood is the change from the tree center (the pith) to the bark, which is frequently referred to as the characteristics of juvenile and mature wood. Covered here and in the previous section on juvenile wood are the radial patterns in wood properties of different species. Because the patterns can vary significantly for different groups of trees, the discussion will be divided into radial patterns for specific gravity of conifers, specific gravity of hardwoods, and other wood properties of both hardwoods and conifers.

As is true of so many wood properties, the literature is voluminous and controversial. There is no possible way we could cite even a significant portion of it. Therefore, we have summarized some representative publications to illustrate the trends. Then, for each section there is a summary, which constitutes our best assessment of the general trends as covered in the literature represented briefly in the tables as well as from our own experience. It is the task of the reader to tie his or her own knowledge into these generalized summarizes. We are, in fact, trying to bring some order to what is now chaos on a widely studied and important subject.

There are a few general and irrefutable facts:
(1) There are radial trends in wood properties for all species of trees. (2) The radial trends vary greatly in magnitude and type, being of key importance in some species and of little importance in others. (3) The radial patterns are reflected in the concept of juvenile and mature wood which is a crude, but useful, way of understanding the within-tree patterns of variation.

Related to the magnitude of radial trends is the number of annual rings from the tree center or sometimes the distance from center, especially when annual rings are not evident. Some species have a very rapid change in wood properties (Fig. 3.5) and in others it is much more prolonged and gentle. Although only a generality, the radial change appears to be more abrupt, and sometimes more severe, in the short-lived, initially fast-growing species than in the longer-lived ones. Especially for the very fast-growing species in the tropics and subtropics, this information can be of considerable importance.

Another frequently misunderstood fact is that the rate of change from juvenile to mature wood is not the same for different characteristics, as previously explained. For example, specific gravity usually changes and stabilizes more rapidly than cell length. Thus, the juvenile zones for specific gravity and tracheid length are not the same for the two characteristics within a given tree.

Commonly, the juvenile zone is defined by the specific gravity pattern and not by that of the other wood properties. There are several causes for the radial patterns, such as the the relationship to the age of the cambium or to the proximity to the crown; these were discussed earlier in relation to juvenile wood formation. Another interesting hypothesis put forth by Schniewind (1961) concerns the support function of the bole of the tree. Schniewind explains the advantage of stronger wood at the periphery of the tree. He maintains that the tree pattern closely follows mechanical efficiency and conductive-storage functions.

3.5.1 Radial Trends in Wood Properties of Conifers

When working with conifers, it is immediately evident that many have a common pattern of radially increasing specific gravity while others do not, and that these must be described separately. It is also evident that not all reports agree totally within each group. These divergences from the norm will be pointed out. There is no disagreement, however, that radial variation exists. As Lantican and Hughes (1973) stated for *Pinus caribaea* "with the exception of latewood radial cell width, all features examined (cell width, wall thickness, latewood percent) showed very highly significant variation with increasing age from the pith." These changes are systematic, not random. An understanding of the radial patterns is essential for an assessment of utilization patterns and best age of harvest.

The patterns occur for characteristics other than density, such as moisture content. Most trees, except older ones containing heartwood, have a high moisture content at the center, which decreases toward the bark. This has been reported for many species, including members of the genus *Populus* (Curro 1960). The variation is very large in cell length and considerable in cell and lumen dimensions and in cellulose yields. It is safe to say that *nearly* all wood properties show a radial change from pith to bark.

3.5.1.1 The Hard Pines – Specific Gravity

A number of studies related to the radial trends in the specific gravity of conifers are summarized in Table 3.9.

The hard pines have a rather uniform pattern of low specific gravity at the pith, a rapidly increasing specific gravity through the juvenile period followed by a series of annual rings that have an essentially constant specific gravity, albeit fluctuating from year to year. Overmature trees-frequently show a reduction in specific gravity near the bark. An extreme example of this was reported by Harris (1971) for the tropical *Pinus caribaea, P. merkusii* and *P. oocarpa* where outside a small zone of corewood, in which latewood was poorly developed, there was a rapid change to multiple bands of very dense wood.

There are exceptions to the normal pattern; for example, Hughes and Mackney (1949) report for *Pinus radiata* that the basic density increases linearly from the center through 22 annual rings. Several researchers cited in Table 3.9 also either state, or imply, that radiata pine has such a continuously increasing specific gravity, at least to 30 years of age, although others report the normal trend of rapid increase near the tree center, leveling off toward the bark. A linear trend was also reported for *Pinus resinosa* by several workers, such as Jayne (1958). For *Pinus patula* and *P. taeda* in South Africa this trend occurred regardless of growth rate according to Banks and Schwegmann (1957). A linear relationship was also found for *Pinus caribaea* by Brown (1971), but several others report for this species the more normal pattern of rapid increase from the center, followed by relative uniformity toward the bark. The *general rule* then is that in the hard pines there is a rapid increase in wood density from the pith outward, then usually a short transition period, followed by a leveling off period toward the bark. This normal pattern is upset if there are deposits of resins at the heart of the tree, especially after heartwood formation (Falkenhagen 1979). Usually, the change from the tree center outward is greatest in latewood and least in earlywood (Peterson 1968). It is important, however, to recognize that there are variations.

Table 3.9. Variation in specific gravity of conifers from the tree center to the bark (juvenile and mature woods)

Species	Reference	Comments
Pinus taeda	Fernow 1896	Regardless of height, specific gravity is low near the pith and increases toward the bark
Pinus patula *P. radiata*	Scott and Stephens 1947	The wood from the periphery of the stem is 11 lbs/ft^3 heavier than that from the center of the stem
Pinus radiata	Hughes and Mackney 1949	Basic density increases nearly linearly from pith to bark (330 to 430 kg/m^3) in 22-year-old trees
Pinus resinosa	Pillow 1952	Heavier wood was formed away from the tree center. Longitudinal shrinkage was much greater near the pith than the usual 0.1 to 0.3% of green length in wood closer to the bark
Picea abies	Hildebrandt 1954	The highest wood densities are usually found near the tree center; then there is a decrease, after which there is an increase toward the bark
Pinus radiata *P. taeda*	Banks and Schwegmann 1957	A radial density pattern was found in both fast- and slow-grown trees; the increase in density was similar, and independent of the rate of growth
Chamaecyparis obtusa	Hirai 1958	Heavy wood occurs at the tree center; it becomes lighter until ring 25 and then heavier again
Pinus resinosa	Jayne 1958	Density increased from tree center to cambium
Picea sitchensis	Jeffers 1959	There was a significant regression of specific gravity with age of ring, with a reduction from the cambium inward and a rise in specific gravity near the pith
Pseudotsuga menziesii	McKimmy 1959	Specific gravity of the latewood increased with successive zones from the pith. Use of Mork's definition to define latewood was necessary only for the first ten rings
Pinus contorta	Schütt 1959	Both specific gravity and tracheid length were higher in the outer wood than in interior wood; this was more evident for specific gravity
Pinus resinosa	Cooper 1960	Specific gravity increased with distance from the pith
Pinus brutia	Hilmi 1960	Density rises sharply from the pith outward for 20 to 30 rings
Picea glauca	Keith 1961	There was essentially no tendency for specific gravity to increase outward from the tree center, independent of ring width
Pseudotsuga menziesii	Littleford 1961	Specimens less than 15 rings from the pith had lower density and strength properties compared to those 15 to 29 rings from the pith. More than 15 rings from the pith, crushing strength, bending strength, stiffness, and specific gravity remain constant or decrease slightly
Larix leptolepis *L. decidua*	Pearson and Fielding 1961	Earlywood density decreased steadily during the first 15 rings and then was uniform
Pinus caribaea	Schmidt and Smith 1961	Contrary to other reports, no pith to bark trend was found for specific gravity in Queensland
Pinus densiflora	Sakai and Uegaki 1962	Specific gravity increased from the pith outward
Pinus elliottii	Sellers 1962	Number of rings from tree center was the chief determining factor for specific gravity, the wood from the outside being the heaviest

Table 3.9 (continued)

Species	Reference	Comments
Pseudotsuga menziesii	Wellwood and Smith 1962	Juvenile wood is formed to the 15th ring. Trees with high specific gravity near the pith are above average in these characteristics at advanced ages. Specific gravity was governed primarily by the square of the number of rings from the pith
Pinus radiata	Harris 1963b	Specific gravity of core wood is significantly correlated with outer wood ($r = 0.717$), but is not considered to be close enough to allow an accurate distinction between the outer wood and core wood
Pinus radiata	Reid 1963	The center core consists of a large pith as great as 0.75 in. in diameter. Wood density increases rapidly for five or more rings, then more gently to the 15th ring from the pith
Tsuga heterophylla	Krahmer 1966	Specific gravity was highest near the pith and decreased for 2 in. and then remained more or less static
Pseudotsuga menziesii	Smith et al. 1966	Number of rings from the pith influenced both earlywood and latewood more than variations in annual growth caused by climatic differences
Pinus radiata	Harris 1967	Latewood and earlywood densities vary from core wood to outer wood
Pinus banksiana	King 1967	The specific gravity of 3-year-old material is not useful in predicting the density of mature trees
Pinus densiflora	Tajima 1967	There were ten rings in the core wood, at both the base and top of the tree. Density is high in the first ring, falls to the fifth, and increases to the 15th where it becomes constant. Outer wood density is 15 – 24% greater than the core wood density
Pinus resinosa	Peterson 1968	The cambium primarily produced thin-walled cells the first 10 years; afterwards latewood made up an increasing proportion of the annual increment
Picea abies *Larix decidua*	Petrik 1968	The differences in yield and strength of pulps between growth groups might be explained by different juvenile to mature wood ratios
Pinus sylvestris	Dorn 1969	Cambial age had a very significant effect on specific gravity
Pinus taeda	Gilmore and Pearson 1969	Juvenile wood had a specific gravity of 0.48, mature wood 0.57 in the southeastern USA
Pinus resinosa	Larson 1969a	Fertilization enlarges the core of juvenile wood
General	Larson 1969b	Juvenile wood has a high proportion of earlywood and a lower proportion of transition-latewood tracheids
Picea abies	Stairs 1969	Juvenile wood was strongly correlated with mature wood. Juvenile wood does not necessarily cause lowered pulpwood yields
Pinus elliottii	de Guth 1970	In Argentina, all five characteristics studied, including wood density (except tracheid diameter), increased from pith to bark
Pinus caribaea	Brown 1971	Density increases linearly with increasing distance from the pith
Abies concolor	Wilcox and Pong 1971	Specific gravity, moisture content, toughness, and shrinkage were lowest near the pith and were constant in the outer third of the radius

Table 3.9 (continued)

Species	Reference	Comments
Pinus clausa	Manwiller 1972	It will be hard to change the normal growth pattern of low specific gravity near the pith and higher specific gravity near the bark
Pseudotsuga menziesii	Megraw and Nearn 1972	Fertilizer treatments had no effect on the specific gravity of juvenile wood, but the juvenile growth period was extended
Pinus caribaea	Brown 1973	Density increased with increased distance from the pith. Trees showed greater variation radially than vertically
Pinus radiata	Burdon and Harris 1973	The density gradient from pith to bark can be reduced by clonal selection; the first five growth rings give a good indication of the density of the outer rings
Pinus caribaea *P. merkusii* *P. oocarpa*	Harris 1973a	All three species had a core of low density wood containing poorly developed latewood. Density gradients are steep from juvenile to mature wood, terminating at the twelfth growth ring. Mature wood of *P. caribaea* has a whole series of what appears to be false rings in the latewood
Pinus palustris *P. elliottii*	Schmidtling 1973	Selection for specific gravity did not greatly influence the yield of juvenile wood
Pinus taeda	Zobel et al. 1973	Juvenile wood occurs within a core of about ten annual rings from the pith
Pinus elliottii	Foelkel et al. 1975	Slash pine begins to form mature wood at the age of 7–9 years. Variability of latewood in the radial direction was more marked than the variability of earlywood
Larix leptolepsis	Isebrands and Hunt 1975	The juvenile wood zone persisted for about 10 years from the pith based upon percent of latewood and extractives
Pinus caribaea	Harris 1977b	Wood specific gravity at the tree center was the same wherever the trees were grown, but the increase across the radius was much greater at lower latitudes. Juvenile wood ceases at about the tenth ring
Pseudotsuga menziesii	Keller and Thoby 1977	Wood density in the first ten rings is highly correlated with density outside the tenth ring
Picea abies	Olesen 1977	Basic density decreases with increasing ring width from the pith to the 8–10th ring. The level then rises and becomes constant between the 15th and 20th ring. Innermost juvenile wood is heavier; outer juvenile wood is lighter than mature wood
General	Kellison and Zobel 1978	Wood specific gravity in man-made plantations is 2–4% less than in naturally regenerated stands because of a larger proportion of juvenile wood
Pinus caribaea *P. elliottii*	Kromhout and Toon 1978	There is a high density at the pith which falls for three to five rings after which it rises to reach its maximum at the bark. The cause of the high density near the pith is resin infiltration. Slash pine has a severe gradient from 374 kg/m^3 near the pith to 596 kg/m^3 near the bark
Pinus caribaea	Barrichelo and Brito 1979	Specific gravity increases rapidly from the rings near the pith and then remains constant to the bark based upon whole rings. When earlywood and latewood are separated, there is no radial pattern of specific gravity

Table 3.9 (continued)

Species	Reference	Comments
Pinus caribaea	Falkenhagen 1979	Specific gravity changed from high at the center, decreased at 3 or 4 years from the center and then increased to the bark. The high values at the center were due to resin infiltration
Picea abies	Lewark 1979	On both seedlings and rooted cuttings in Germany, specific gravity was high near the pith, fell to about the tenth year, and then rose again
Pinus radiata	Bunn 1981	The predictable pattern for basic density is lowest near the pith, it increases outward to about the 25th ring, and then remains constant. Wood density can vary as much as 150 kg/m^3 from pith to mature outer wood
Pinus elliottii *P. caribaea* var. *hondurensis* var. *caribaea* var. *bahamensis* *P. oocarpa*	Greaves 1982	The radial density gradient is steeper in var. *hondurensis* than in the other two varieties, but the steepest gradient is in *P. elliottii*. *P. oocarpa* had a very small radial density gradient
Picea glauca	Taylor et al. 1982	Juvenile wood (rings 1 through 10) has higher specific gravity than mature wood. At every height in the tree, rings 1 – 5 had a high specific gravity, which decreased sharply to rings 5 – 10
Pinus radiata	Bamber and Burley 1983	In Australia, the amount of latewood increases from about 5% near the pith to 30% at the 20th ring from the pith; after which it remains constant
Picea sitchensis *Pinus radiata*	Brazier 1983	Density of sitka spruce in New Zealand varies from 300 kg/m^3 in the core to 400 kg/m^3 in mature wood. In radiata pine in New Zealand density varies from about 300 kg/m^3 in the center to 450 kg/m^3 in mature wood
Pinus caribaea	Cown et al. 1983	There was both a radial and vertical gradient, but the specific gravity was higher than quoted by others. It did not have the extreme outer wood values reported by some authors
Pinus caribaea	Plumptre 1983	Juvenile wood lasts 5 – 8 years; there is a large gradient in density to the mature wood
Pinus banksiana	Roddy 1983	The specific gravity of juvenile wood (first ten rings from pith) was 0.39 and less than mature wood
Pinus banksiana *Larix laricina*	Yang 1983	The number of sapwood rings shows a strong correlation with tree age
Pinus taeda	Zobel et al. 1983	Specific gravity varies from 0.32 at the center to 0.55 near the bark in 30-year-old trees
Pinus taeda	Bendtsen 1984	The modulus of elasticity of juvenile wood ranged from 47 to 63% of mature wood. Age of demarcation varied by the characteristic
Pinus patula	Ladrach 1984	Mature wood begins to form in rings 4 to 8; the change to mature wood is gradual
Picea glauca	Wang and Micko 1984	The radial pattern of density is complex. The density is slightly higher closest to the pith than further out, but lower than that of the mature wood. This differs from some previous studies which showed juvenile wood to have the highest density
Pinus taeda	Kellison et al. 1985	The specific gravity of loblolly pine juvenile wood is 25% less than that of mature wood. Juvenile wood has limited utility for solid wood products

Table 3.9 (continued)

Species	Reference	Comments
Pinus taeda	Megraw 1985	Specific gravity increases from the center to the outside at all heights, but is greater at the base of the tree than higher up
Pinus contorta	Fries 1986	Juvenile wood density is not a good estimator of mature wood density
Pinus oocarpa *P. patula* *P. kesiya* *P. caribaea*	Ladrach 1986c	The change from juvenile to mature wood occurs at 5 to 6 years for *P. oocarpa* and *P. patula*, while it may be 8 years or more in *P. kesiya* and *P. caribaea*
Pinus oocarpa	Ladrach 1987	Mature wood begins to form between the fourth and sixth rings

3.5.1.2 Other Conifers – Specific Gravity

Table 3.9 also contains examples of conifers other than pines.

Members of the Cupressaceae (*Chamaecyparis, Cupressus*, and others) usually have a high specific gravity near the tree center, which diminishes for a few rings from the pith, and then becomes greater again toward the bark (Hirai 1958). Some members of the Pinaceae (*Abies, Picea, Tsuga*) sometimes show this pattern (Jeffers 1959; Krahmer 1966) or a pattern of very little change from the pith to the bark (Keith 1961; Taylor et al. 1982). The *general rule* for fir, spruce, and hemlock is to have a high specific gravity near the tree center with a decrease for a few rings followed by a leveling off or a moderate increase toward the bark. Members of the Cupressaceae tend to have a high density wood at the tree center, a decrease in density for a period, and then an increase. There are, as always, exceptions to these general rules. The growth patterns from pith to bark are normal and will be difficult to change, as stressed by Manwiller (1972).

Several genera of conifers, such as *Larix* and *Pseudotsuga*, have a wood pattern similar to that of the hard pines with a low density at the tree center, a rapid increase in the juvenile zone, and a leveling off toward the bark (McKimmy 1959; Wellwood and Smith 1962; Isebrands and Hunt 1975).

3.5.1.3 Radial Changes in Conifer Wood of Properties Other than Specific Gravity

Some examples of radial variation in wood properties of conifers other than specific gravity are shown in Table 3.10.

Most reports on radial changes of wood properties other than specific gravity in conifers relate to tracheid length. The usual patterns for all species is to have short tracheids near the tree center followed by a rapid increase through the juvenile zone then a leveling off. However, some species show only a slight leveling, and tracheid length can be as much as four of five times greater near the bark than near the pith.

Table 3.10. Change in conifer wood properties other than specific gravity from tree center to bark

Species	Reference	Comments
Pinus radiata	Dadswell 1958	Tracheid length increases for 12 to 15 years from the pith and then becomes relatively uniform. The change in cell length may be three to four times
Pinus caribaea	Schmidt and Smith 1961	Tracheid length increased with age and distance from the pith to a relatively constant length at about 19 years. Over 91% of the mean tracheid length was associated with the initial length. There was no radial pattern for spiral grain
Pseudotsuga menziesii *Tsuga heterophylla*	Smith et al. 1961	In Douglas-fir, 77% of the variation in tracheid length and 60% in hemlock are related to the number of rings from the pith. The larger the fibers near the pith, the longer are the fibers in the whole tree
Pseudotsuga menziesii *Thuja plicata*	Wellwood and Smith 1962	Tracheid lengths of Douglas-fir were controlled primarily by ring number from tree center, which accounted for 65% of the variation
Pinus densiflora	Sakai and Uegaki 1962	Tracheid length increased from the pith outward
Conifers	Noskowiak 1963	Spiral grain is largest at the tree center, then decreases and sometimes reverses and can become larger near the bark in older trees
Picea mariana	Rydholm 1965	Tracheid length was 0.9 mm in the 1st ring, 2.7 mm at the 10th ring, 3.3 mm at the 20th, and 3.7 mm at 75 to 100 years
Pinus taeda	Zobel et al. 1968	Spiral grain was highest near the tree center. Generally, in the mature wood spiral grain was not large enough to affect utilization
Pinus taeda	McMillin 1969	Earlywood ash content decreased with increasing distance from the pith; latewood ash content was independent of distance from the pith
Pinus caribaea	Andrews and Hughes 1973	The largest source of variation in tracheid length was a systematic trend from pith to bark
Pseudotsuga menziesii	Erickson and Arima 1974	Fibril angle decreased from over 30° near the pith to 7° at the 30th ring; tracheid length was highly correlated with fibril angle. Lignin percent was highest nearest the pith and decreased for about 20 rings; both holo- and alpha-celluloses increased until about the 20th ring
Picea abies	Olesen 1977	Tracheid length increases with ring number from the pith
Pinus caribaea	Barrichelo and Brito 1979	Tracheid length and width sharply increased until the fifth or sixth ring; from the seventh ring, tracheid dimensions increased slowly to a constant value
Pinus caribaea	Whyte et al. 1980	There was no clear trend of spiral grain with position in the tree stem
Pinus radiata	Harris 1981	The microfibril angle decreases from 40° to 10° from pith to bark. Outerwood has somewhat higher rates of cellulose to lignin than corewood
Pinus radiata	Cown and McConchie 1982a	Resin content decreased with distance from the pith
Picea abies	Olesen 1982	Tracheid width increases steeply with distance from the pith in trees grown from seed; in vegetatively propagated trees, narrow tracheids are not found in juvenile wood if cuttings come from mature branches

Table 3.10 (continued)

Species	Reference	Comments
Pinus caribaea	Plumptre 1983	Tracheid length increases from pith to bark, from about 2 mm near the pith to 6 mm near the bark at 14 to 20 rings
Pinus taeda	Zobel et al. 1983	Tracheid length varies from under 2 mm at tree center to 4 to 6 mm near the bark
Pinus patula	Ladrach 1984	Tracheids in mature wood are always longer than in juvenile wood
Picea glauca	Wang and Micko 1984	Tracheid length increases with increasing ring width from the pith outward. Frequently, length stabilizes in the mature wood
Pinus taeda	Megraw 1985	The fibril angle is large near the pith and is less closer to the bark

There are scattered reports of other wood property changes with radial position such as microfibril angle (Erickson and Arima 1974; Harris 1981; Megraw 1985). The fibril angle is highest near the pith and usually diminishes rapidly toward the mature wood nearer the bark. The same applies to spiral grain which often is large near the tree center and even reverses direction toward the bark (Noskowiak 1963; Zobel et al. 1968b; Harris 1988). However, this patterns does not always hold. Whyte et al. (1980) report no trend of spiral grain with the position in the tree for *Pinus caribaea*.

The ratio of alpha- and holocelluloses also varies; in Douglas-fir, both increased from pith to bark (Erickson and Arima 1974). In loblolly pine, Zobel et al. (1966) showed that holocellulose was higher near the tree center.

As expected, the resin content, found primarily in the heartwood, is highest near the tree center (Cown and McConchie 1982b), but the rate of inception of heartwood formation varies greatly from tree to tree and often is not directly related to the age of the ring (Hillis 1987).

3.5.2 Radial Trends in Hardwoods

3.5.2.1 Specific Gravity

Some papers reporting on radial trends in wood properties of hardwoods are listed in Table 3.11.

Several different patterns for the variation of specific gravity from pith to bark are found in the hardwoods. Some species, such as *Swietenia macrophylla, Liquidambar styraciflua, Liriodendron tulipifera*, and others, have an increase in specific gravity from the pith outward (Briscoe et al. 1963; Hunter and Goggans 1963; van Eck and Woessner 1964; Webb 1964; Herpka 1965; Sluder 1970). Of particular importance are the eucalypts, which generally show a small increase from the tree center outward (Ferreira 1972; Hans et al. 1972; Skolman 1972). However, Taylor (1973b) found a decrease in wood density for *E. grandis* for the first few years and then an increase at breast height level, but little change at high-

Table 3.11. Variations in several wood properties from pith to bark of some hardwood species

Species	Reference	Comments
Quercus robur	Eichorn 1895	Fiber length increased until an advanced age
Fagus crenata	Kaneski 1938	Increase in fiber length continued for over 50 years from tree center
Eucalyptus regnans	Bisset and Dadswell 1949	Fiber length varied from 0.6 mm near the pith to 1.35 mm in the outermost growth rings in one tree. Increase was rapid for ten rings, then more or less constant
Fraxinus spp.	Bosshard 1951	There was a steady increase in fiber length to 40 years and more
Populus japanogigas	Inokuma et al. 1956	Fiber length increased outward from the tree center
Eucalyptus regnans	Dadswell 1958	Cells in mature wood are twice as long as those in juvenile wood; this increase occurs in both fibers and vessels. Vessel diameter also increases with distance from the pith. The juvenile period is about 20 years for cell length
Populus euramericana	Scaramuzzi 1958	There was very little variability in volume proportions of various types of wood elements across the radius of the tree; ray volume showed the most variation
Euro-American poplar hybrids	Currò 1960	Basic density increased from pith to bark
Populus spp.	Göhre 1960	There was a central core of heavy wood followed by lighter and then again by heavier wood
Eucalyptus camaldulensis	Chudnoff 1961	Mechanical properties vary greatly from pith to bark; a central brittle-heart zone has very low strength properties
Quercus falcata	Hamilton 1961	Fiber length increased from pith to bark but specific gravity, latewood percent, fibrous tissue, hardness, and toughness decreased
Liriodendron tulipifera	Thorbjornsen 1961	The specific gravity increased from inner to outer wood, being 0.42 at 5, 0.45 at 15, and 0.46 at 25 years. No significant relationship was found between fiber length at 11 and 31 years.
Populus tremuloides	Valentine 1962	The major portion of the variation in specific gravity across the radius of the tree is related to distance from the pith
Liriodendron tulipifera	Barefoot 1963	Age from the pith has an effect on specific gravity and toughness
Swietenia macrophylla	Briscoe et al. 1963	Specific gravity increases outward from the pith
Tectona grandis	Kedharnath et al. 1963	Fiber length is short at the tree center and increases outward. There are nine rings of juvenile wood
Liquidambar styraciflua	Webb 1964	Specific gravity was lowest and fiber length shortest in the region near the center of the tree. Both increased toward to bark, but specific gravity leveled off at about 15 rings and fiber length at 25 rings
Salix alba	Herpka 1965	Specific gravity and fiber length increase from the center toward the bark.
Numerous hardwoods	Knigge and Koltzenburg 1965	A number of investigators could find no relationship between the proportion of cell types and distance from the pith, but others (in *Quercus* and *Fagus*) found more vessels near the bark

Table 3.11 (continued)

Species	Reference	Comments
Populus deltoides	Farmer and Wilcox 1966a	Wood near the cambium was denser than that near the center of the stem but there was a weak correlation between juvenile and mature woods, making early prediction of specific gravity unreliable.
Fagus sylvatica	Koltzenburg 1966	Fiber length and vessel diameter increase with ring age up to 30 years; there is no change in fiber diameter or fiber wall thickness
Liquidambar styraciflua	Johnson and McElwee 1967	Fiber length at the 30th ring was always greater than at the 15th ring
Prunus serotina	Koch 1967	After a slight increase, there is a marked decrease in specific gravity from near the pith toward the bark through the 30th ring; the values are 0.512 near the pith and 0.487 near the bark
Acer saccharum	Sajdak 1967	There is high specific gravity near the pith, followed by a decrease, and then an increase
Liquidambar styraciflua	Hunter and Goggans 1968	There is a definite zone of juvenile wood within the first ten rings from the pith
Liriodendron tulipifera	Koch et al. 1968	There was an increase in specific gravity from the pith outward except at stump height
Gmelina arborea	Lamb 1968	Density increases gradually from the pith outward, from 25 to 34 lbs/ft^3. Fiber length increases rapidly from the pith to the end of the third year (0.6 to 1.2 mm) and then increases only slightly
Liriodendron tulipifera	Taylor 1968	Heartwood formation is not governed by the age of xylem (distance from the cambium); specific gravity increases from the first 20 rings, levels off, and then decreases with ring number
Platanus occidentalis	Taylor 1969a	Specific gravity increases with ring number in both the upper and lower bole
Carya illinoensis	Taylor 1969b	Specific gravity increases from pith to bark at all heights in the tree
Liriodendron tulipifera	Sluder 1970	Specific gravity was lowest at the pith, increased rapidly to ring 25, increased slightly to 45, and then decreased slowly to the bark
Liriodendron tulipifera	Gilmore 1971	Juvenile wood is lighter than mature wood
Eucalyptus saligna	Santos and Nogueira 1971	Maximum fiber length was reached at the tenth ring The juvenile stage lasts 10 years
Celtis occidentalis	Taylor 1971	Specific gravity decreases linearly with ring number from the pith; fiber length increases with age from rings 1 to 25; outer rings have more vessel volume with a reduction in fibrous tissue
Populus tremuloides	Einspahr et al. 1972	Specific gravity was high near the tree center, lower from three to five rings, and then increased toward the cambium
Eucalyptus grandis	Ferreira 1972	Specific gravity increased from pith to bark
Eucalyptus grandis	Hans et al. 1972	Density varied from 420 kg/m^3 near the pith to 470 at the bark in 7-year-old trees; fiber lengths increased from 0.87 to 1.01 mm
Eucalyptus robusta	Skolman 1972	There is an increase in specific gravity with distance from the pith; there was a difference of 0.3 in specific gravity between inner and outer wood

Table 3.11 (continued)

Species	Reference	Comments
Eucalpytus grandis	Taylor 1973a	There is a strong correlation and little variation between specific gravity of juvenile wood and mature wood. Fiber length of juvenile wood is not strongly related to that of mature wood
Eucalyptus grandis	Taylor 1973b	Specific gravity decreases for the first few years from the pith, then increases at breast height; at a height of 10.7 m, distance from the pith had little effect and at 22.9 m, there was a decrease from the center outward. Both fiber length and fiber diameter decrease with distance from pith
Salix nigra *Quercus phellos* *Platanus occidentalis* *Carya illinoensis* *Celtis laevigata*	Taylor and Wooten 1973	In all species there were large increases in fiber length from the pith, followed by a more gradual increase until a maximum was reached
Eucalyptus spp.	Hillis and Brown 1978	In the eucalypts, the density varies more with increasing age or distance from the pith than it does with height. In *E. grandis*, the minimum density (370 kg/m^3) was found near the pith and the maximum (495 kg/m^3) at the periphery of the tree
Eucalyptus globulus	Anonymous 1979	Wood density and cellulose content of juvenile wood are low, but increase with age; increase in cellulose content does not begin until trees are mature, while increase in density starts in young trees
Eucalyptus grandis	Brasil et al. 1979	Wood specific gravity increased from the center of the tree outward
Plantanus occidentalis	Land and Lee 1981	Specific gravity increased from 0.44 to 0.47 over the first 20 to 25 growth rings from the pith
Eucalyptus spp.	Zobel 1981	Juvenile wood in the eucalypts is mild and has only a minor effect on the final product
Eucalyptus regnans	Frederick et al. 1982	Basic density usually increased from the pith outward at all heights
Eucalyptus spp.	Ferraz 1983	Some species (*E. alba*) are constant in wood density from pith to bark, others (*E. grandis*) have a small increase, and others (*E. saligna*) have a large increase
Triplochiton scleroxylon	Oteng-Amoako et al. 1983	Fiber length increased 30% from pith to mature wood; specific gravity increased slightly with ring age
Populus tremuloides	Yanchuk et al. 1983a	Wood density is high near the pith, decreases, then increases again in the mature wood zone after about the 15th ring
Diffuse-porous hardwoods	Wheeler 1987	Fiber wall thickness increases outward from the pith which causes an increase in specific gravity
Ring-porous hardwoods	Wheeler 1987	In many species the specific gravity is higher in the juvenile wood than in mature wood

er levels in the tree. Other species have a high specific gravity near the tree center; for example, *Acer saccharum* (Sajdak 1967; Einspahr et al. 1972) and *Populus tremuloides* (Yanchuk et al. 1983b). This pattern is usual in ring-porous species, such as the oaks. Other species have little change across the radius as shown for *Alnus rubra* by Harrington and DeBell 1980.

In summary, all possible patterns of wood density variations appear in the hardwoods. The middle to high density diffuse-porous hardwoods generally follow a pattern of low specific gravity near the pith and then an increase, followed by a slower increase or leveling off toward the bark. The low density, diffuse-porous woods, such as *Populus*, seem to have a somewhat higher density at the pith, although some have a uniform density from pith to bark. The ring-porous hardwoods tend to have a high density at the center, which decreases and then increases to some extent toward the bark.

3.5.2.2 Other Wood Characteristics

Some properties other than specific gravity are also shown in Table 3.11.

Most reports on radial patterns in hardwoods deal with fiber length, and most agree that fibers are shortest near the center of the tree. Sometimes the fiber length differences are considerable, as much as 2 to 1, but usually the change is slight. An example of a small change is the tropical hardwood *Triplochiton scleroxylon* from Ghana in which there was an increase of fiber length of less than 30% from pith to mature wood (Oteng-Amoako et al. 1983). This is the usual pattern for hardwood trees with a storied structure. A few reports indicate a reduction in fiber length toward the bark. Changes in ray or vessel volume or other characteristics have also been shown to occur across the radius, but these changes are usually not major.

Knigge and Koltzenburg (1965) found a rapid increase in cell length the first 10 to 20 years in hardwoods, followed by a leveling off. This pattern was also present in *Populus* (Jayme et al. 1943; Scaramuzzi 1955; Boyce and Kaiser 1961, and others), *Eucalyptus* (Bisset and Dadswell 1949), *Liriodendron* (Thorbjornsen 1961), and *Carya* (Pritchard and Bailey 1916). Other genera, such as *Quercus, Fagus*, and *Fraxinus* have a continuous increase in fiber length for 40 to over 100 years, depending on the species. Vessel length continued to increase with ring number from tree center, according to Knigge and Koltzenburg 1965, who report that the changes in cell characteristics occur at all heights in the tree, and that they depend on the distance from the pith.

The radial changes in wood properties other than specific gravity could be described as confusing. Nearly every possible pattern occurs. However, it is important that the pattern for each species be determined for the site on which the trees are grown. The most consistent is fiber length, which usually increases from pith to bark.

3.6 Variation from the Base to the Top of the Tree

Table 3.12 lists the changes in a number of wood properties from the base to the top of the tree.

Many trees have wood properties that vary at differing heights in the tree. For those species in which juvenile wood greatly differs from mature wood, a change in wood properties with height is automatic since the proportion of juvenile wood

Table 3.12. Variations in several wood properties from the base to the top of the tree for conifers and hardwoods

Species	Reference	Comments
Conifers		
Pinus taeda *P. echinata* *P. palustris* *P. elliottii*	Chidester et al. 1938	This early report emphasizes the increase of earlywood from the butt to tree top, with high wood density at the base
Pinus radiata	Hughes and Mackney 1949	Wood density decreased from the base to the top; infiltration of extraneous materials at the butt near the pith caused high density
Pinus elliottii	Lindgren 1951	Specific gravity decreased progressively from the base to the top with butt bolts averaging 25% higher than top bolts. Moisture content of butt bolts averaged only about two-thirds that of top bolts
Picea excelsa	Nylinder and Hägglund 1954	Pentosan content rose strongly with increased height in the stem; oven dry density rose only slightly with a rise in the stem
Chamaecyparis obtusa	Hirai 1958	Wood is heavy at the tree base, lighter in the middle and heavy at the top
Pinus resinosa	Jayne 1958	Density decreased from the base upward
Picea sitchensis	Jeffers 1959	There was a significant decrease in specific gravity with distance from the butt of the tree to the top
Pinus resinosa	Cooper 1960	Wood formed in the bole below the living crown was heavier than wood formed within the living crown, and the density decreased with increasing height
Pinus brutia	Hilmi 1960	Specific gravity decreases upward in the stem (outside the core) from 0.59 at 5 ft to 0.55 at 11 ft, and to 0.50 at 24 ft
Picea mariana	Risi and Zeller 1960	The influence of stem height upon specific gravity was very small
Pinus elliottii	Sellers 1962	Specific gravity decreased with increasing height
Pinus contorta	Tackle 1962	Specific gravity decreases noticeably from the stump to 20 ft, but changes little beyond that
Picea abies	Stern 1963	Among 1 000 trees in 12 different stands, four stands showed no change in specific gravity with height, while the rest showed an increase
Tsuga heterophylla	Krahmer 1966	Specific gravity was highest at the stump level, decreased rapidly to 8 ft, and then decreased gradually to the crown region
Pinus resinosa	Maeglin 1966	Taller trees had a specific gravity that decreased from the tree base to the crown
Pinus resinosa	Baker and Shottafer 1967	Wood specific gravity is considerably higher at the base of the tree than at the top
Pinus pinaster	Polge and Illy 1968	Wood specific gravity is positively correlated with height ($r = 0.68$)
Pinus taeda	Gilmore and Pearson 1969	At the 3-ft level, juvenile wood had a specific gravity of 0.481 and mature wood 0.527; at 33 ft the values were 0.407 and 0.435, respectively. The overall decrease of specific gravity with height was considerable
Pinus taeda	van Buijtenen 1969	The difference between the butt log and the top log in any one tree is usually greater than the largest difference between any two butt logs of trees of the same species growing under similar conditions
Pinus caribaea	Brown 1971	Wood density showed a decreasing linear trend with increasing height
Pinus resinosa *P. banksiana* *Larix laricina* *Picea glauca*	Provin 1971	There was a clear decrease in specific gravity with height in pines and in the larch. Specific gravity decreased only to 16 – 24 ft; this was followed by a steady increase to the top in spruce, fir, and hemlock

Table 3.12 (continued)

Species	Reference	Comments
Picea mariana *Abies balsamea* *Tsuga canadensis*		
Abies concolor	Wilcox and Pong 1971	Highest values for moisture content, specific gravity, tangential and radial toughness, and tangential and radial shrinkage were in the butt log and decreased with height
Pinus taeda	Zobel et al. 1972	Topwood has properties similar to juvenile wood at the tree base except that the tracheids are somewhat longer
Pinus caribaea	Brown 1973	Density showed a decreasing trend with increasing height, but this varied greatly from tree to tree. Some trees showed no trend at all
Tsuga heterophylla *Picea sitchensis*	Farr 1973	Specific gravity in the stem of hemlock decreased to mid-height, then increased to the top. Specific gravity of spruce increased with increasing tree height
Picea abies	Olesen 1973	There is no difference in basic density between different heights in the same tree, with the exception of the top which consists of juvenile wood
Picea mariana *Abies balsamea* *Pinus contorta*	Heger 1974	Factors governing wood structure in stems of conifers depend more on the relative height in the stem than on species or size
Pinus taeda	Zobel 1975a	Topwood is primarily juvenile wood, while basal logs of older trees have a high proportion of mature wood
Pinus caribaea	Kromhout and Toon 1978	The highest density was in the base of the tree, the lowest in the top portion
Pinus caribaea	Falkenhagen 1979	Vertical density gradations were not pronounced
Pinus radiata	Corson and Foster 1981	The energy to achieve a given freeness level of pulp decreased with a decrease in density from the butt to the top logs
Pinus caribaea	Cown 1981	There was a decrease in wood density with height in the stem
Pinus patula	Palmer et al. 1982a	Usually the upper log was the least dense within a tree
Picea glauca	Taylor et al. 1982	Wood in top logs has a specific gravity and tracheid length that are equal to or exceed those in butt logs
Pinus radiata	Kibblewhite and Lloyd 1983	Slabwood pulps had substantially higher tear and lower burst than did top-core wood pulps
Pinus banksiana	Roddy 1983	Specific gravity was 0.42 at breast height, 0.40 at 5-m height, and 0.37 at the 5-cm top diameter
Pinus banksiana *Larix laricina*	Yang 1983	The number of sapwood rings steadily declined with increasing tree height
Picea glauca	Wang and Micko 1984	Tracheids from the top log are longer than those from the lower heights
Pinus taeda	Megraw 1985	There is a sharp decrease in the specific gravity of juvenile wood from the base of the tree to 3 m; above 5 m the specific gravity is constant regardless of height
Hardwoods		
Populus japanogigas	Inokuma et al. 1956	Fiber length showed a constant decrease in the upward direction; specific gravity exhibited no trend
Euro-American poplars	Currò 1960	Basic density increased from butt to top, while moisture content decreased
Populus spp.	Göhre 1960	The wood density in the top part of the stem was greater than at the base
Eucalyptus camaldulensis	Chudnoff 1961	Mechanical wood properties do not vary at different heights

Table 3.12 (continued)

Species	Reference	Comments
Quercus falcata	Hamilton 1961	There is a zone of less strength in the middle of the bole
Swietenia macrophylla	Briscoe et al. 1963	Specific gravity was high at the base, dropped to a minimum at 8 ft, and increased near the crown
Liquidambar styraciflua	Webb 1964	Fiber length was largest at the stump and decreased steadily with increasing height. Both specific gravity and interlocked grain reached a maximum at the stump, decreased to 10 to 20 ft above ground, and then increased slightly
Nyssa silvatica	McElwee and Faircloth 1966	There was essentially no difference in wood density with stem height
Acer rubrum	Saucier and Taras 1966b	In 1-year-old sprouts up to 17.3-ft-tall, specific gravity did not vary between internodes, but fiber length did, fibers becoming progressively shorter from base to top
Eucalyptus regnans	Dargavel 1968	Basic density increased up the tree for one third to two-thirds of the merchantable height
Liriodendron tulipifera	Koch et al. 1968	Specific gravity at stump height was considerably greater than further up the stem
Gmelina arborea	Lamb 1968	Specific gravity decreased from 0.41 to only 0.39 with increasing tree height
Liriodendron tulipifera	Taylor 1968	Specific gravity decreases sharply to 10 ft and then decreases slightly, reaching its lowest values between 30 and 50 ft, after which it remains constant
Platanus occidentalis	Taylor 1969a	Specific gravity is the same at all heights sampled
Carya illinoensis	Taylor 1969b	Specific gravity is significantly less at breast height than at 10 ft further up the bole
Liriodendron tulipifera	Gilmore 1971	The heaviest wood is near the ground and becomes lighter up the stem until a plateau is reached
Celtis occidentalis	Taylor 1971	Specific gravity is higher at breast height than at 10 ft and above; fiber length increases with height, but there is no pattern of variation of vessels and rays with height
Populus tremuloides	Einspahr et al. 1972	Specific gravity was high near the tree base, decreased for some distance, and then leveled off
Eucalyptus grandis	Ferreira 1972	Specific gravity increased from the base to the top of the tree
Eucalyptus robusta	Skolman 1972	There is an increase in specific gravity with stem height
Eucalyptus grandis	Taylor 1973b	Fiber length is not affected by stem height, but the cell diameter is less at 22.9 than at 10.7 m. Specific gravity decreases from 1.5 to 4.6 m and then increases
Eucalyptus grandis *E. camaldulensis*	Franklin and Meskimen 1975	Wood density decreased from the base of the tree to the 10-ft level, and then increased sharply to 25 ft. Above this level, *E. grandis* varied little, but *E. camaldulensis* decreased rapidly
Eucalyptus propinqua	Brasil et al. 1977	Wood basic density increases from the base of the tree to half-height and then decreases. The dense wood at the middle of the tree is caused by reaction wood
Plantanus occidentalis	Land and Lee 1981	There was no consistent change in specific gravity with stem height
Eucalyptus regnans	Frederick et al. 1982	Basic density increased up the tree, whereas moisture content decreased
Eucalyptus grandis	Barrichelo et al. 1983	Basic density is highest at the base of the tree up to 50–75% of the commercial stem height
Populus (hybrids)	Stevens et al. 1983	The basic density decreases from the base of the bole until a height of 3 m and thereafter increases with height

Table 3.12 (continued)

Species	Reference	Comments
Platanus occidentalis	Land et al. 1983	Specific gravity decreased and moisture content increased with increasing stem height
Populus tremuloides	Yanchuk et al. 1983b	Wood density is high at the base of the tree, decreases to a minimum at midheight, and increases again near the top
Eucalyptus grandis *E. saligna*	de Souza et al. 1986	Basic density decreases up to 25% of the merchantable height and then increases to the top

in the stem increases extensively from the base to the top. For example, van Buijtenen (1969) stated that the difference between the butt and top logs of *Pinus taeda* is usually greater than the largest difference between the butt logs of trees of the same species growing under the same conditions. For several species and genera, Heger (1974) found that factors governing wood structure of conifers depend more on the relative height in the stem than on species or tree size.

In his discussion of the southern pines Lindgren (1951) states that the much higher specific gravity and lower moisture content of freshly cut butt logs compared to top logs have a major effect on pulp yields; much more pulp would be obtained from the butt logs based on either volume or weight. This has been shown by the senior author for above-ground heights and specific gravities for 15-year-old *Pinus caribaea* from tropical areas: 3.0 ft (0.55); 6.5 ft (0.52); 10.0 ft (0.51); 13.0 ft (0.48); 20.0 ft (0.49). As expected, the basal logs have the highest specific gravities. Differences in specific gravity are mostly a direct result of the proportion of top wood, which consists of juvenile wood (Zobel 1975b). But it is more than that! Megraw (1985) shows a rapid decrease of specific gravity of the juvenile wood from the base of the tree to 3 m height in loblolly pine; in the same species, Gilmore and Pearson (1969) report the commonly observed situation where there were large variations in the specific gravity of both juvenile and mature woods themselves from the top to the base of the tree. In those species in which juvenile wood does not greatly differ from mature wood, there is usually only a mild change from base to top.

Usually the hard pines show a considerable decrease in specific gravity with increasing height (Jayne 1958; Brown 1971; Provin 1971). These differences can be considerable; for example, Hilmi (1960) found a specific gravity of 0.59 at 5 ft, 0.55 at 11 ft, and 0.50 at 24 ft for *Pinus brutia*, and Roddy (1983) noted for *Pinus banksiana* a change from 0.42 at breast height to 0.40 at 5 m height and 0.37 at the 10-cm-top diameter (Fig. 3.12).

Members of the genus *Picea*, however, show only a slight change with height (Nylinder and Hägglund 1954; Jeffers 1959; Stern 1963; Provin 1971; Taylor et al. 1982). Some conifer species, such as *Chamaecyparis obtusa* (Hirai 1958) or *Tsuga heterophylla* (Krahmer 1966) have heavy wood at the base of the tree, lighter in the middle, and heavier at the top. The usual trend is to have longer tracheids at the base of the tree and shorter ones near the top (Fig. 3.13). A few reports have been made on wood properties other than tracheid length with respect to

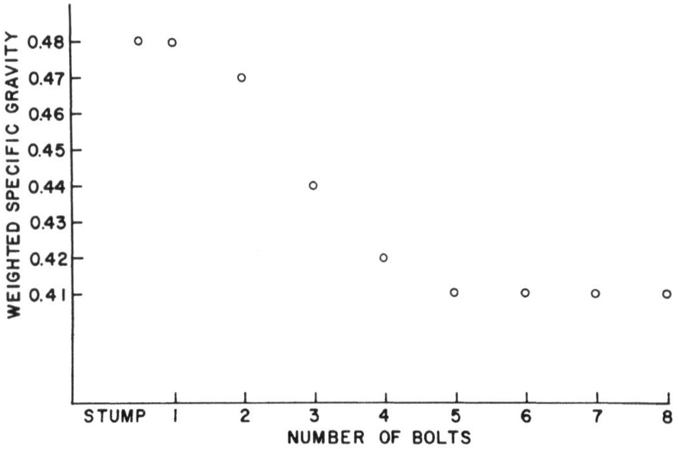

Fig. 3.12. The effect of juvenile wood on specific gravity up the bole (5-ft bolts in 20-year-old hard pines) shows how the top bolts all have a low specific gravity because they all consist of juvenile wood

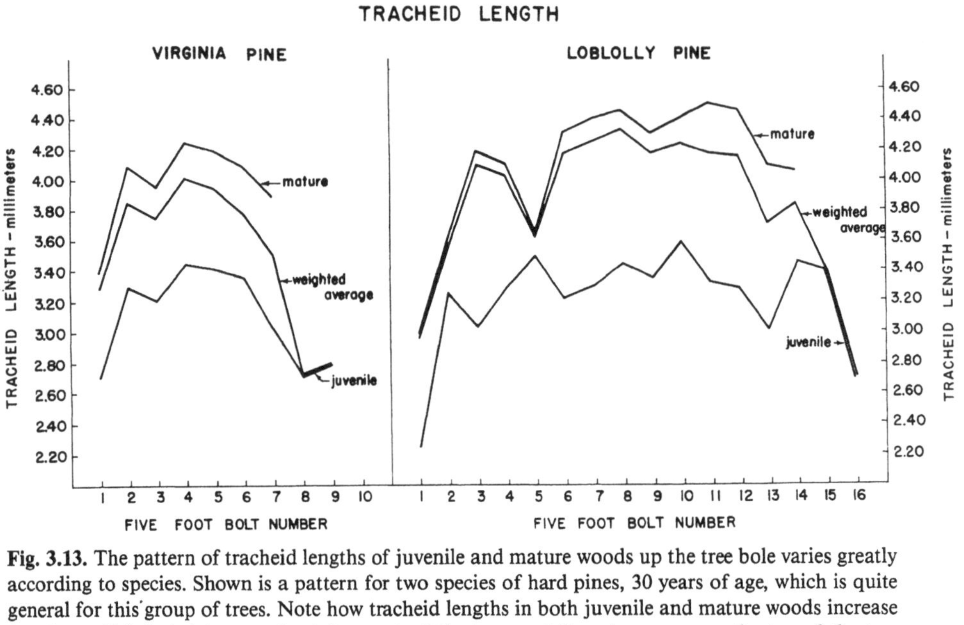

Fig. 3.13. The pattern of tracheid lengths of juvenile and mature woods up the tree bole varies greatly according to species. Shown is a pattern for two species of hard pines, 30 years of age, which is quite general for this group of trees. Note how tracheid lengths in both juvenile and mature woods increase for about 10 ft in height, stay level for most of the tree, and then decrease near the top of the tree

height but mostly the variation was not great. Andrews (1986) found that the upper log of a Douglas-fir had a higher cellulose and lower lignin content compared to the butt log.

Moisture content also varies greatly with height in the tree. For example, in 30-year-old loblolly pines, on the average, the moisture content will be about 130% in the top logs (1.3 lbs of water for every pound of dry wood) to about 80%

moisture in the basal logs (0.8 lb of water for every pound of dry wood). Juvenile wood usually has a much higher moisture content than mature wood (Fig. 3.9).

The effects of height on wood properties is equally as variable in hardwoods as it is in conifers (see Table 3.12). Perhaps the most common pattern is to have only minor changes with height, but some species, such as *Swietenia macrophylla* (Briscoe et al. 1963), *Liriodendron tulipifera* (Taylor 1968), *Populus tremuloides* (Einsphahr et al. 1972; Yanchuk et al. 1983a), and *Liquidambar styraciflua* (Webb 1964), showed a high density at the base, a decrease for some distance up the tree, followed by an increase toward the merchantable top. Some reports show an increase in density with height in the eucalypts and *Populus* (Göhre 1960; Dargavel 1968; Skolman 1972; Taylor 1973b). The same pattern as in conifers, with the highest density at the base and the lowest at the top, was evident in some hardwoods, but most diffuse-porous hardwoods have little variation in specific gravity from the base to the top. The fiber length patterns vary but the most common is to have slightly longer fibers at the tree base than in the top or for the fibers to be essentially of the same length at all heights.

In summary, one can find almost all possible patterns of longitudinal variation in either the conifers or the hardwoods. For example, Okkonen et al. (1972) made the following generalizations in a study of the relationship of wood specific gravity to height in 28 commercially important timber species. In 17 species the specific gravity decreased with an increase in height, in five it increased, in three it decreased for a time and then increased, and in three no specific change was observed. They warned wood users of the importance of determining the wood properties of upper logs.

For specific gravity of the hard pines, Douglas-fir, and larch the wood pattern is heavy at the tree base and lighter at the tree top. For some other conifers, the pattern is reversed. Tracheid length is usually shortest at the tree top. For hardwoods, even though patterns are evident, their magnitude is generally small and differences in both specific gravity and fiber length with height do not have a major effect on the utilization of the wood. Many species show such slight variation with height that a breast height sample gives a rather accurate estimate of the wood properties of the whole stem.

To understand fully the changes of wood properties with height in the tree it is necessary to study the trends in juvenile and mature woods separately. When this is not done, the different proportions of juvenile wood at different heights will largely determine the trends. Unfortunately, there are only a few cases where data were obtained separately for juvenile and mature woods.

3.6.1 Breast Height to Total Tree Correlations

Table 3.13 was prepared to show a few of the studies relating breast height to total tree values.

A major problem always faced in sampling the wood of a tree is to determine where the sample should be taken to give meaningful results, representing the whole tree. Usually the sampling must be non-destructive and it must be done in a convenient location to minimize cost and effort. Under all conditions, sampling

Table 3.13. The relationship between breast height and total tree wood properties

Species	Reference	Comments
Populus spp. *Betula* spp. *Quercus* spp. *Picea* spp. *Pinus* spp.	Götze 1959	There is a good correlation between the density of wood laid down in youth and the density of whole trees
Pinus taeda	Zobel et al. 1960a	Good breast height to total tree correlations were found. The best correlation in density was between mature wood at breast height and total tree values
Abies spp. *Cryptomeria japonica*	Kâno 1961	The bulk density of whole stems can be estimated by means of small increment cores at breast height
Pseudotsuga menziesii	Knigge 1962	Increment cores at breast height are suited to show the differences in average density of stems and stands
Pinus contorta	Tackle 1962	Specific gravity at breast height determined by increment cores gave a satisfactory estimate of whole tree specific gravity values with $r = 0.754$. Disks gave the same results as increment cores
Swietenia macrophylla	Briscoe et al. 1963	Whole tree specific gravity can be estimated from wood at the center of the tree, with $r = 0.618$
Abies grandis	Stage 1963	Both specific gravity and dry weight of the whole tree can be estimated from increment cores at breast height
Pinus radiata	Warren and Mirams 1963	A correlation coefficient of 0.808 was found between whole tree basic density and the basic density of the outer 20 growth rings at breast height. The correlation between whole tree basic density and breast height density was 0.763. Even a small sample at breast height is a good indication of whole tree values
North American Conifers	Mitchell 1963	The breast-height core to whole tree specific gravity correlations did not vary significantly with site or geographic location
Pinus radiata	Harris 1965	The outer wood specific gravity at breast height can be confidently used to predict the average specific gravity of the whole stem ($r = 0.743$); average specific gravity at breast height gives a better estimate ($r = 0.846$). Specific gravity of core wood at breast height gives a good estimate of core wood throughout the stem but does not give a good estimate of the entire stem ($r = 0.365$)
Pseudotsuga menziesii	Wahlgren 1965	Merchantable specific gravity can be acceptably measured from a single increment core taken at breast height to a maximum depth of 10 in.
Pinus taeda	Zobel et al. 1965	Breast height values of specific gravity were a good indication of total merchantable tree specific gravities. Correlation coefficients were $r = 0.88$ for total tree to total breast height values
Pinus resinosa	Maeglin 1966	The correlation between the core wood and total tree specific gravities is better in plantation trees than in natural stands
Pinus strobus *Pinus resinosa* *Abies balsamea* *Picea glauca* *Picea mariana* *Picea rubens* *Tsuga canadensis* *Larix laricina*	Wahlgren et al. 1966	Regression equations were developed for all eight species between the increment core specific gravity at breast height and specific gravity of the merchantable volume. Each species had a characteristic pattern with height which affects the regressions

Table 3.13 (continued)

Species	Reference	Comments
Liquidambar styraciflua	Johnson and McElwee 1967	There was a good correlation in specific gravity between 10-year-old trees and mature trees
Eucalyptus alba *E. saligna*	Ferreira 1968	There is a definite relationship between basic density at breast height and tree basic density
Eucalyptus regnans *E. calophylla* *E. deglupta*	Rudman et al. 1969	Breast height sampling is inadequate to represent the whole tree; higher locations (15 ft) are acceptable, but different heights are required for fiber length and density
Pinus serotina	Taras and Saucier 1970	There was good correlation between a single increment core at breast height and total tree specific gravity
Pinus taeda	Wahlgren and Yandle 1970	A simple model developed to estimate tree specific gravity fitted the data better than the empirical breast height to total tree methods commonly used
Pinus ponderosa	Echols 1971	The correlation between basic density of increment cores at breast height and whole trees was 0.843
Eucalyptus grandis	Ferreira 1972	There is a high, positive, and significant correlation between specific gravity of the tree and breast height values
Picea sitchensis	Farr 1973	Whole tree specific gravity was significantly related to increment core specific gravity at breast height
Pinus elliottii *P. taeda*	Higa et al. 1973	Specific gravity of wood at breast height (1.3 m) is correlated with that of young trees, 9- to 13-year-old, grown in Brazil; r = 0.93 for slash and r = 0.97 for loblolly pine
Pinus caribaea	Cown 1981	Good correlations were obtained between the wood density of increment cores and whole tree wood densities
Pinus radiata	Cown and McConchie 1982a	The regression of the density of the outer wood at breast height was significantly related to topwood at the 5% level
Eucalyptus regnans	Frederick et al. 1982	The plot mean basic density could be estimated within 2%, using the disk value at breast height
Pinus banksiana	Roddy 1983	The correlation coefficient between breast height and total tree specific gravities was 0.82
Eucalyptus grandis	Barrichelo et al. 1983	Basic density of the whole tree is highly correlated with the increment core or wedge at breast height. There is no correlation between cellulose or pulp yield at breast height and total tree values
Pinus caribaea	Cown et al. 1983	Estimation of whole tree values from increment cores was excellent, confirming results of others
Eucalyptus spp.	Brito et al. 1984	The density of the wood of the whole tree was correlated with the density of a disk taken at breast height
Pinus patula	Ladrach 1984	The difference between breast height and total tree specific gravities is largest in young trees. In general, the whole tree averages 0.02 less in specific gravity than the breast height value. Tracheid length of the whole tree is less than that obtained at breast height
Pinus caribaea	Nariyoshi et al. 1986	The positions at 25 and 50% of the commerical height were most representative for estimating specific gravity

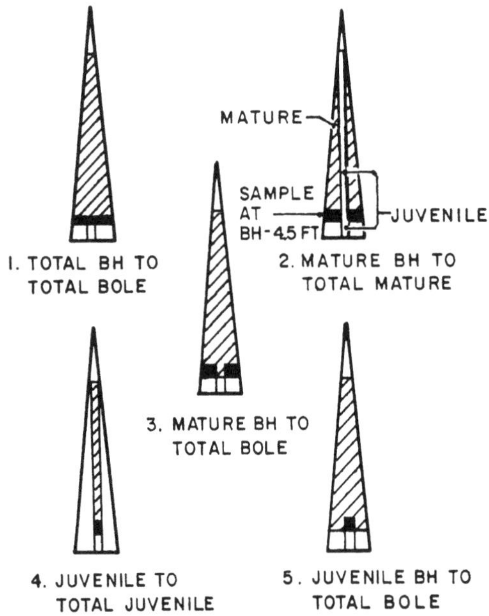

Fig. 3.14. Some simple way of estimating whole tree values must be used. The usual method is to relate breast height to total tree values. Shown are five different methods that have been used by several researchers. *Number 3* (i.e., mature breast height to total tree) usually gives the highest correlations

for wood properties is difficult, time-consuming, and expensive. Therefore, it is necessary to use the simplest and easiest sampling method that has reasonable accuracy.

The simplest method is to take a wood sample at breast height and relate it to the wood properties of the whole tree. This is the most commonly used method and has been relatively well developed. All kinds of methods have been tried, as shown in Fig. 3.14. Many studies indicate that the most accurate relationship for wood density is obtained when mature wood at breast height is related to total tree values.

It is evident that the ratio of juvenile wood to mature wood differs in an increment core compared to a wedge or a full cross-section. Whether this is serious enough to negate the use of increment cores has been well studied; use of large (10 mm) diameter increment cores gives good comparative estimates of wood properties. Small increment cores (4.5 mm) are marginally suitable for specific gravity, but are not usable for tracheid lengths which frequently exceed 4 mm; their use results in mostly cut cells. Because of the built-in inaccuracy of weighting the amount of juvenile and mature woods using increment cores, some have tested this concern and found that cores gave as good, or nearly as good, correlations as wedges at breast height. It is sufficient here to cite Tackle (1962) who reported for *Pinus contorta* that disks gave the same results as increment cores. We obtained the same outcome for the southern pines, which is shown graphically for moisture content for the three major southern pines in Fig. 3.15.

Breast height sampling in conifers is usually quite satisfactory in predicting whole tree values with correlation coefficients around 0.8. Frequently, tables are constructed which enable the prediction of total tree and thus acreage weight val-

Fig. 3.15. There is concern about the accuracy of using increment cores rather than wedges to estimate wood properties. Despite the inaccurate weighting of juvenile and mature woods using increment cores, results are similar. This is shown for the moisture content in three species growing together in Georgia despite the known difficulty of obtaining accurate moisture content determinations from increment cores

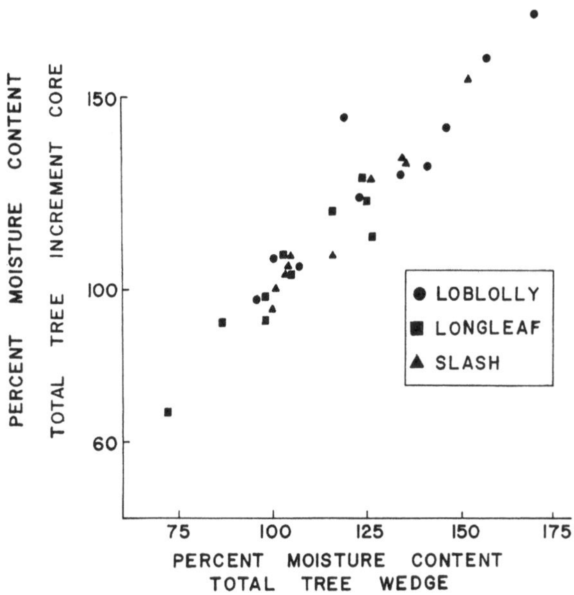

ues from breast height sampling (Zobel et al. 1965; Roddy 1983). This enables one to estimate projected tonnage of wood yields in addition to the normal volume yields, the common practice in the past to indicate productivity of a forest. Tons of usable fiber have much more meaning than volume alone, and many forestry organizations are using this to measure inventories or growth.

The accuracy of predicting whole tree specific gravity values from breast height values is really remarkably good for both conifers and hardwoods (Fig. 3.16). Essentially, all the works cited in Table 3.13 showed a moderate to strong trend of breast height to total tree values; a few examples of correlation coefficients for six species are shown in Table 3.14.

A number of researchers have reported using only the mature wood for comparison, rather than using the full cross-section of the stem. This has reduced the problem related to differential proportions of juvenile wood in trees of different ages. It frequently happens that the breast height to total tree relationship is better when only the mature wood at breast height is compared to the whole tree values (Zobel et al. 1960c) (Figs. 3.14 and 3.16). Problems are minimal when trees of the same age are sampled.

The use of only a few outer rings at breast height is sometimes satisfactory. For example, Harris (1965) and Warren and Mirams (1963) obtained r values of 0.74 and 0.81 according to this method with radiata pine. Warren and Mirams stated that "even a small sample at breast height is a good indication of whole tree values." Less satisfactory is the use of only juvenile wood to predict total tree values. Zobel et al. (1960c) found that the correlation coefficients using juvenile wood were much smaller than those for whole cores or mature wood only. Similar results were reported by Harris (1965) who found a correlation of only 0.37 when juvenile wood was used compared to 0.85 when the whole increment core was employed.

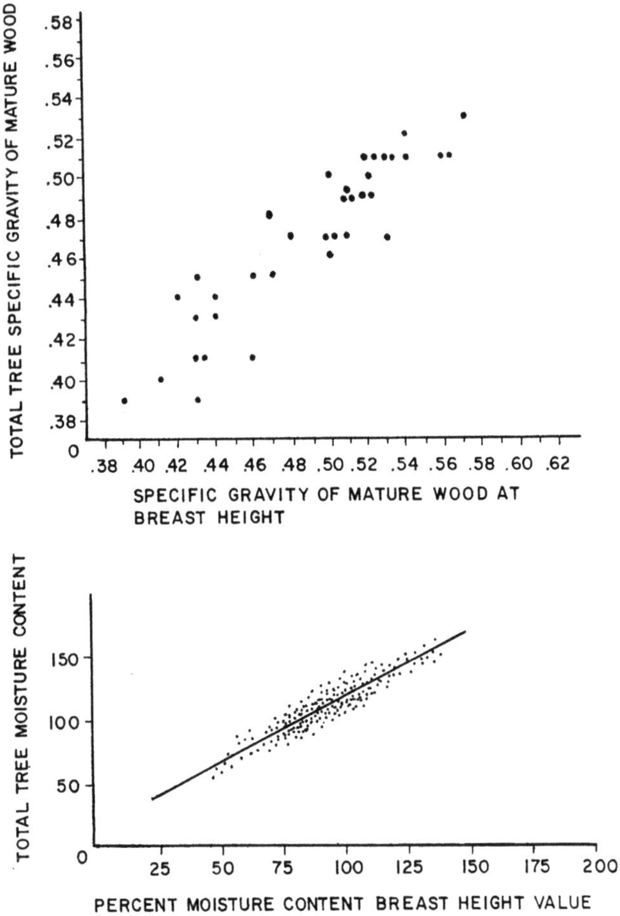

Fig. 3.16. It is important to obtain reliable breast height to total tree values. This is shown in the graphs for wood specific gravity and moisture content. Note the very good relationships between the breast height and total tree values

Table 3.14. The correlation coefficients of breast height to total tree specific gravity for a few pine species

Species	Correlation
Pinus radiata	0.85, 0.76
P. contorta	0.75
P. elliottii	0.93 (13-year-old)
P. taeda	0.97 (13-year-old)
P. banksiana	0.82
P. ponderosa	0.84

A number of researchers feel that the crude empirical method of correlating increment core values with total tree values could be improved. Some studies have been made to test this; for example, Cown and McConchie (1982a) developed regression equations, and Wahlgren and Yandle (1970) used a simple model to estimate tree specific gravity. They found that their models fitted the data better than the usual empirical methods.

One would expect an even better correlation between breast height and total tree values in hardwoods than in conifers because of the mild juvenile wood and small longitudinal variations. This is, in fact, frequently true, and the breast height density value is often nearly the same as the whole tree density. There are some reported exceptions, however. Even though several workers report a high correlation between breast height and tree densities in *Eucalyptus grandis* and other eucalypt species, Rudman et al. (1969) report that for the three species they tested breast height sampling was inadequate to represent the whole tree. They found a 5 m height sample was fairly representative but different heights were needed for wood density and fiber length.

3.7 The Effect of Total Tree Age on Wood

There is much concern about the influence of tree age on wood properties. According to Zobel et al. (1965), the age of a stand has a direct effect on wood properties because it affects the percentage of juvenile wood. A few studies related to stand age are shown in Table 3.15. One study in the tropics on *Pinus patula*

Table 3.15. The effect of total tree age on wood properties

Species	Reference	Comments
Pinus resinosa	Pillow 1952	Wood from young trees had much greater longitudinal shrinkage (especially in the juvenile wood) than the usual 0.1 to 0.3% of green length. Older trees had heavier wood
Populos euramericana	Scaramuzzi 1958	There is no variation in proportion of different wood elements by volume with increasing age of trees
Pinus resinosa	Cooper 1960	There is little relationship between age of tree and wood specific gravity
Pinus radiata	Harris 1965	Shorter rotations will cause a change toward a higher percentage of core wood and lower specific gravities
Hardwoods in General	Knigge and Koltzenburg 1965	The proportion of types of cells in hardwoods are not clearly related to the age of the samples
Pseudotsuga menziesii	Wahlgren 1965	The reciprocal of tree age was by far the most important, accounting for 16% of the specific gravity variation
Pinus taeda	Zobel et al. 1965	There is a strong relation between age of stand and proportion of juvenile to mature wood; the correlation between stand age and specific gravity is 0.64. Some of the dry weight differences associated with age of stand are due to the deposition of resinous materials in the heartwood

Table 3.15 (continued)

Species	Reference	Comments
Fagus sylvatica	Koltzenburg 1966	The relative weight of wood decreased in trees up to 60 years of age
Pinus taeda	McMillin 1969	The ash content of wood is related to tree age. The ash content of juvenile wood from short rotation trees is different from that of juvenile wood in long rotation trees
Picea abies	Stairs 1969	There was a lack of relationship of specific gravity with age, as reported in other spruce studies
Pinus taeda	Zobel 1970	Young thinnings (11-year-old) had a wood density of 24 lbs/ft^3, and older trees had 29 lbs/ft^3. Moisture content for young trees was 160 and 100% for older trees
Pinus caribaea	Palmer and Tabb 1971	Wood from old trees produces pulp with high tearing and low bonding strengths. In Malaya, wood from 6-year-old trees gave the best bonding, while wood of 12-year-old trees gave the best tearing strength
Pinus taeda	Zobel et al. 1972	Specific gravity of the merchantable bole of loblolly pine was 0.43 at 17 years, 0.45 at 25 years, 0.47 at 30 years
Eucalyptus spp.	Petroff 1973	Trees 3 to 4 years of age produced paper with lower tensile and bursting strengths and higher tearing strength than trees 5- to 10-years-old
Pinus radiata	Cown 1974	Older stands (of the same diameter) had 10% higher density and 10% longer tracheids than young stands
Pinus oocarpa	Foelkel et al. 1975	Many wood and pulp properties increased with tree age, including cell length, cell wall thickness, wood density, pulp yields, and tensile, burst, and tear strengths
Eucalyptus globulus	Farrington et al. 1977	Consumption of both wood and alkali per ton of pulp decreased with increasing age up to year 6, as did pulp strength
Pinus caribaea	Ferreira 1978	There was an increase in wood specific gravity with tree age up to 14 years
Eucalyptus saligna *E. grandis*	Ferreira et al. 1979	Wood specific gravity increases with plantation age
Pinus caribaea	Cown 1981	There is an increase in wood density with tree age
Pinus taeda	Talbert and Jett 1981	An increase of the specific gravity of juvenile wood with tree age was found
Pinus radiata	Cown and McConchie 1982a	Between the ages of 25 and 45 years, sawlog densities increased by 10 to 15 kg/m^3. As trees age, heartwood formation becomes more extensive, and moisture content decreases. Sawlogs showed a decrease in green weight with age from 940 to 810 kg/m^3 and a 10% increase in density from 390 to 403 kg/m^3. Logs from young trees are heavier because of a higher moisture content but have a lower basic density
Eucalyptus regnans	Frederick et al. 1982	Trees more than 13 years old had greater basic density and lower moisture content than younger trees

Table 3.15 (continued)

Species	Reference	Comments
Pinus patula *P. radiata*	Palmer et al. 1982a	The lightest 10-year-old samples of *P. patula* had a wood density of 325 kg/m^3 and the heaviest 20-year-old samples had a density of 440 kg/m^3. Tearing strength of pulp increased, and bursting and tensile strengths decreased with increasing age
Eucalyptus grandis	Barrichelo et al. 1983	The wood basic density increases as stand age increases
Betula pendula	Nepveu and Velling 1983a	The effect of tree age on shrinkage and basic density is very significant
Triplochiton scleroxylon	Oteng-Amoako et al. 1983	Young plantation trees compared to 10-year-old natural trees had: fiber length 1.29 vs. 1.42 mm, specific gravity 0.279 vs 0.337, and ring width 11.1 vs 8.8 mm
Eucalyptus tereticornis	Bhat and Bhat 1983	Basic density of 1-year-old trees is 2.6% higher than that of 8–9-year-old trees, while fibers are 22% shorter
Pinus banksiana	Roddy 1983	There was only a weak relationship of specific gravity with tree age for trees over 30 years old
Pinus taeda	Bendtsen 1984	At age 40, plantation trees contained about 25% juvenile wood
Pinus elliottii	Howell et al. 1984	The pulp yield of dry wood per ton green wood increased regularly with 10-, 19-, 28-, and 44-year-old trees; yield of oven dry pulp followed the same trend. Tall oil and turpentine yields decreased through age 28, then increased slightly
Pinus patula *Cupressus lusitanica*	Ladrach 1984	The wood of young pine trees is lighter than the wood of older trees due to the higher percentage of juvenile wood. This relationship with age does not hold for cypress
Eucalyptus grandis	Vital et al. 1984	Log density increases each year with the age of the tree by 25 kg/m^3
Eucalypts	Delwaulle 1985	In the Congo, specific gravity increases with age at about 0.025 per year (a hybrid, *E. saligna*, increased from 0.49 at age 4 to 0.58 at age 6.5)
Eucalypts	Higgins 1985	The age of the tree has a large influence on basic density, vessel size, and extractive content
Pinus oocarpa	Ladrach 1987	Specific gravity increased with age, from 0.35 at 4 years to 0.46 at 11 years

(Ladrach 1986b), the results of which are shown in Table 3.16, is illustrative of the effect of both tree age and elevation on wood properties. Differences are large; the actual age effects will vary by species and geographic location where the trees are grown.

It is axiomatic that any influence of total tree age on wood properties is closely related to the proportion of juvenile wood. This is shown for loblolly pine in Tables 3.17 and 3.18. Older conifer trees often have higher specific gravities (Wahlgren 1965; Cown 1974; Ferreira 1978), while this difference is much less for hardwoods (Jett and Zobel 1975). In a rather complete study of age effects, Cown and McConchie (1982b) found for *Pinus radiata* that in trees over 40 years of age green density decreased from 1025 to 815 kg/m^3, basic density increased from

Table 3.16. Wood quality of *Pinus patula* by age, site, and altitude[a]

Tree age	Specific gravity			Tracheid length (mm)			Dry wt/green ton (kg)		
Elevation (m)	1800	2200	3000	1800	2200	3000	1800	2200	3000
7	0.33	0.32	0.30	3.9	3.8	3.2	369	353	322
10	0.37	0.36	0.33	4.2	4.1	3.9	412	396	364
15	0.42	0.41	0.39	4.7	4.6	4.4	477	461	429
20	0.45	0.44	0.42	5.2	5.1	4.9	537	521	489
25	0.47	0.46	0.44	5.8	5.6	5.4	594	578	547
30	0.48	0.47	0.44	6.3	6.2	5.9	651	635	603

[a] After Ladrach (1984).

Table 3.17. Percentage of juvenile wood by dry weight and volume in the merchantable bole of loblolly pine stands of different ages[a]

Age of stand (years)	Amount of juvenile wood	
	Dry weight (%)	Volume (%)
15	76	85
25	50	55
40	15	19

[a] After Zobel and Blair (1976).

Table 3.18. Percentage of juvenile wood in stands of various ages growing throughout the range of loblolly pine

Age of stand	Type of stand	Number of trees	Average juvenile core diameter (inches)	Years of juvenile wood	Percentage of juvenile wood in total volume
15	Plantation	25	4.3	7	35
18	Plantation	25	4.5	9	38
18	Natural	25	5.2	9	47
23	Natural	43	3.7	6	19
28	Natural	26	3.2	6	12
30	Plantation	25	4.0	7	22
33	Natural	62	3.2	7	12
35	Natural	25	3.9	8	14
38	Natural	43	3.3	8	10
43	Natural	41	3.3	8	10
48	Natural	34	3.1	8	8
53	Natural	57	3.1	9	7

Table 3.19. The effect of age of stand on the green and dry weight of loblolly pine[a]

Age of trees (years)	Green wt./100 ft^3 (lbs)	Dry wt./100 ft^3 (lbs)	Ratio green wt/dry wt.
18	6230	2696	2.31
30	5880	2759	2.13

[a] After Zobel and Talbert (1984).

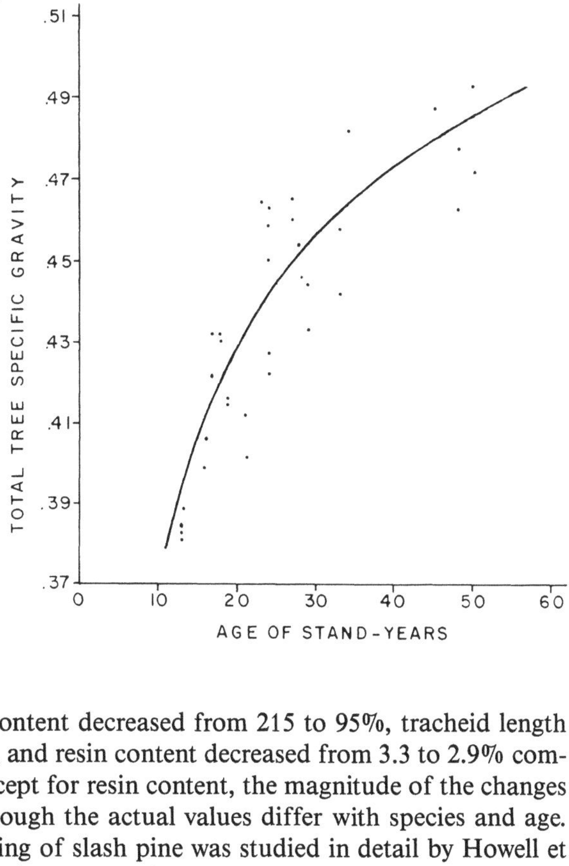

Fig. 3.17. As trees become older, the percentage of juvenile wood is reduced, and the specific gravity increases. At older ages the effect of tree age upon yield and quality also becomes less

325 to 420 kg/m^3, moisture content decreased from 215 to 95%, tracheid length increased from 2.5 to 3.5 mm, and resin content decreased from 3.3 to 2.9% compared to 25-year-old trees. Except for resin content, the magnitude of the changes reported are quite usual, although the actual values differ with species and age.

The effect of age on pulping of slash pine was studied in detail by Howell et al. (1984). Trees 10, 19, 28, and 44 years of age were studied and developed into prediction models for specific gravity, moisture content, bark percentage, tons of dry wood contained per ton of green wood, chip size, and other data. In summary, they found that pulp yields for the 10-year-old sample were considerably lower than for older samples, but they continued to increase only slightly beyond 19 years.

One complication with age correlation and wood density is that older trees frequently contain heartwood and that there are variations in moisture content (Table 3.19). Heartwood is often accompanied by resin or other extractives (along with a lower moisture content), which may greatly increase specific gravity. Although reported by numerous researchers, it is sufficient to quote Cown and

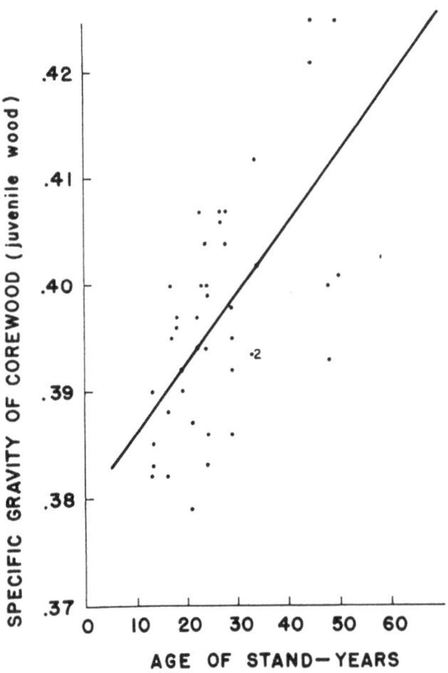

Fig. 3.18. Once laid down, juvenile wood should not change in specific gravity as the trees age, if the resinous deposits are removed. However, juvenile wood of older trees has a higher specific gravity than that from younger trees, as shown. See the text for a discussion of this phenomenon

McConchie (1982a) who pointed out that as trees age, heartwood becomes more extensive, and moisture content decreases. A real assessment of wood properties on these kinds of trees can only be made after extraction with organic solvents, which remove the resins.

Some pine species, such as *Pinus virginiana*, do not normally produce much resin in the heartwood, but the moisture content is drastically reduced, making the wood much lighter in the green condition. For example, a young Virginia pine may have over 110% moisture content (based on dry wood weight), while an older tree may have less than 30% moisture content at the tree center after heartwood formation has taken place.

It is evident that as trees become older (and the percent of juvenile wood becomes less) that the effect of tree age becomes less (Fig. 3.17). For example, for *Pinus banksiana* over 30 years of age, Roddy (1983) found only a weak relationship between specific gravity and age. The relationship with age was also weak in *Pinus taeda* for trees 20 to 40 years of age (Talbert and Jett 1981). This was also reported for *Pinus resinosa* of higher ages by Cooper (1960). In species with only mild juvenile wood, the specific gravity to age relationship is weak or absent; for example, *Picea abies* shows little variation (Stairs 1969).

Age also has an effect on spiral grain, as reported by Harris (1984, 1988). He states that the "economic pressures to reduce the ages of trees at time of felling have resulted in relatively higher proportions of the stem than formerly, consisting of young wood with significant spirality."

An unusual phenomenon, reported by Zobel et al. (1972) and Talbert and Jett (1981), is that the juvenile wood of older loblolly pine trees has considerably higher specific gravity than juvenile wood of young trees, even after resin extraction. Since all but the ray cells in this area of the tree are inactive in the older trees, there appears to be little chance for additional cell material to be deposited. Yet, the steady increase in specific gravity of juvenile wood with age is a fact that should be recognized, but generally is not (Fig. 3.18). The reasons are unknown; it has been suggested that it results from internal tree stresses, sometimes referred to as the compaction theory. No one has studied this or confirmed an increase in the density of juvenile wood with age in other species, but it was clear and the increase was quite large in the two studies on loblolly pine reported above. Changes in ash content of juvenile wood were also associated with tree age (McMillin 1969).

Chapter 4

Wood Property Variation as Related to Tree Form and Reaction Wood

4.1 Introduction

One of the best and easiest ways to alter wood properties is to manipulate tree form. This effective approach to controlling variation in wood is frequently overlooked. Tree form is so useful because it can be manipulated both genetically and silviculturally, and changes in wood properties can be obtained relatively rapidly.

It was stated in Chapter 1 that essentially anything that will affect the growth of a tree can have an effect on wood properties. This is also true for tree form; anything the forester does to alter the form of a tree can also affect the quality of the wood that the tree produces (Burdon and Thulin 1965). Extensive literature has been developed relative to tree form and wood, many of which are summarized in Anonymous (1962) and will not be repeated here. In this chapter we only attempt to list a few of the many references available, because the results are so well known or are covered in other summary or bibliographical publications.

As used in this book, tree form refers mainly to the straightness of the tree stem and to the number, size, and angle of the limbs produced (Fig. 4.1). Bole straightness has a major effect on both the quality and yield of solid wood products and on fiber products. It affects wood qualities through the grain of the wood as well as through the reaction wood associated with crooked tree boles. This influences the methods used in the manufacturing process and their results. Limb characteristics affect the number, size, and quality of the knots and the amount of abnormal grain along with the reaction wood associated with branches and knots. Some like Sanwo (1983) have related wood properties to crown class in the stand (see Chap. 7). For example, in teak (*Tectona grandis*) he showed that crown class had only a small effect on specific gravity, with the dominant trees having a specific gravity 0.06 higher than the subdominant trees.

4.2 Wood Properties Associated with Poor Tree Form

The form of a tree influences both solid wood products and fiber products made from it. The variability in wood between poor and well-formed trees affects different products in various ways; therefore we will separate the discussion in relation to the effects of tree form on solid wood and on fiber products.

Poor tree form can be the result of crooked or leaning stems; any time a tree does not grow straight or vertically, it forms reaction wood (Zobel 1980b, Timell 1986). For example, in *Pinus caribaea* a positive relationship has been established between sawn board recovery and stem straightness (Smith 1973). Branching habit such as large limbs, steep-angled limbs, or abnormal growth like "foxtailing", in

Fig. 4.1. Tree form, straightness, and limb characteristics have a major effect upon wood qualities. Degraded stands such as the one shown will produce very little desirable wood. The best way to improve tree form is to straighten the trees and produce small, flat-angled limbs by silvicultural and genetic manipulation

which the tree produces limbs on only part of the tree bole or no limbs at all (Whyte et al. 1980b), is also part of tree form (Sect. 4.4.3). Other types of abnormal branching such as the virgate form of *Picea abies*, which has a pronounced reduction in the number of buds and shoots with horizontal or declining branches, has been studied for its wood properties by Ericson (1960b). He concluded that this unusually formed tree has slightly higher wood density than normal spruces because of a higher latewood content. He states "...the extreme branching habits of virgate spruce obviously have no influence on the wood properties...".

The size of limbs and associated knots are major factors in determining the quality of the resultant timber. For example, Malan and Toon (1980) found that knots in *Pinus caribaea* and *P. elliottii* grown in South Africa are well distributed and small, and their influence on wood quality is minimal. This contrasts with the large knots and limbs and poor wood quality in *Pinus radiata, P. patula, P. montezumae*, and *P. kesiya*. In *Picea abies*, Lewark (1981b) reported that knottiness is positively correlated with tree size. He states that although a volume increase of up to 25% may be obtained from selection, the changes in knottiness will reduce the wood quality and thus the potential gain from the added growth. Knots not only influence strength but they also affect workability, appearance, gluing properties, and paintability. The resin often found around knots create utilization problems, as does the abnormal grain angles which are present (Plumptre 1983). The adverse effect of knots relative to mechanical stress-grade yields was covered by Grant et al. (1984).

4.2.1 Solid Wood Products

Tree form influences qualities of solid wood products such as boards and plywood not only by stem straightness, limb size and number, and limb angle, but also by bark inclusions and chemcial deposits related to nonnormal wood and knots.

The effect of poor tree form on solid wood products is readily evident. For example, Cown et al. (1984) describe how conversion factors decreased with increasing sweep. Straight logs yielded 45% of the outturn in clear grades compared to 34 and 28% for moderately and severely swept logs, respectively. Sweep also reduced the length of clear boards. Crooked trees are difficult to saw or to peel efficiently, and the boards and plies produced are usually not stable, since they have an unusually high longitudinal shrinkage, resulting in cracking or warping. This was stated by Kellison et al. (1985) who reported that straight trees increased lumber recovery because there is less board deflection from straight trees. The effects and occurrence of compression wood and branch wood in radiata pine was well covered by Bamber and Burley (1983). For example, they report that compression wood comprises 30–45% of the wood in this species and it was not confined to leaning trees. About 37% of the branches consisted of compression wood. Compression wood was especially common in juvenile wood.

The quality considerations for solid wood products are of two types. Surface characteristics are of major importance in wood used for panelling and furniture, while the properties of stiffness and strength are important for wood used for construction and framing. Both types of quality are sensitive to wood property changes resulting from tree form differences.

Boards made from stems with large limbs have large knots which weaken them; this is especially so when numerous branches occur together in a cluster, which commonly occurs in tropical as well as some temperate climate pines. The grain deviations associated with knots result in weak boards. As reported for radiata pine by Bamber and Burley (1983), the angled grain from near the branch is 40 times weaker than straight-grained wood, resulting in low strength properties when knots are included in boards. If the limb angle is steep, the knots are even more undesirable and result in boards with low strength or in lumber that is difficult to finish.

In plywood, large knots, especially those with steep angles, cause considerable degrade. Crooked logs are difficult to peel and the plies formed have differential shrinkage in and around the knots. In *Pinus sylvestris*, Boutelje (1965) found that longitudinal shrinkage on the upper side of the knot was 1.5%, while on the underside of the knot it was 3.7%.

Additional defects from large limbs occur from ingrown bark around the knot, and quite frequently there are chemical depositions in association with the knots (Boutelje 1965; Rydholm 1965; Fig. 4.2), especially resin. These prevent the

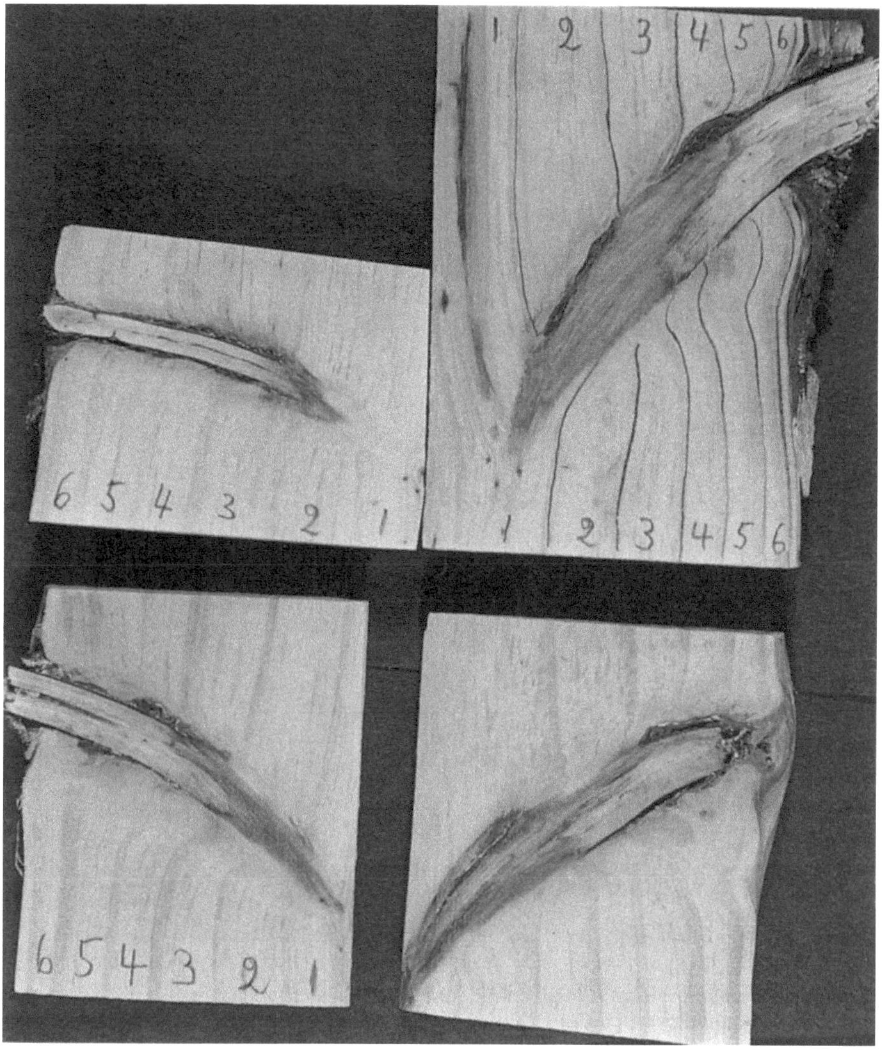

Fig. 4.2. Not only is knot wood itself undesirable but often there are resin deposits and ingrown bark associated with the knots along with compression wood. Such defects are undesirable for all wood products

production of a smooth finish, inhibit the proper use of natural stains and even of some paints, and cause problems in liquor penetration during pulping.

It is safe to say that if good quality solid wood products are to be obtained, trees must be straight and have small, flat-angled limbs. Any method to improve tree form, whether it be genetic or silvicultural, will greatly reduce wood variability and improve its properties. Much improvement is possible, as outlined in Chapter 7 which deals with the results of silvicultural manipulations of wood by methods such as stocking control and pruning. In addition to pruning and long

Fig. 4.3. The size and number of knots can be controlled by genetic manipulation. Shown are three clones (uninodal, multinodal, and intermediate) of *Pinus radiata* grown in Zimbabwe. Each tree type will have differing wood properties related to limb (and knot) characteristics (Zobel and Talbert 1984)

rotations, in certain species such as *Pinus radiata* it is possible to genetically develop trees with long, clear internodes between whorls; such work is well advanced in New Zealand (Chap. 8; Fig. 4.3).

Sometimes perfectly normal or apparently well-formed trees can have wood undesirable for solid wood usage. Trees that grow crooked at a young age sometimes "straighten up" by differential growth and formation of reaction wood to produce a well-formed mature tree that, in fact, can have a considerable defect in the stem center caused by crook when the tree was young. Such trees are sometimes referred to as having "wandering pith." Straightness of a young tree will be affected by wind, snow, and ice: all trees have some reaction wood no matter how straight they appear to be. Eccentricity of the bole in regard to the pith was quite pronounced in the straight and cylindrical yellow-poplar trees investigated, according to Taylor (1968).

Trees that have been pruned can appear to be superior externally at older ages, but the center will not only contain the knots from the pruned branches but bark pockets and resin depositions where the pruned limb has healed (Rydholm 1965). Even in apparently normal trees, one side can grow faster than the other, and abnormal wood can result which is inferior for solid wood products (Illy and Polge 1967).

4.2.2 Fiber Products

As explained in greater detail in the following section of this chapter, poor tree form can have a major effect on fiber products. Most problems are related to the production of reaction wood with its many effects on the manufactured product; because of its importance and interest to the pulp industry, a short, separate section dealing with reaction wood follows below.

Knots have an effect on fiber production through the bark and chemical impurities that develop in conjunction with knots (Rydholm 1965). Also, since knot wood is denser than stem wood and has a high resin content in the conifers, it is difficult to penetrate with pulping liquors, and the partially pulped knots are

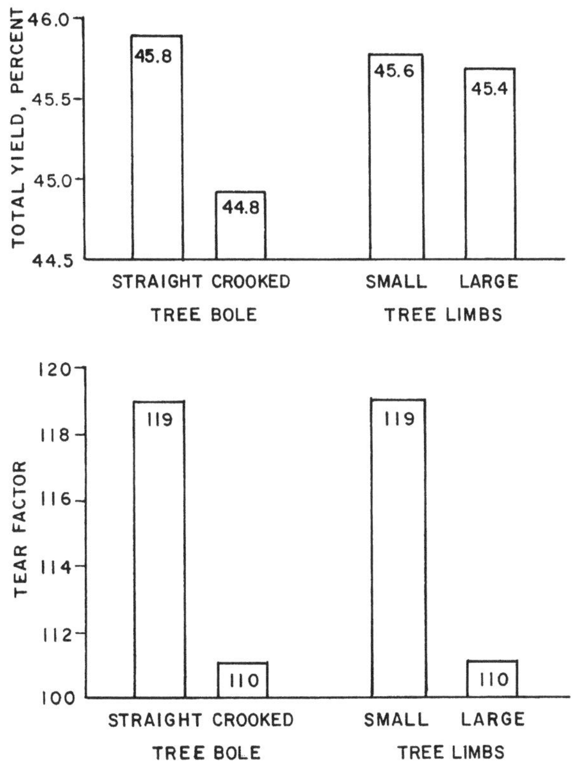

Fig. 4.4. Tree straightness and limb size affect yield and quality of the paper produced. Shown is a summary of a very large study involving straight, small-limbed trees; straight, large-limbed trees; crooked small-limbed trees; and crooked, large-limbed trees of young loblolly pines. For details see the text and Blair et al. (1974)

present in the screenings and as unpulped residues. The result is an increase in power consumption and costs, especially for groundwood or thermomechanical pulping. Knots sometimes must be recycled several times to become useful. Branch size affects the final product not only by knot volume and associated compression wood, but also by the proportion of included bark and resin depositions.

Contrary to the belief of some investigators, tree form is of major importance in determining the quality of fiber products. We have been seriously challenged about attempts to produce straight, small-limbed trees for pulp products. There are now quantitative data that show the importance of tree form for fiber products. For example, in an intensive pulping study of the effect of bole straightness and limb characteristics on pulp yield and quality of loblolly pine, Blair et al. (1974) found that both percentage yield and tear strength were considerably better in paper from straight than from crooked trees, while those trees with large and steep-angled limbs produced paper with poor tear strength and lower pulp yields (Fig. 4.4).

4.2.2.1 The Effect of Limbs Upon Fiber Products

In the past, the use of limb wood was primarily for fuel. But as total tree chipping becomes more prevalent, wood qualities of branches become of increasing importance. The percentage of chips that come from limbs can be very large in certain species, especially when the trees are open-grown. There has been a strong reaction against the use of limb wood, mostly because of its high bark percent and because of the high proportion of reaction and knot wood present. The quality of branch wood will continue to become of increasing importance as better harvest utilization is practiced. This kind of wood contributes considerably to the variability of the quality of wood used as chips.

The properties of limb wood have been assessed by several investigators. Some of these relate to the possibility of using branches to assess the wood quality of the boles of the trees; this would be a most useful nondestructive sampling procedure if it works (Fig. 4.5). In 35-year-old cherry (*Prunus serotina*) and red oak (*Quercus rubra*) Hamilton et al. (1976) found a low but statistically significant relationship between branch and bole wood specific gravity. Limb specific gravities were higher than bole specific gravity and decreased toward the end of the stem. For another hardwood, *Platanus occidentalis*, Land et al. (1983) found that the specific gravity of limb wood was greater than that of bole wood.

In a study of branch to bole wood of spruce, Holst (1960) reported that a high correlation was found between the specific gravity of adjacent branch samples of the fourth, sixth, and eighth whorls and the stem specific gravity. More distant branch samples had specific gravities that were only slightly correlated with that of the stem. For *Cryptomeria*, Ando and Hirozami (1961) found that the base of the branch had a higher bulk density than the stem, but the ends of the branches had a lesser density. They noted that the value at the middle portion of the branch was close to that of the stem. A correlation between bulk density of stems and branches was significant only in the older age classes (29–35 years of age).

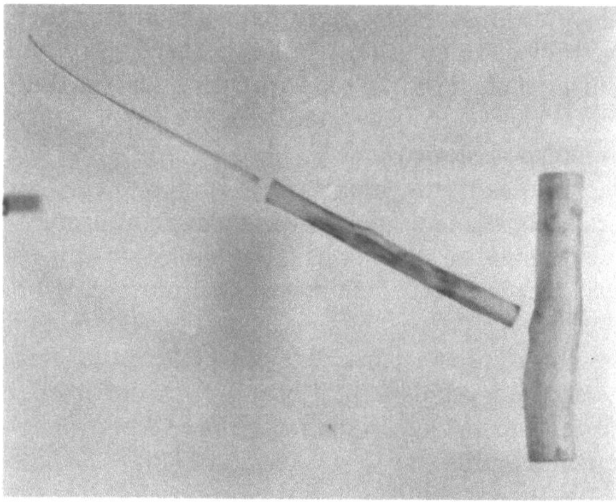

Fig. 4.5. Considerable research has been done to determine the relationship of limb to bole wood characteristics in a tree. Shown are the sampling points tried. Overall, such methods of limb sampling have not been successful, and the bole of the tree itself must be assessed if reasonable results are to be obtained

Similar assessments were made for black pine (*Pinus nigra*) by Lee (1971) and for Douglas-fir by McKimmy and Ching (1968). In pine, there was no correlation between tracheid length of branch and bole wood but an r = +0.76 was found for specific gravity. In Douglas-fir a very high correlation was found between the specific gravities of branch and bole woods. McKimmy and Ching (1968) found that the specific gravity of branch wood is usually higher than that of the bole. This relationship seems to be quite general as was reported by Watanabe et al. (1962) for *Pinus densiflora*. For *Pinus pinaster* in France, Polge and Illy (1968) showed that there was such a high correlation between branch and bole woods of young trees that branches could be used to obtain a good estimate of bole wood. In a study of several conifers, Schultz-Dewitz (1969) reported that branches have shorter and narrower tracheids than the bole wood. The authors concluded that the inclusion of branch wood in the furnish within the natural amount would be feasible in kraft pulping.

The general concensus seems to be that branch wood can be used to assess bole wood, but the relationship is often not as strong as desired.

Cell size and lumen diameter were very small in branch wood compared to stem wood (50 to 60% of stem wood sizes) in *Pinus densiflora* (Tajima 1967). Branch wood also had lower holocellulose and alphacellulose yields and slightly higher hemicellulose yields than stem wood.

Some studies have been made on the use of limb wood for pulp. For example, Law and Koran (1982) found that barked white spruce branch wood was comparable in yield to bolewood; unbeaten pulp from branchwood had higher bulk and stretch and comparable tensile and burst strengths to that from bole wood. Branch wood also had lower tear strength than bole wood. When beaten, pulp from branches gave similar tensile and stretch values as bole wood, while burst and tear were less; when bark was included there was an increased chemical consumption, and sheet cleanliness was unsatisfactory. In a summary of the potential use of radiata pine limbs, Bamber and Burley (1983) state "branches are a potential source of fibre for papermaking while the usefulness of radiata pine branch wood for paper does not appear to have been examined. It is likely from its anatomical characteristics – that it would be markedly inferior to stem wood and that its characteristics would resemble paper made from trunk wood containing compression wood."

Because of the heavy limbs and branched tree form of hardwoods, and because of the movement toward greater utilization of forest residues, the utilization of limbs of hardwoods is increasing rapidly. For example, Law and Lapointe (1983) reported on chemimechanical pulping of branches of white birch (*Betula papyrifera*) and trembling aspen (*Populus tremuloides*). The branches of the birch and aspen contained less fiber and more vessel elements than the corresponding boles; the fibers and tracheids of the branches were 30%–50% shorter and smaller in diameter than those from the boles. In bisulfite pulping, branches produced 5%–7% less pulp than the boles; the pulps were also weaker.

The major effect of knots on pulpwood is related to the knot itself as well as the related reaction wood (Watanabe et al. 1962). Knot wood is considerably higher in density than stem wood; from 1020 to 1080 kg/m^3 in spruce (Nylinder and Hägglund 1954). Baker (1968) recommends that samples be taken 7–9 in. (18–23 cm) above the bole from the knot to obtain wood that has normal properties, including specific gravity. Upon pulping, knots sometimes produce an excess of fines and shives, causing a high amount of rejects. Quite frequently the knot wood must be recycled, sometimes several times during the pulping process before satisfactory delignification occurs. In a large study involving trees with differing limb sizes (Blair et al. 1974), the chief defect of paper that resulted from more and larger knots was its low tear strength.

Limb angle is important since steep-angled limbs that tend to grow strongly upward have larger knots along with a large amount of associated reaction wood. Knot wood and its associated reaction wood occur in an approximate ratio of 1:7

Table 4.1. Knot wood and compression wood in 5-ft bolts of 11-year-old loblolly pines[a]

Bolt No.	Average branch diameter (inches)	Branch angle (deg.)	Volume bolt with:			Percentage of bolt with knot and compression woods
			Knot wood	Compression wood	Knot and compression woods	
1	0.4	69	0.3	5.4	5.7	21
2	0.7	62	0.7	7.0	7.7	28
3	0.8	59	0.8	6.3	7.1	25
4	0.9	56	1.0	6.5	7.5	25
5	1.1	52	1.5	8.3	9.8	28
6	1.1	53	1.6	7.3	8.9	23
7	1.1	54	2.3	11.1	13.7	22
8	1.0	53	2.3	10.5	12.0	17
9	0.9	51	2.6	11.1	13.7	8
10	0.7	51	1.9	7.4	9.3	3

[a] Adapted from von Wedel et al. (1967).

in 7- and 8-year-old loblolly pine (von Wedel et al. 1968). The last merchantable bolt near the limby top of the tree, contained 2.3% knot volume together with 8.9% by volume of associated compression wood. The proportion of knot wood and compression wood associated with knots is shown in Table 4.1. Note the percentage increase of knot plus compression woods with increasing stem height except for the top bolt which has very small limbs.

In their study of radiata pine, Bamber and Burley (1983) reported in some detail on wood of this relatively large-limbed species. Branch wood shows properties that differ from stem wood, having higher density but lesser tracheid length and tracheid diameter. The knots contain more resin and are harder to surface than normal stem wood when wood is finished.

4.3 Reaction Wood

Reaction wood is of major importance in causing variation in wood properties. It was studied and described a long time ago, as discussed by Dadswell (1960). For example, he states "reaction wood formation usually involves eccentric radial growth of the stem; in hardwoods this is on the upper side of leaning stems whereas in softwoods it is on the lower side" (Fig. 4.6).

There are several excellent summaries of reaction wood, such as the recent one on compression wood by Timell (1986), and it would be repetitious to go into great detail about reaction wood in this chapter. Therefore, we will cover reaction wood somewhat superficially, only going into enough detail to indicate its importance and effects upon wood variation plus a little about its control. Naturally, therefore, the literature coverage of reaction wood is very incomplete.

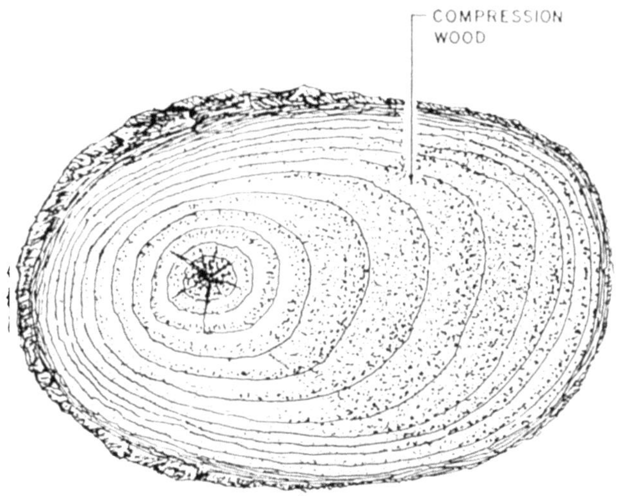

Fig. 4.6. Stem eccentricity is common when reaction wood is formed. It usually occurs on the upper side of the stem for hardwoods and on the lower side for conifers, as shown

When a tree leans or twists, it produces wood with characteristics that differ from the normal which is called reaction wood (Scurfield 1973). Reaction wood is of two types: when produced in a softwood it is called *compression wood*, while wood from crooked or leaning hardwoods is called *tension wood* (Dadswell et al. 1958). Both types of reaction wood are generally considered to be undesirable for most forest products. Although the generalization about kinds of reaction wood in conifers and hardwoods is usually true, there are exceptions. For example, hardwoods such as *Buxus sempervirens*, whose wood consists primarily of tracheids, form compression wood instead of tension wood (Höster and Liese 1966).

There is controversy as to whether reaction wood should be considered to be abnormal wood. For example, Shelbourne and Ritchie (1968) state that reaction wood varies greatly from normal wood in specific gravity, cell dimensions, and chemical constituents. Similarly, Barefoot (1963a) considers reaction wood to be abnormal. However, other authors feel that reaction wood is always present to some extent and it should be considered as a normal variation of wood. Clarke (1939) states that, in fact, reaction wood is not abnormal but is only an extreme form of normal wood. Similarly, Rendle (1956) refers to compression wood as a natural defect of softwoods because it is so common. As Dadswell (1958) stresses, reaction wood can result from any number of external influences and the changes in structure that it causes are superimposed on the usual changes occurring in normal wood.

Despite the widespread study of reaction wood, some researchers feel that its importance to wood and product quality has been overemphasized. For example, Shelbourne[1] states that all but the more extreme forms of reaction wood have a minimal effect on utilization or product quality. He reports that wood and pulp technologists in New Zealand do not seem to be greatly concerned about compression wood in radiata pine. As described below, others consider reaction wood to be of great importance for the quality of both solid wood and fiber products in both conifers and hardwoods.

The causes of reaction wood are complex and somewhat controversial. The common causes cited are that reaction wood occurs when a stem is not straight or vertical and that it is also formed in association with limbs. It can also result from constant prevailing winds from one direction (Burton and Smith 1972) or from a one-sided crown. The causative mechanism usually is related to gravitational responses of the tree, and their effect on the distribution of auxins and turgor pressure.

Boyd (1977) claims that tension wood results from imposed tensile stresses and that compression wood is a response to compressive stresses. Although this is still debated, the formation of reaction wood appears to be a gravitropic response, related to intrinsic growth direction and auxin distribution. It has been found that reaction wood can sometimes be produced in straight trees, as observed by the authors. This was also reported by Shelbourne et al. (1969).

Reaction wood varies among trees that are of equal straightness and from species to species. For example, Plumptre (1983) reported that there was more

---

[1] Personal communication, Forest Research Institute, Rotorua, New Zealand.

compression wood in *Pinus caribaea* var. *caribaea* than in the other two varieties of *P. caribaea*, even though the trees of this variety are straighter. In cases like this, trees may be straighter because they respond more strongly to deviations from straightness and thus produce more reaction wood which straightens the tree.

Sometimes reaction wood is caused by an apparent physiological upset or as a result of an injury. Fire or pest attack appear to be causative mechanisms in apparently straight trees. For example, Amman (1970) and Smith (1967) reported that compression wood (Amman calls it redwood) was common in Fraser fir (*Abies fraseri*) and grand fir (*Abies grandis*) caused by the balsam wooly aphid (*Adelges piceae*); Doerksen and Mitchell (1965) reported similar production of compression wood in the western firs (*Abies grandis, A. lasiocarpa*, and *A. amabilis*). The authors state that the mechanical and pulp properties of the wood from aphid-infested trees differ significantly from wood of trees free from attack. This has become a major problem in South Africa where the wood of *Pinus taeda* has so much compression wood following aphid (*Cinara cronartii*) attack that planting of this species has been stopped and is only now starting again since aphid-tolerant loblolly pines are available (van der Sijde et al. 1985).

The causes of reaction wood are not the subject of this book, but the fact that reaction wood causes wood variation makes reaction wood relevant to discussion. The reader must be aware that tree form is generally related to the amount of reaction wood, and that if wood variability is to be reduced, reaction wood must be controlled. Therefore, its general causes and control must be appreciated. Improving tree straightness, reducing limb size, and improving limb angle are ways of reducing the amount of reaction wood and improving wood unformity and utility. Additionally, there appears to be genetic control of a hormonal imbalance, as well as the imbalance resulting from thinning, pruning, and insect and/or disease infection (Shelbourne 1966, van der Sijde et al. 1985).

4.3.1 Compression Wood – Conifers

Conifers produce compression wood on the underside of the leaning stem when a tree is not straight (McElwee and Zobel 1962); it is also produced in association with limbs and sometimes from a physiological upset. It is difficult to refer to compression wood as a given type of wood. According to Timell (1983), compression wood is a vital tissue in the gymnosperms, and has probably existed throughout the evolution of the conifers. As emphasized by Shelbourne and Ritchie (1968), there are three recognizable grades of compression wood; the severe type has the most effect on the quality of wood products.

Compression wood has been widely studied and its general causes and specific effects are well known; these are summarized in the reviews by Low (1964), Westing (1965, 1968), and Timell (1981 a, b). The ultimate coverage of compression wood is the three-volume treatise by Timell (1986). He discusses every conceivable aspect of compression wood from its causes and development to its effect on the final product. The world literature on the subject is thoroughly reviewed. One chapter of Bamber and Burley's book on the wood properties of radiata pine (1983) includes the properties, occurrence, and effects of compression wood.

Compression wood is generally considered to be inferior for both pulp and solid wood products.

One fact not commonly understood is that, in addition to lean in a tree or in association with limbs, compression wood is often related to the slope of the terrain on which the trees are grown and to the direction of prevailing winds. This was reported for *Pinus caribaea* from Puerto Rico by Boone and Chudnoff (1972). In an attempt to show the effects of wind stresses, Burton and Smith (1972) guyed trees to prevent swaying. They found that the tied tress had a reduced specific gravity and less compression wood than the trees that were influenced by the wind.

Compression wood often has tracheid characteristics and a chemical composition different from those of normal wood but this is not always so. Even though compression wood has shorter cells, Goddard and Strickland (1964) found only a small positive correlation ($r = 0.08$) between the straightness and tracheid length in slash pine. This is probably the result of the relatively small amount of mild compression wood related to this relatively straight species. A series of papers by Wardrop and Dadswell (1950, 1952) covered the cell wall organization, characteristics, and formation of compression wood. A most interesting work by Timell (1983) summarized the origin and evaluation of compression wood from a study of fossils and primitive extant gymnosperms. It was less definite and somewhat different in early forms of gymnosperms such as *Ginkgo biloba* and not even evident in some of the earliest forms of gymnosperms.

There can be varied (and considerable) amounts of compression wood in a tree. For example, in a good lowland site Carmichael (1969) found Canadian conifers contained from 5.3% to 26.1% by volume of compression wood. Generally, the largest amount of compression wood was found at the base of the tree. When comparing amount of compression wood with tree straightness, Zobel and Haught (1962) found that 6% of the volume of straight loblolly pine consisted of compression wood, average trees had 9%, and crooked trees of marginal value contained 16% compression wood. One very crooked cull tree had 67% of its volume as compression wood. Generalized for pine, compression wood varies from 5.4% to 57.1% of the commercial volume (Low 1964).

Many more examples of the amount of compression wood could be cited; the key point is that compression wood often makes up a significant part of the merchantable volume of leaning, crooked and heavy-limbed trees. The best way to reduce the percentage of compression wood is to improve tree form.

There has been considerable discussion on the properties of wood formed on the side of the tree opposite to compression wood. Although some authors (Tajima 1967) refer to it as tension wood, that terminology is not correct since tension wood only occurs in hardwoods. Most persons find that opposite wood is very similar to normal wood. For example, Timell (1973a) reported that opposite wood has exactly the same lignin content, cellulose, and hemicelluloses as corresponding normal wood. However, Timell (1973b) found differences in cell structure and concluded that normal coniferous wood can be regarded as intermediate between opposite wood and compression wood. Larson (1969b) feels that opposite wood may not be normal because it sometimes has a higher cellulose and lower lignin content than normal wood. The concept that opposite wood constitutes a third kind of wood within the tree adds even more to the picture and confusion of wood variability in forest trees.

Compression wood formation is the method whereby trees that are not straight are "pushed" into an upright position. As a result of this, Shelbourne (1966) found that some of the straighter trees among the southern pines had the largest amount of compression wood. According to Allen (1985), "...trees having particularly poor values for compression wood should be rejected from breeding populations... ".

4.3.1.1 *Characteristics of Compression Wood*

There are many references that indicate how compression wood differs from regular wood. Any broad generalizations about compression wood are difficult to make because it varies from mild to severe and varies greatly with growth conditions, tree species, tree age, and individual tree. Most of the adverse characteristics of this type of reaction wood are related to the more severe form. Reports in the literature usually do not make clear what type of compression wood is being discussed. A few of the most important characteristics of severe compression wood that affect wood uniformity are listed below:

1. Compression wood is generally assumed to have a higher specific gravity than normal wood. For severe compression wood this can vary from 15% to 40% above normal as stated by Low (1964), or twice as heavy according to Timell's (1981a, b, 1986) recent, comprehensive reports. In *Pinus wallichiana*, Jain and Seth (1979) found that where false rings or compression wood occurred, there was a localized increase in specific gravity. The change is slight on the border of the false ring compared to that of a normal ring. Contrary to the above reports, Shelbourne and Ritchie (1968) did not find any relationship between amount of compression wood and specific gravity in 11-year-old loblolly pine.

2. Compression wood has a higher lignin content according to Timell (1981a, b, 1983) who reports 30% lignin in normal wood and 39% in compression wood of conifers. Koch (1972) shows for loblolly pine that normal wood has 28% lignin, while compression wood has 35%.

3. Tracheids of compression wood are shorter than normal wood (Wardrop and Dadswell 1950, Shelbourne and Ritchie 1968). Fibril angle is much flatter, usually 30°–50°.

4. Compression wood cells tend to be rounded, leaving voids at the corners where the cells come together. They usually have helical cavities in the cell wall, resulting in cell fractioning during manufacture.

4.3.1.2 *Effects of Compression Wood*

Although compression wood can often produce a poor final product, the relatively small overall quantity harvested from a good forest stand has a minimal adverse effect. When excessively crooked or limby forests are produced as a result of poor silviculture, ice, wind, or use of genetically inferior stock, or when thinnings are made of the poorest formed trees, the amount of compression wood becomes large enough to cause an appreciable loss in quality of the final product.

Considerable amounts of compression wood can have a major effect on the quality of the final product. Since compression wood is lower in cellulose (30%) than normal wood and has a higher specific gravity, both yields and quality of paper produced will be different than from normal wood. Because of the differences in cell morphology and chemistry, a negative correlation usually occurs between compression wood and pulp yield. For example, Ladell et al. (1967) found an $r = -0.745$ between pulp yield and percent compression wood in black spruce (*Picea mariana*). They calculated that the reduction in pulp yields from compression wood in this species would result in the loss of about 1 ton of usable fiber per acre.

Working with *Pinus radiata*, Watson and Dadswell (1957) found that compression wood produced a paper with low strength properties and gave low yields of pulp with a high lignin content. They concluded that the poorer strength properties from compression wood are the result of changed fiber dimensions and structure rather than because of an undesirable chemical composition. For example, the short tracheids found in compression wood result in a low tearing strength, and the bulky compression wood fibers do not respond well to beating and produce paper with low burst and tensile strengths.

In a very early study on loblolly pine, Pillow and Bray (1935) found that in using the sulfate process the characteristics most affecting the quality of pulp made from compression wood are the shorter fibers and greater fibril angles compared to normal wood. Individual compression wood fibers disintegrate easily on mechanical treatment. Since compression wood has a lower cellulose and higher lignin content, the yield of pulp is lower, and consumption of chemicals is higher than for normal wood. The adverse quality effect of compression wood is particularly important for mechanical pulp, according to Dadswell et al. (1958). The flat fibril angle and low cellulose content of compression wood tracheids cause them to have a low tensile strength which contributes to poor quality pulp and paper. The high lignin content produces problems in bleaching.

The presence of compression wood can cause degrade in solid wood products because of its effect on shrinkage and related seasoning defects. The flat fibril angle in compression wood tracheids result in excessive shrinkage along the grain and reduced shrinkage in the radial and tangential directions. Wood normally shrinks very little longitudinally (generally 0.1% to 0.3%), but compression wood can have shrinkage in excess of 3%. Thus, rather than the normal shortening of a 16-ft board by 0.2 in. upon drying, the board with severe compression wood will shrink 5.7 in. or more (Wooten 1968). The real problem occurs when part of a board is compression wood and part is normal wood. The differential shrinkage results in all kinds of defects, such as warping (Wooten 1968). Even though longitudinal shrinkage increases with increasing amounts of compression wood, careful seasoning using kilns can reduce degrade, according to Boone and Chudnoff (1972).

Most strength properties of products made from wood are fairly closely related to specific gravity. However, even though compression wood usually has a high specific gravity, strength properties of products made from it often do not reflect the qualities that would be expected from a given specific gravity. In a comparison of the strength of boards made from compression and normal woods, Banks (1957) reported that the modulus of elasticity and stiffness were less, while hardness and side compression were stronger for compression wood than for normal wood with the same specific gravity. Since compression wood is usually denser, strength property differences from compression and normal woods from the same tree are often not too large. However, instability of the boards made from compression wood is usually severe.

The properties of compression wood are particularly important in short rotation, tropical conifers. Many plantations have poor tree form with excess crook, forking, and foxtailing (i.e., the tree does not produce normal side limbs). The effect of foxtailing on wood qualities has been partially covered in Section 4.4.2. Its importance is debated; some report that normal wood is formed (Woods et al. 1979), while the authors have observed a large amount of compression wood in foxtailed trees. Often the wood of severely foxtailed trees is soft and weak because it is spotted with needle traces, and because the amount of foliage maintained on older trees is insufficient for suitable lignification. The needle trace pattern can result in a rather attractive grain pattern when machined correctly.

Not everyone agrees on the seriousness of compression wood. Shelbourne and Ritchie (1968) found that the mild type of compression wood had very minor importance. For *Pinus caribaea* Plumptre (1983) reports that compression wood is not such a serious defect if the wood is cured correctly. Certainly, modern technology has been able to somewhat reduce its adverse effects.

4.3.2 Tension Wood – Hardwoods

Hardwoods produce tension wood on the upper side of the bole when a tree is not straight and on the upper side of the limbs. As defined by the International Association of Wood Anatomists, "reaction wood is formed typically on the upper side of branches and leaning or crooked stems of dicotyledonous trees are characterized by lack of cell wall lignification and often by the presence of an internal gelatinous layer in the fibres" (Barefoot 1963a). Tension wood produces tensile forces that bring the tree into a more normal, upright position (Wardrop 1961). To test this Wardrop bent limbs upward from their natural position, and found that tension wood was formed on the bottom of the limb so it would return the limb to its normal position. Although gelatinous fibers are typical of tension wood (Arganbright and Bensend 1968), not all leaning trees produce them. Some species, such as *Tilia cordata* and *Liriodendron tulipifera* do not produce typical reaction wood (Scurfield 1973), although Barefoot (1963a) reports its presence in about 30% of the *Liriodendron* trees he examined.

Tension wood has been studied almost as extensively as compression wood. The most extensive reviews are those by Hughes (1965) and Wicker (1979). Information is also found in papers such as those by Wardrop and Dadswell (1955), Wardrop (1956), and Timell (1969). Rather than having heavily lignified cell walls as in compression wood, tension wood is characterized by greatly reduced cell wall lignification in the zone of gelatinous fibers. The reasons for tension wood formation have been extensively discussed, but it is not well known. It appears to be a gravitropic response similar to compression wood, and it results from hormonal stimuli. Rendle (1955) categorizes tension wood as a natural defect of hardwoods just as he does for compression wood in conifers.

4.3.2.1 Characteristics of Tension Wood

Some of the most important characteristics of tension wood that make it different from normal wood are:

1. The vessels are fewer and smaller, and the fibers are usually slightly longer than those in normal wood.
2. The lignin content in the overall tension wood zone can vary from low to about normal, while the cellulose content is high; the cellulose is highly crystalline (Bentum et al. 1969). The wood outside the gelatinous layer has about the same lignin and hemicellulose contents as normal wood. It is the addition of the pure cellulose gelatinous layer that produces lower lignin and hemicellulose contents by weight (Timell 1969).
3. Gelatinous layers are formed which are the main diagnostic feature of tension wood (Timell 1969). This layer is nearly pure cellulose and is not lignified.
4. Specific gravity of tension wood is often greater than that of normal wood (Kaeiser and Boyce 1964).

4.3.2.2 Effects of Tension Wood

Some foresters minimize the importance of tension wood and choose to emphasize other causes for wood variation. However, it can have a major effect on the quality of the final product. For example, Goebel et al. (1960) reported excessive splitting in hickory trees that had gelatinous fibers, probably from excessive shrinkage along the grain. Longitudinal shrinkage of tension wood is much larger than that of normal wood. Tension wood also causes warping and sometimes "collapse" in boards. It also causes instability in solid wood products on drying. Gelatinous fibers also cause an excess amount of radial shrinkage (Barefoot 1963a). Dadswell (1985) states that shrinkage is less than 1.5% which is considerably lower than the 3%–5% sometimes found in compression wood but 1.5% is still a lot of shrinkage, when compared to 0.1%–0.2% for normal wood.

The effects of tension wood on fiber properties is debated. For example, Knigge and Koltzenburg (1965) cite five papers on *Fagus* and *Populus* where no influence of tension wood was found on fiber length, two papers where fibers were longer, and one where they were shorter. They state that much more detailed and more controlled studies need to be made; this conclusion is still largely correct in 1988, because most of the work in this area was done before the 1970s.

In a study on *Populus*, Parham et al. (1976) found that tension wood produced papers with lower strength properties but with higher bulk, porosity, opacity, and brightness than normal wood. In fast-growing hybrid poplars, as little as

1.8% and as much as 35% of tension wood was found by Stevens et al. (1983), who concluded that "one can suppose that the phenomenon always occurs in one degree or another in poplar boles."

According to Perem and Clermont (1961), tension wood often results in weaker pulps because it easily "overcooks"; however, it grinds easily and may be pulped by the semichemical process to yield pulps having good strength properties. In a study of two *Eucalyptus* species, Watson (1956) found that pulp yields were much higher for tension wood than for normal wood but the tension wood pulps had inferior strength properties, especially those depending on fiber bonding. The main factors causing the low strength were the low hemicellulose content and thick cell walls of the tension wood.

In summary, tension wood, like compression wood, causes variability in wood quality that can affect the final product. It is more complex in its formation and effect than compression wood but can be controlled to a considerable extent, both silviculturally and genetically, by improvement of tree form in hardwoods.

4.4 Methods to Improve Tree Form

4.4.1 Better Forest Management

The effects of various forest management practices on tree form and wood properties are considered in detail in Chapter 7. However, a few will be mentioned here so that the reader is aware that improvement of wood through the production of better formed trees by silvicultural manipulation is possible.

The most obvious method of improving tree form is through silvicultural techniques such as maintaining a stocking which allows the trees to grow normally, and thinning at the best time for the development of good tree form. Such prompt silvicultural action is particularly needed for many of the hardwoods which tend to produce a rounded top with large, heavy limbs if open-grown. Proper stocking can overcome this problem to some extent, although all tree species tend to produce excessively large limbs at low stocking levels. Trees that are spaced too closely and thus are tall and skinny, are particularly susceptible to bending by wind, ice, and snow. Proper stand density is essential for producing a well-balanced tree with a good form (Fig. 4.7). Another silvicultural practice to improve wood properties used with some species is to prune the trees to obtain the maximum production of clear wood.

One of the least understood forest management practices to ensure better quality trees is correct planting. Trees planted with U or J roots produce a tree that may grow well but will be unstable. As the trees become larger, high winds and rain or snow will cause the trees to lean, leading to a forest with sometimes as much as 40% or 50% reaction wood in the merchantable bole. Poor planting is a widespread problem and often results in forests with an excess of undesirable wood.

Another forest management practice that sometimes results in poorly formed trees is the heavy application of nitrogen fertilizers. It has been found that trees caused to grow very rapidly "artificially" are often crooked. In contrast, however,

Fig. 4.7. Good forest management is required if satisfactory wood is to be produced. For example, the overdense natural stand of loblolly pine (*left*) will produce greatly different wood than the stand grown at normal spacing (*right*)

genetically fast-growing trees are usually no more crooked than slower growing trees.

4.4.2 Use of the Proper Seed Source

An easy way to improve tree form is to use the best geographic seed source. This was covered in Chapter 2, and only a few examples are listed here. Ill-adapted provenances are often crooked and forked. In tropical pine plantings they frequently have the foxtailed form (Leding and Whitmore 1981) and are often multistemmed or "basket-topped". The foxtailing habit is discussed in Section 4.5.3 in Zobel et al. (1987). A number of studies have been made on the effect of the foxtail habit on wood properties. Since the needle traces go to the pith (center) of the tree, wood from foxtailed trees have numerous small holes giving the wood a dippled appearance when finished. Much of the wood from foxtail trees is juvenile so it has the characteristics of this kind of wood. In *Pinus caribaea*, Plumptre (1983) found that the specific gravity was normal but there was a lesser gradient from the center of the tree to the bark. Similarly, Whyte et al. (1981) found that foxtail trees had normal wood densities. In Venezuela, Woods et al. (1979) found no difference at breast height but lower specific gravity at the 5-m height. Others report that foxtail trees produce wood properties different from normal. We have observed that in older foxtailed trees there is much compression wood and that the wood density is less than that of normal trees.

It has been found that seed sources from the higher latitudes and higher elevations usually produce straight, small-limbed trees with relatively horizontal branches, but these well-formed trees are often slower growing so their use is limited (Zobel and Talbert 1984). For Scots pine, Remröd (1976) reported that the length and diameter of branches are smaller in northern provenances; these also have limbs at a right angle between the branch and the trunk. There can be large differences from different stands growing in the same general environment. For example, for Douglas-fir, Campbell (1965) found considerable differences in stem sinuosity among different stands within each of the tree farms assessed, although there generally was little overall difference among the farms. In ponderosa pine, La Farge (1974) reported large differences in stem form and taper among different seed sources.

The better form of trees resulting from the use of the best-adapted sources can have a major effect upon wood quality. The source consideration for improved tree form is particularly critical for the tropical pines, where vast acreages have been planted which are producing wood of marginal value because of very poor tree form. Poor tree quality from use of a poor provenance can be somewhat improved by intensive selection and development of a land race. This has proven to be successful in many species. Tree form responds exceedingly quickly when land race development is employed. Therefore, very poorly formed plantations, such as *Pinus caribaea* in the Orinoco Basin of Venezuela, can be greatly improved with selection of the occasional excellent tree from the overall poorly formed plantations already established.

Provenance is also of great importance in *Eucalyptus*. For example, Karami (1973) found that the good stem form of the improved *E. grandis* from Uganda was also present in the Wingham provenance from Australia. He points out that the same improved form could have been obtained earlier if the good Wingham provenance had been used rather than waiting the many years it took to develop the Uganda land race from unknown sources.

4.4.3 Breeding for Branching Characteristics and Bole Straightness

The inheritance of wood properties is discussed in Chapter 8. However, this short section on the genetics of tree form is included to illustrate the potential for bettering wood properties by improvement of tree form. Fortunately, inheritance for tree bole straightness is strong enough so that it is possible to breed successfully for straighter trees (Shelbourne and Stonecypher 1971). The genetic control of straightness appears to become stronger as trees grow older.

Straight trees usually have the advantage of containing less severe compression wood, although some families tend to produce more mild compression wood than others quite independently of tree straightness (Shelbourne et al. 1969). Although not explained, the inheritance pattern was relatively strong, with some families being much more prone to the formation of mild compression wood than others.

Fig. 4.8. Although branch size is often reported to have a low inheritance, differences are sufficient to make breeding for improvement worthwhile. For example, seven families of pine showed considerable differences in branch size, as illustrated. Trees are of the same age, growing at the same site

Table 4.2. A few examples of the inheritance of tree straightness

Species	Reference	Inheritance pattern
Pinus taeda	Perry 1960	Straight parents produced straight progeny
Pinus taeda	Goddard and Strickland 1964a	Large differences occur in the inheritance of crookedness. Tracheid length was only slightly affected by straightness
Pinus elliottii	Gansel 1965	The number of crooks/ft had $H^2 = 0.38$; degree of crook had $H^2 = 0.34$
Larix spp.	Keiding and Olsen 1965	Large differences in straightness were observed among clones
Pinus taeda	Shelbourne 1969	$h^2 = 0.39$ for stem straightness, a heritability high enough to obtain good gains quickly
Pinus sylvestris	Ehrenberg 1970	There was a good inheritance for straightness
Dalbergia sisoo	Vidacović and Ahsan 1970	Heritability of crookedness varied from 0.42 to 0.65

This tendency has not been exploited by tree breeders. The propensity of compression wood to be genetically controlled was also reported by Burdon (1974) for *Pinus radiata* where differences in amount of compression wood were observed among families that appeared to be equally straight. The authors have both observed that some trees with the same amount of lean develop wood with much larger amounts of reaction wood than others.

Many studies have shown a positive response to breeding for straightness. A few are listed in Table 4.2.

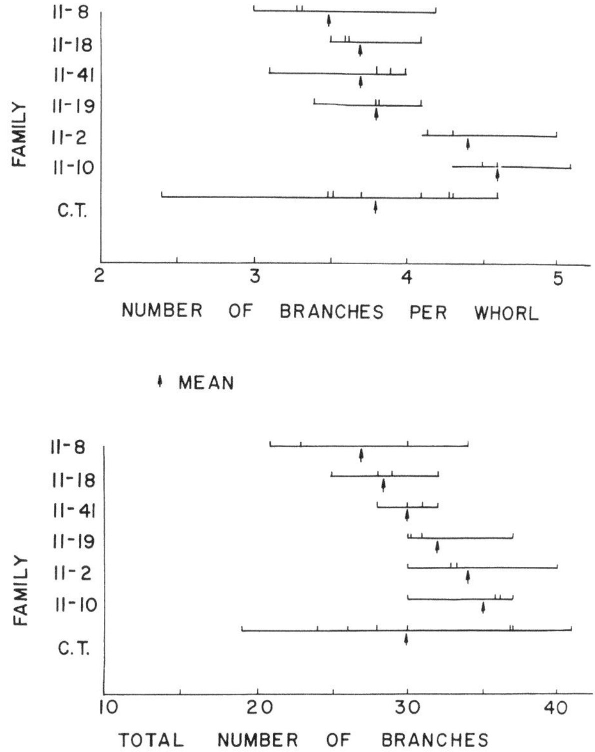

Fig. 4.9. The number of limbs is of great importance in wood quality determination. Both the total number of branches and branches per whorl vary by family as indicated in the graphs for young pine. The average for the family (*arrow*) and individual tree variation (*horizontal line*) within the family are illustrated. *C. T.* is the commercial check

Inheritance of most branch characteristics is less clear-cut than bole straightness, even though many studies have been made of the genetic influence on the size and angle of branches (Fig. 4.8). Most researchers agree that branch angle is strongly inherited (Ehrenberg 1970; Bailey et al. 1974), and branch size and branch clusters have intermediate inheritance (Fig. 4.9), while branch length has a very low genetic control (Strickland and Goddard 1965b; Bannister and Burdon 1969). Nearly all researchers agree that branching is worthy of inclusion in a selection and breeding program, although Campbell (1965) emphasizes the limited gains to be achieved from genetic manipulation of branch characteristics in Douglas-fir. In contrast, Ehrenberg (1970) showed that narrow-crowned and small-limbed trees produced progeny similar to the parents.

In sycamore, Ferguson et al. (1977) found usable inheritance patterns for branch characteristics. In some species, such as radiata pine, the number of limb whorls is of great importance; Fielding (1960) found the broad sense heritability for this characteristic to be strong, about 0.7. The value of breeding for number of branches per whorl and number of branch whorls to improve product quality has been recognized for a long time. A few studies relative to the inheritance of limbs have been summarized in Table 4.3.

In their discussion on the effect of branches on wood properties, Bamber and Burley (1983) report on the inheritance of branch size in radiata pine. They quote

Table 4.3. A few studies on inheritance of branch characteristics, including forking of the tree bole

Species	Reference	Inheritance pattern
Picea glauca	Holst 1960	Acute branch angle is strongly inherited
Pinus radiata	Reid 1963	There is a close correlation between branch diameter and its angle to the vertical
Pinus kesiya	Shelbourne 1963	Great differences were found in inheritance in crown form
Pinus elliottii	Gansel 1965	H^2 for size of branches = 0.40; for crown width H^2 = 0.43, relatively high heritability values compared to most studies
Cryptomeria japonica	Arita 1967	Crown slenderness is genetically set and used in vegetative propagation
Pinus pinaster	Polge and Illy 1967	There was a high heritability for branch characteristics
Salix alba	Kristiníc 1968	Branch number is controlled more by environment than by genetics
Quercus rubra	McGee 1968	Of the seedlings from one source 41% were multistemmed while two sources had only 4% multistemmed trees
Acer saccharum	Howe 1969	Stem forking was strongly inherited
Pinus contorta	Franklin and Callaham 1970	Branch whorls and forking were genetically controlled
Pinus banksiana	Polk 1972	Branching habit is strongly inherited
Populus spp.	Nelson et al. 1979	Spacing affects branches but genetic control is also reasonably strong
Picea abies	Karki 1980	Narrow tree crowns and fine branch characteristics are inherited
Picea abies	Lewark 1981b	Knottiness characteristics, except branch angle, are directly related to height and tree diameter. Increased volume from selection will lead to a loss in quality due to large branches
Pinus radiata	Bamber and Burley 1983	Spacing and stocking are most important for knot characteristics. Branch size is not strongly inherited; branch angle has a reasonably strong genetic component
Populus tremuloides	Jefferson and Yanchuk 1985	Branch diameter has a low heritability; branch angle has a high heritability

Fig. 4.10. Limb angle has a major effect on knot size and associated abnormal wood. Shown (*below*) are knots from six trees from a flat-angled, small-limbed family and (*above*) eight trees from a steeper-angled, larger-limbed family. Note the relative amount of knot wood and defects such as resin and included bark

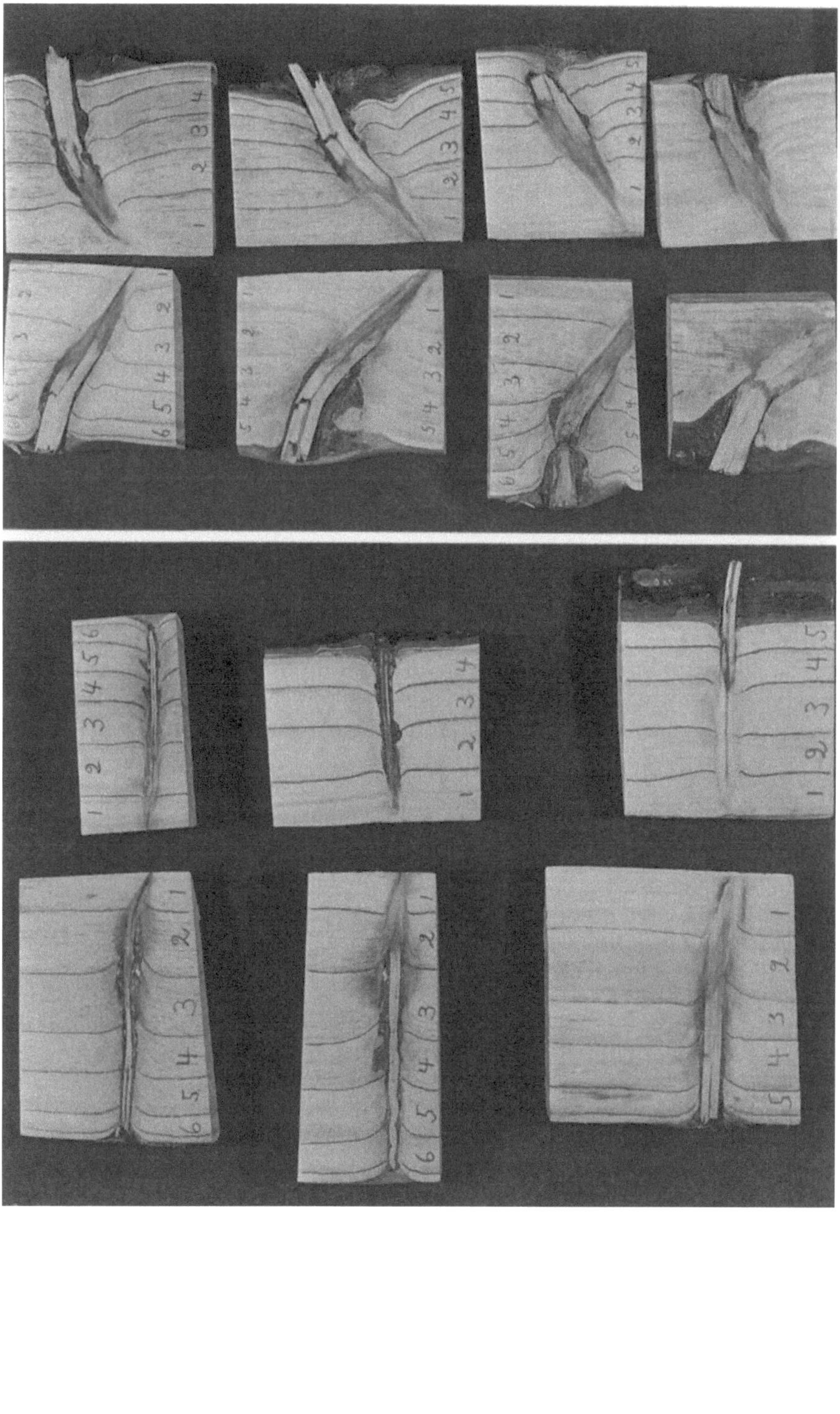

several persons who feel that spacing and stocking are most important for knot characteristics. Branch size was not considered to be strongly inherited in radiata pine, although there is a strong enough genetic component so that meaningful improvements in branch size can be attained. Branch angle, which has a resonably strong genetic component, is perhaps more important than branch size per se in affecting wood properties. Branch size and branch angle tend to be correlated (Fig. 4.10).

4.4.3.1 Summary of Form Improvement Through Genetics and Silviculture

The easiest way to obtain better and more uniform wood is to develop trees with better form. This can be done either by silvicultural or genetic manipulation of tree straightness and crown form. (Silvicultural methods are covered in Chapter 7 and genetic methods in Chapter 8.) The genetic changes can be made either by choosing the most suitable geographic seed sources or through the use of individuals with the desirable tree form. This was summarized by Martin (1984) as: "...genetic selection will result in straighter trees and more disease-free trees with fewer and smaller knots...".

Straightness is the easiest to control genetically. For example, the North Carolina State Cooperative was able to straighten the relatively crooked loblolly pine (*P. taeda*) enough in one generation of breeding so that little additional improvement will be needed in subsequent generations. Crown form is more influenced by environment than straightness, but branch angle has a strong genetic component (Fig. 4.10) and branch size has enough genetic control to ensure considerable gain from genetic manipulation. Limb length is difficult to control genetically in individual trees but can be changed by the choice of the proper geographic source along with silvicultural manipulation with respect to stocking and thinning.

Considerably larger gains in tree form can be made genetically when vegetative propagation is used rather than seed generation (Zobel et al. 1983). Stands that are very homogeneous in limb form, bole characteristics, and wood properties can be obtained. The value and limitations of using vegetative propagation to develop desired trees have been discussed by Zobel and Talbert (1984) and Zobel et al. (1987).

Chapter 5

The Effect of Growth Rate on Wood Properties

5.1 Introduction

Perhaps no aspect of wood quality variation and its control has been more discussed and debated than the effect of growth rate on wood properties (Fig. 5.1). For example, concern was expressed for several species of southern pines many years ago by Paul (1927) that free and fast growth would produce wood that would have characteristics unsuitable for quality products. The need and future for fast-growing trees was emphasized by Bingham (1983). It is evident that forestry is moving in that direction and the kind of wood produced by harvesting fast-growing trees at young ages will have to be accepted. As Zobel (1980b) states: "the proportion of trees from fast grown plantations of both conifers and hardwoods will continue to increase until it will predominate in the next quarter-century. Therefore, the industry must learn to use it effectively; just complaining about how different it is from what it has been using is not good enough. Wood from fast-grown plantations is not necessarily 'bad wood' but certainly it is different, especially in the conifers." In his discussion of variation in wood quality, Hillis (1983) emphasizes that the effect of compound interest on final timber growing costs will place future emphasis on rapid growth and shorter rotations. In the future it seems inevitable that the forest industry will have to defer to the economic realities and accept rapid growth *unless* it can be proved that the consequences are so marked as to reduce the present net value of the forest crop.

The controversy over the effect of growth rate on wood was discussed in 1972 by Larson, who dealt with the concern that improved forest management practices may result in fast-grown wood of poor quality. He mentions how difficult it is to define exactly what constitutes fast growth and resultant poor wood quality. One type of wood may be acceptable and of high quality using one set of standards and unusable under another. Larson goes on to say that ring width, ring structure, and ring uniformity, along with internal wood properties, determine wood quality; each can be changed with a variation in growth rate. In 1973, Zobel states: "there is a lot of misinformation on the effects of growth rate on wood qualities. ... High and low density woods often are found in equally fast-growing trees." He goes on to discuss how environment can affect both growth rate and wood density. For example, an environment that is so good that a pine tree grows continuously (i.e., has no dormant stage) produces low wood density which frequently has many false rings.

Despite the very widespread interest and study over the years there still is much controversy, and a search of the literature will yield publications which can be used to support nearly any chosen point of view. As just one example, when growth was represented by height, Allen (1977) reported a positive correlation of

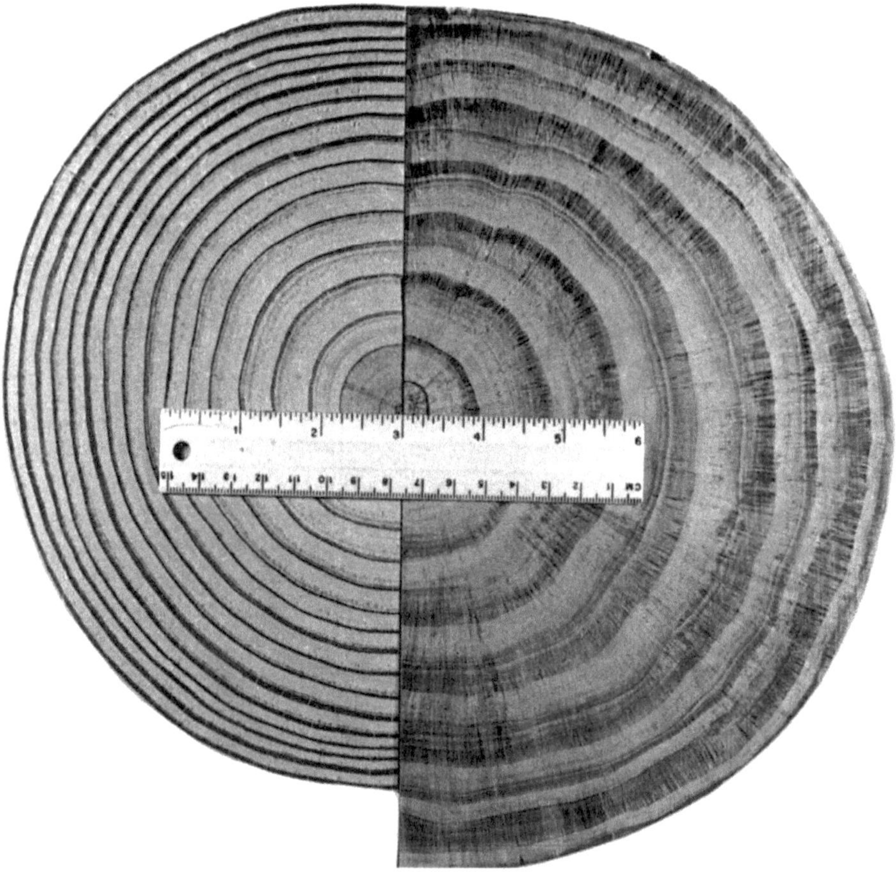

Fig. 5.1. Better silviculture and genetic improvement results in rapid tree growth. There is concern about the quality of wood produced. Shown is an 8-year-old, fast-grown pine compared to a 17-year-old normally grown tree. The cause for concern about wood properties is obvious

height growth to specific gravity in slash pine but when growth was represented by diameter, no correlation was found in the closely related loblolly pine (Matziris and Zobel 1973). In plantations in Colombia, Ladrach (1986) found that fast growth did not affect the wood density of the several hard pines he tested.

As another example, Smith et al. (1961) reported that in *Pseudotsuga menziesii* specific gravity was negatively correlated with growth rate, while Wellwood and Smith (1962) found that there is no relationship between growth rate and specific gravity in Douglas-fir. Many of the opposing views are shown in the tables in this chapter.

It is evident that any relationship between growth and wood properties depends on the species and the site concerned as well as on how growth is expressed, i.e., height, diameter, basal area, or volume. Volume and basal area, the ultimate in growth expressions, are too seldom used. Volume, however, is not satisfactory in young stands because the diameter and height measurements are too difficult to make accurately, and volume tables for small trees are rather inaccurate, especially when bark is included.

Generally, the effects of growth rate are related to cell dimensions, specific gravity, or moisture content. However, Smith and Ker (1960) point out the adverse properties of wide growth rings per se, and that wide rings are not suitable for some solid wood products (see Fig. 5.1). This problem is becoming increasingly important as juvenile wood becomes more abundant. Manwiller (1972) also stresses this fact and the necessity of maintaining a constant ring width if quality solid wood products are to be produced.

As has frequently been mentioned, growth rate is only one of many factors that can influence wood quality (Mitchell 1963). In fact, anything that affects crown development and thus growth rate may affect wood properties (Larson 1972), making an assessment of the growth to wood property relationship particularly difficult and confusing. According to Larson (1962), "most of the confusion in the literature results from attempts to relate wood quality directly to environmental influences. A more logical approach would be to relate wood quality to growth and development of the crown."

There has recently been a whole series of publications relating wood quality to fast-grown plantations (Bendtsen 1978; Zobel 1981; Bamber and Burley 1983; Zobel 1984; Megraw 1985). All emphasize that there will be an increase in the proportion of juvenile wood and the wide-ringed wood will produce products that differ from those made from old growth forests. All also agree that, as this occurs, acceptability of fast-grown wood will increase as technology improves and product standards change.

An additional source of confusion relates to the cause of faster growth. Trees caused to grow more rapidly by silvicultural manipulation or by fertilizers will often have their wood affected differently than genetically fast-growing trees compared to genetically slow-growing ones. The genetic relationship of growth rate related to wood properties often appear to be quite independent in some species.

As warned several times in this book, it is not acceptable to relate wood properties to ring width with rings of different ages (Chap. 3). Yet, this has frequently been done in the past and is still being done, leading to false and controversial ideas about the effect of growth rate. For example, in *Pinus patula* in South Africa, Scott and Stephens (1947) state that the heavier pieces of wood were of slow growth with narrow annual rings and the light pieces had wide rings and "it was assumed that the rate of growth was largely responsible for the differences in weight...". No clarification was made as to whether juvenile wood with its wider growth rings was involved.

5.1.1 Patterns in Wood Production Resulting from Growth Rate Differences

It is immediately evident to one who works with the relationship of growth rate to wood that there are no patterns that always hold. Tree age and especially the age of the cambium are very important. Young trees are especially responsive to environmental and growth factors that influence wood, as emphasized by Reck (1969).

Differences can occur within a species depending on where it is grown. For example, in the southeastern United States both the present authors studied loblolly pine. Zobel (1960b) found essentially no relationship of growth rate to wood specific gravity in the eastern part of the species range, while in the western part of this range, van Buijtenen (1969) reported a strong negative correlation between growth rate and specific gravity.

Some patterns are quite general; for example, tracheids and fibers usually are shorter in faster growing trees (Bannan 1967). This has been explained by Bannan through the relationship of the length of the cell to the period of time the cambial initials exist before anticlinal divisions produce new cells. When growth is rapid, the initials divide before they have a chance to reach their potential length. Since it is generally accepted that the size of a cell is largely determined by the size of the initial cambium cell from which it is formed, the description by Bannan is understandable. This relative effect of growth rate on cell length was described by Bailey (1920) and Priestley (1930) in which the cambial initials are shorter because of earlier transverse divisions in the initials that occur with rapid diameter growth.

Other wood properties are affected by growth rate. An example is growth rate and fibril angle in Douglas-fir. Also, accelerated growth increased the lignin content and decreased both holocellulose and alpha-cellulose contents (Erickson and Arima 1974). In northern red oak, Maeglin (1974) found that the proportion of fibers increased, while ray and vessel proportions decreased with greater ring width. In the overall literature, however, there are only a limited number of studies relating growth rate to wood properties other than specific gravity and cell length.

5.1.1.1 Growth Rate-Wood Property Grouping of Forest Trees

Although the effect of growth on wood varies greatly among species, it tends to be similar within certain groups of species. Because of this, and for easier understanding, the discussion will be divided into the following groups of trees:

1. Hard pines;
2. Other conifers, including spruce, fir, Douglas-fir, hemlock, soft pines, and members of the family Taxodiaceae;
3. The ring-porous hardwoods;
4. The diffuse-porous hardwoods.

Although the separation of forest trees into these four categories is obviously arbitrary, it does permit a less confusing picture of the relationship of growth rate to wood properties than would occur if one tries to generalize for all forest trees, as has so often been done. In fact, a lack of recognition of the differing responses among the four categories is one of the major causes of the current controversy. This was indicated in an early summary on the relationship of environment to wood properties by Goggans (1961), who stated that, based on the evidence then available, growth rate has a minor or insignificant effect on wood specific gravity. Goggans warned that this relationship applied primarily to conifers that contain distinct latewood zones and might not hold for other species. Yet, numerous times in the literature one sees Goggans' summary paper cited as supporting the thesis that growth rate and specific gravity are not related, with no recognition of the species or group qualification he made.

5.1.2 Literature on Growth Rate and Wood Properties

The published information relative to wood properties and growth is so voluminous that it will not be possible to cite more than a small fraction of it. A complete coverage of the subject is especially difficult because many references are incidental statements in papers published on related subjects. Nearly every person who works on various aspects of wood properties includes growth rate as one contributing factor. Because of this incidental mention regarding growth, many references are difficult to locate.

During the period 1950 to 1970, several major summary papers were published relative to growth rate and wood properties; these included Paul and Smith (1950), Spurr and Hsuing (1954), Goggans (1961), von Pechmann (1958). In the mid-1950s a heated debate on the effect of growth rate took place between Aldridge and Hudson (1955) and Hiley (1955). Aldridge and Hudson believed that rate of growth was the factor that controls wood specific gravity and the resultant strength in conifers, while Hiley stated that the strength of conifer wood depends, not on the breadth of the annual ring, but on the age of the cambium which produced it. In later articles, Aldridge and Hudson (1958, 1959) continued to assert that Turnbull's radial density hypothesis, which holds that conifer wood was denser toward the bark, a theory espoused by Hiley, is wrong and that fast-grown timber will not be strong. In their 1959 article they stated that "the first single guide to quality in softwoods is rate of growth."

Similar arguments continue to this day. It is clear that sometimes an article is written by a person who is promoting a favored hypothesis, and that the author is not considering other viewpoints. It is not unusual for bias to become more serious and obvious in the growth to wood quality controversy as the subject becomes more emotional. Unfortunately, sometimes ideas have been "proven" in the literature by citing only those results that support the particular author's views. There will be an attempt in this book to cite faithfully opinions that may be in conflict.

It will not be possible to discuss data from each paper cited, but general concepts will be summarized from selected papers. Tables are included listing results from a number of studies made with different species in various areas to help the reader develop a meaningful picture of the complexity and controversy associated with the relationship between growth rate and wood properties. Most emphasis will be on specific gravity, with less on cell dimensions. A number of papers deal with gross wood morphology and growth rate, including the amount of reaction wood. For example, Cown and McConchie (1981) found that a form of mild compression wood was associated with accelerated growth.

5.2 Growth Rate and Wood Specific Gravity in Conifers

Many attempts have been made to summarize the relationship between growth rate and specific gravity in conifers. For example, Spurr and Hsuing (1954) and Goggans (1961), along with several others, reviewed a large number of early studies concerned with the relation of growth rate to wood properties. Some researchers have concluded that in conifers there is little relationship between ring width and specific gravity at a given position in the tree and for a given age of the ring. However, others have stated equally positively that fast growth results in a low wood specific gravity; this is especially true in conifers other than the hard pines. As discussed in Chapter 3, any effects of growth rate must be tempered with age. As stated by Bendtsen (1978): "...if the specific gravity of rapid- and slow-growth (hard) pine trees of similar diameter is measured, the slow growth trees will generally have a higher specific gravity because of the consistent pattern of increasing density from pith to bark associated with age. If the age effect is ignored, growth rate can be wrongly interpreted as the causal factor of differing specific gravity. Actually, if the specific gravity of the rapid-growth trees is compared to that of the slow-growth trees over the same number of annual rings, counting from the pith, they tend to be similar."

The complexity of making generalized statements relative to growth rate and specific gravity was described for loblolly pine (*P. taeda*) by van Buijtenen (1963). When 6-year-old progeny of mother trees which had been selected for both high and low specific gravities were test-planted in three locations, results showed that: (1) when all trees were considered, regardless of parentage or planting location, a small positive correlation was found between growth rate and specific gravity, (2) when averages of whole replications were used, there was a strong positive correlation between growth rate and specific gravity, and (3) there was a moderate negative correlation when based upon progeny averages. According to van Buijtenen (1963), these results reflect the differing effects of environment and genetics in their relation to growth and specific gravity. The results obtained are shown in Table 5.1, which also indicates the effects of growth rate on tracheid length.

The feeling that fast growth will result in low specific gravity wood results from the question about the ultimate use of the photosynthetic substances produced by the tree crown. If a given amount of photosynthate is produced, will it be used to make more new cells in faster growth or will it be used to produce thicker cell walls in cells already formed? The feeling expressed by most foresters is that a fast-growing tree will not have the reserve food necessary to produce thick, secondary cell walls. Undoubtedly, this relationship exists to some extent, but it is likewise true that some of the fastest growing trees within a species also have the highest specific gravity wood and thickest cell walls.

The actual situation appears to be that if the necessary conditions for good growth are present, the growth pattern and growth rhythm of some trees are such that they can make full use of the resources available to produce rapid growth plus thick-walled cells, while others cannot. When resources for tree growth are limiting, as is the usual case, the manipulations by the silviculturist can result in a better balance of photosynthetic resources so that the potential growth patterns

Table 5.1. The relationship of growth rate to specific gravity and tracheid length for loblolly pine from Texas and the Atlantic coast[a]

Relationship	Correlation coefficient	
Rings per inch × specific gravity (Atlantic coast)		0.133
DBH × specific gravity, based on individual trees (Texas)	Plantation 1	−0.123
	Plantation 2	−0.006
	Plantation 3	+0.121
DBH × specific gravity, based on progeny averages (Texas)	Plantation 1	−0.426
	Plantation 2	−0.110
	Plantation 3	−0.075
Width of 5 rings (24−29 years) × tracheid length at 30 years (Atlantic coast)		0.259
Width of 5 rings (9−14 years) × tracheid length at 15 years (Atlantic coast)		−0.336

[a] For details, see Zobel et al. (1960b) and van Buijtenen (1963).

of individual trees can be better expressed, as described in Chapter 7. Since most actions of the silviculturist are to increase growth rate, wood specific gravity may be affected (Fig. 5.2).

Several studies, such as the one by Haigh (1961), have shown that the growth rate to wood density relationship can be considered only within given provenances. He reported that a negative correlation between growth rate and specific gravity was evident in Douglas-fir when provenance was not considered, but the correlation became less when the effect of different provenances was taken into account. Different provenances may have differing growth rates and they also can have differing specific gravities (Dorn 1969). Dorn emphasizes the importance of determining the cause of growth rate differences, i.e., whether the wood properties are due to geographic source or to growth rate per se. This can only be resolved by studies where trees are grown at different rates within each provenance. Additionally, he found a significant interaction between growth rate and provenance, showing that growth rate affects specific gravity differently in some provenances than in others.

A recurring question about the effect of growth rate on conifer wood relates to the relative control of specific gravity by the environment compared to control by genetics. This question can only be answered satisfactorily with well-designed experiments. Many reported results make it appear that genetic control is strong, since author after author state with certainty their conclusions that the wood differences were related to genetics (von Pechmann 1958; Reck 1969; Barker 1973).

Fig. 5.2. One major result of thinning is to promote faster growth. There is concern about the effect on wood properties of such actions. An example is the twice-thinned pine stand shown. It is of importance that faster growth caused by thinning may affect wood differently than rapid growth resulting from fertilization

The aspects of genetic control of specific gravity are discussed in detail in Chapter 8.

The determination of the relative importance of environmental or genetic control of wood properties is also difficult because the relationship is often influenced by genotype×environment interactions. Thus, an individual tree of a species (or a species as a unit) may produce one type of wood in a given environment and similar wood in another environment, while another tree (or species) will produce a different kind of wood in the second environment. Such interactions cause uncertainty as to the kind of wood that will result when similar trees are grown under differing environments. Luckily, it is common for specific gravity to have a minimum of genotype×environmental interaction within what might be termed "normal" environments; this was reported for southern pines by Barker (1973) and for sycamore by Land et al. (1983). This is not true for all conifer species nor for all environmental conditions, and the only sure way to determine whether an interaction exists is to grow trees in the differing environments and then to assess the wood produced.

Looking only at growth rate and specific gravity can be misleading. A number of investigators have reported greater wood weight in faster-grown trees even if they have slightly lower density. For *Pinus patula*, 23 years of age, Banks and Schwegmann (1957) add another consideration. Although the faster-grown trees

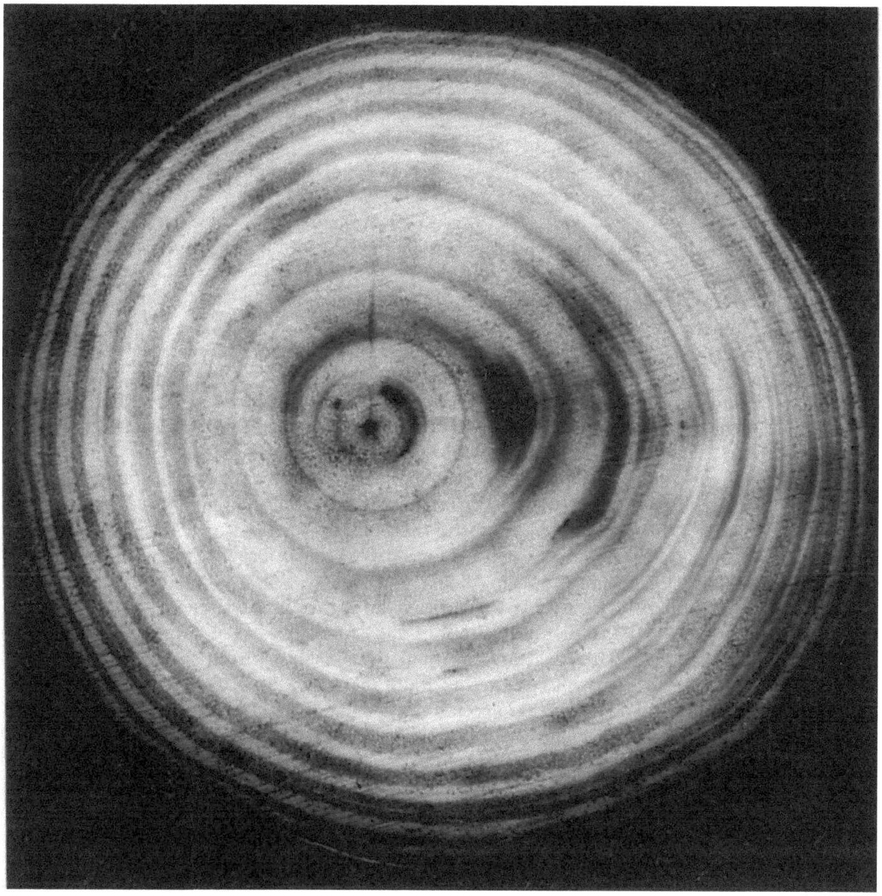

Fig. 5.3. One result of fast growth in conifers can be the production of excess compression wood (see the *darker areas* in the photo). This phenomenon has been reported on several occasions, but its extent is not known

were somewhat lower in wood specific gravity based on increment cores, the average density of the fast grown trees as a whole is higher than the slower growing trees because the dense wood formed toward the end of the rotation constitutes a larger proportion of the whole tree than of the increment core. As pointed out by Westing (1965); the formation of reaction wood is sometimes favored either directly by rapid growth or by the wide spacings used to achieve rapid growth, which subject the trees to greater external stresses (Fig. 5.3).

5.2.1 Growth Rate – Specific Gravity Relationship in the Hard Pines

It is in the hard pines where most controversy about the effect of growth rate on wood occurs, and in which the growth rate-specific gravity relationship is of greatest importance. To indicate the different results and opinions, Table 5.2 was

Table 5.2. Some publications related to growth rate and wood specific gravity in the hard pines[a]

Species	Reference	Comments
P. taeda	Paul 1932	Fast-growing trees had a lower specific gravity than slow-growing trees of the same age
P. patula *P. radiata*	Turnbull and du Plessis 1946	No relationship exists between specific gravity and tree diameter for trees within a stand
P. taeda	Lindgren 1951	Bolts from faster growing trees had the same average specific gravity as those from the slower growing ones
P. elliottii	Kramer and Smith 1956	Only a limited portion of the variation in timber strength could be ascribed to rings/inch
P. taeda	Zobel 1956	A very weak negative relationship exists between specific gravity and growth rate
P. patula	Banks and Schwegmann 1957	Fast-grown trees had only slightly lighter wood when assessed within individual years; overall, there was no relationship between wood density and growth rate
P. elliottii	Larson 1957	Rate of growth exerts a negligible influence on specific gravity
P. elliottii	Perry and Wang 1958	There were no differences in basic density among trees growing two to three times faster than others
P. nigra	Rendle and Phillips 1958	After a certain age, rapid growth can produce wood of reasonably high density
P. koraiensis	Huang and Liu 1959	There was no significant difference in the strength of wood from fast-grown plantations and from slow-grown natural stands
P. resinosa	Cooper 1960	There was no consistent relationship between specific gravity and growth rate or ring width
P. radiata	Fielding and Brown 1960	The average specific gravity had a slight tendency to be positively correlated with rate of growth but there were exceptions
P. ponderosa	Myers 1960	There was a very low correlation between ring width and specific gravity of Black Hills ponderosa pine
P. caribaea	Schmidt and Smith 1961	There was no relationship between ring width and basic density, and tracheid length was slightly negatively correlated with growth rate
P. elliottii	Sellers 1962	The effect of growth rate on specific gravity was small
P. contorta	Polge 1963	A very low correlation was found between growth rate and specific gravity
P. elliottii	Goddard and Strickland 1964b	There was a small negative correlation between growth rate and specific gravity
P. radiata	Nicholls and Fielding 1964	Wood characteristics of clones are not affected significantly by growth rate as determined by the environment
P. echinata	Ralston and McGinnes 1964	Specific gravity of 40-year-old trees is not affected by radial growth rate
P. virginiana	Thor 1964	No relationship was found between growth rate and specific gravity before resin extraction but fast growth results in lower specific gravity after extraction
P. taeda	Geyer and Gilmore 1965	Little of the variation in specific gravity can be related to growth rate alone
P. taeda	Stonecypher and Zobel 1966	A strong negative correlation was found between growth rate and specific gravity in very young trees
P. densiflora	Tajima 1967	No relationship exists between ring width and specific gravity
P. resinosa	Peterson 1968	Growth rate accounted for only a small proportion of the total variation in specific gravity

Table 5.2 (continued)

Species	Reference	Comments
P. pinaster	Polge and Illy 1968	Wide annual rings had a higher specific gravity, an unusual response
P. sylvestris	Dorn 1969	Control of wood properties by limiting growth rate is not recommended; growth rate has more effect in some provenances than in others
Southern pines	Saucier and Taras 1969	The diameter to specific gravity correlations were low and non-significant in all cases
P. taeda	van Buijtenen 1969	A strong negative correlation was found between growth rate and specific gravity in the western part of the species range
P. ponderosa	Voorhies 1969	Growth rate is not related to specific gravity
P. taeda	Zobel et al. 1969a	There was no relationship between growth rate and specific gravity. Some fast-grown families had high density, some low density
P. taeda	Zobel et al. 1969b	From both open and control pollinated families it was obvious that growth rate, measured by diameter or volume per acre, was not closely related to wood specific gravity
P. taeda *P. echinata* *P. resinosa* *P. banksiana*	Larson 1972	Faster growth resulting from wider spacing has little effect on wood specific gravity
P. glabra	Manwiller 1972	Specific gravity did not differ with growth rate
P. patula	Burley 1973	Dominant trees with wider rings had less latewood and lower density than narrow-ringed trees
P. elliottii	de Guth 1973	None of the wood characteristics studied, including wood density, is influenced by the annual ring growth
P. taeda	Matziris and Zobel 1973	A small to negligible correlation exists between height and diameter growth and specific gravity
P. radiata	Nicholls and Brown 1973	No difference in specific gravity was found between large and small stems in double-stemmed trees. Wood properties were not affected by size of stem
Southern pines	Schmidtling 1973	Increase in growth did not change latewood percentage; in loblolly and longleaf pines, the faster growing trees, resulting from better culture, had a higher specific gravity
P. taeda	Burton and Shoulders 1974	Sawlogs from fast-grown trees were unaffected by treatment to enhance fast growth, compared to logs from normally grown trees
P. sylvestris	Unsvaara 1974	Acceleration of growth rate reduces basic density
P. radiata	Nicholls and Wright 1976	Ring width was negatively correlated with minimum and average density
P. elliottii	Allen 1977	A strong positive relationship between height and specific gravity was found, but none between diameter and specific gravity
P. caribaea	Barnes et al. 1977	Fast growth was not necessarily responsible for low wood density on five sites in Zimbabwe. There was only a slight negative correlation between diameter and specific gravity
P. caribaea	Palmer and Gibbs 1977	There was no significant difference in the specific gravity of fast- and slow-grown trees
P. caribaea	Plumptre 1977	Ring width accounted for 12–15% of the variation in specific gravity
Pines in general	Bendtsen 1978	The apparent growth rate and low specific gravity correlation is mostly related to the age of wood from the tree center, not to growth rate

Table 5.2 (continued)

Species	Reference	Comments
P. radiata	Matziris 1979	In Greece, specific gravity has a weak negative relationship with the diameter at breast height and is not related to tree height
P. elliottii	de Guth 1980	There was no correlation between wood density and stem diameter for trees grown in Argentina
P. taeda	Pearson and Gilmore 1980	Mechanical properties of trees selected for fast growth should be similar to those of trees of the same age in the population as a whole
P. radiata	Bannister and Vine 1981	There was a weak negative correlation of specific gravity to radial growth rate
P. radiata	Cown and McConchie 1981	An increase in radial growth is associated with a decrease in both latewood percentage and wood density. A basal area increase of 150% gave a density reduction of 20%
P. elliottii *P. patula* *P. taeda*	Boden 1982	Diameter of the tree was inversely correlated with wood density
P. taeda	McKinley et al. 1982	Selection for growth alone would result in a negative response for wood specific gravity
P. taeda	Taylor 1982	Trees can be rapidly grown without affecting their utilization. Accelerated growth from genetic or silvicultural manipulation will not appreciably affect wood density
P. radiata	Bamber and Burley 1983	Despite the numerous statements to the contrary, specific gravity is usually not related to growth rate unless growth is rendered irregular by silvicultural treatments
P. patula	Ladrach 1984	Specific gravity is the same in trees of different volumes when they are grown on the same site and are of the same age
P. taeda	Megraw 1985	An inherent relationship between growth rate and specific gravity does not exist, although the environmental effect on growth rate may cause denser or lighter wood
P. caribaea	Nariyoshi et al. 1986	There was no significant correlation between average specific gravity and diameter class in any of the ages studied in the Amazon

[a] This list represents only a fraction of the many publications dealing with the subject.

constructed, showing results from some of the many publications which relate growth rate to wood specific gravity in the hard pines. The references in Table 5.2 usually relate to normal growth rate differences, not to those effects from increased growth rate resulting from a stimulation by silvicultural treatments. To assess the effects of silvicultural manipulation, it will be necessary to compare results in Table 5.2 with those in Table 6.1 along with comments in Chapter 7. This should give a clearer picture of the total effect of changes in growth rate from a variety of causes.

5.2.1.1 Summary – Specific Gravity and Growth Relationships in Hard Pines

One observation evident from Table 5.2 is that many of the reference dates are prior to the mid-1970s. During this period there was a great deal of interest in the growth rate to specific gravity relationship, regardless of the cause of the increased growth. In more recent years, there has been a greater emphasis on the causes of variation in wood properties associated with growth rate. Currently, there is much emphasis on wood property variation from changing growth rates caused by silvicultural, genetic, and other manipulations; these are summarized in Chapters 7 and 8.

Studies shown in Table 5.2 indicate that there are all possible relationships between growth rate and specific gravity in the hard pines. Summarizing, 35 of the studies showed no relationship between growth rate and specific gravity, 11 showed a significant reduction in specific gravity with increased growth rate, 9 showed a small negative relationship, and 4 showed a higher specific gravity when the trees grew faster. Although Table 5.2 contains results for only 59 of the hundreds of studies on the hard pines, they are quite representative for this group of trees. It is evident from the literature and our experience that growth rate-specific gravity effects are largest in young trees (Rendle and Phillips 1958; Stonecypher and Zobel 1966). It is also clear how important the method to measure growth is, whether it be height, diameter, or volume (Allen 1977; Matziris 1979).

Although the relationship between growth rate and specific gravity tends to be consistent within a species, this does not always hold. For example, Paul (1932) and van Buijtenen (1969) obtained a negative correlation with growth in *Pinus taeda*, while Zobel (1956, 1970), Burton and Shoulders (1974), Pearson and Gilmore (1980), and Megraw (1985) indicate a lack of relationship of specific gravity with growth rate within the same species.

Although much of the discussion relative to growth rate and wood properties relates to wood density or specific gravity per se, several papers were cited which are concerned with growth and timber strength properties that are closely related

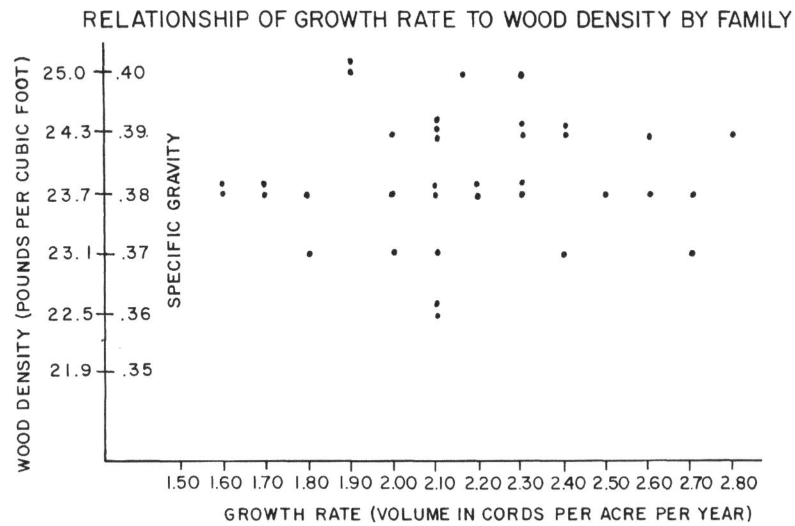

Fig. 5.4. There is usually little or no relationship between specific gravity and growth rate in the hard pines. Such a lack is illustrated for *Pinus taeda* families; each *dot* is an average of 40 trees (Zobel and Talbert 1984)

to specific gravity. For example, Burton and Shoulders (1974) found that growing *Pinus taeda* as rapidly as possible to obtain sawtimber early had no effect on the strength of the wood produced in the fastest growing trees. Conversely, MacPeak et al. (1987) again emphasize that fast-grown trees have a larger core of juvenile wood. Veneer from fast-grown trees gives a lower yield and grade than that cut from slow-grown trees of the same size. Fast-grown trees had only 1% log input compared to 20% from the slow-grown trees.

In summary, it appears that for the hard pines there is generally little or no relationship between wood specific gravity and growth rate of the individual tree (Fig. 5.4). This was clearly stated by Bendtsen (1978) as: "a rather common misconception appears to be that in conifers, particularly those with a pronounced latewood zone, rapid grown wood is inherently low in specific gravity. . ." This must be tempered by results discussed in Chapter 7 where growth has been altered by silvicultural manipulation. In general, although discussion will be centered on genetic aspects of growth and wood properties in Chapter 8, numerous authors state that even when there is a negative relationship, it will be possible to develop fast-growing trees with a high specific gravity because of the relatively low genetic correlation between these two characteristics.

5.2.2 Growth Rate – Specific Gravity Relationships in Other Conifers

Many studies on the relationship of growth rate to specific gravity have been made on conifers other than the hard pines. Species such as Douglas-fir (*Pseudotsuga menziesii*) and the larches (*Larix*) have wood similar to the hard pines and respond to growth rate changes in somewhat the same manner, although negative correlations appear to be more common. Species in the genera *Picea* and *Abies* are considered to have quite different responses to growth; the soft pines are usually placed in this category. The common belief is that species in these genera produce lower density wood with increased growth rate, but this generalization also has many exceptions as indicated in the references in Table 5.3.

The 13 species and 8 genera listed in Table 5.3 sometimes show the expected pattern of lower wood density with faster growth, but there are surprises. For example, most members of the genus *Picea* produce lower specific gravity with faster growth (Olesen 1973), but there are several reports in which growth and specific gravity were not related (Richardson 1959; Stairs et al. 1966; Petrik 1968). Douglas-fir varied, with some reports showing a negative relationship and others no relationship of specific gravity to growth. Although it is generally accepted that the specific gravity of fast-grown soft pines will be low, the five studies cited for *Pinus strobus* in Table 5.3 show no relationship in four and the surprising finding by Lee (1974) that fast growth sources had higher specific gravities than the slower growing ones. Based upon these studies, the accepted relationship for the soft pines may not be correct. Although a negative relationship is common for genera such as *Abies* and *Tsuga*, there are even reports for these genera that wood properties are not closely related to growth rate (Hawthorn 1961).

Most of the studies on conifers other than the hard pines are concerned with *Pseudotsuga* and *Picea*. Many species in *Abies, Picea, Juniperus, Cupressus,*

Table 5.3. Some publications related to growth rate and wood specific gravity in conifers other than the hard pines

Species	Reference	Comments
Picea abies	Klem 1942	Specific gravity gradually decreases, and percentage of earlywood increases with enlargement of the annual ring
Picea abies	Hildebrandt 1954	Dominant (fast grown) trees have the lowest wood densities
Chamaecyparis obtusa	Hirai 1958	There is a tendency for the slower growing trees to have a higher wood density
Pseudotsuga menziesii	McKimmy 1959	Some negative relationship exists between fast growth and specific gravity; percent latewood influenced specific gravity more than growth rate
Picea sitchensis	Richardson 1959	Specific gravity is not related to growth rate
Picea glauca	Holst 1960	Slow growing trees had a higher wood density
Pseuodotsuga menziesii	Mozina 1960	Wood density was influenced only to a small extent by annual ring width
Picea mariana	Rise and Zeller 1960	There was no detectable relation between ring width and specific gravity
Picea abies	von Pechman and Wutz 1960	A large increment increase is related to a decrease in wood density
Pseudotsuga menziesii	Haigh 1961	Wood specific gravity was less in faster growing provenances but when provenance differences were removed, specific gravities were similar for all growth rates
Abies spp. *Tsuga* spp.	Hawthorn 1961	Growth rate is not an accurate indicator of specific gravity in either genus
Picea glauca	Keith 1961	Veneer logs showed a positive correlation between ring width and specific gravity
Pseudotsuga menziesii	Kennedy 1961	The single factor exhibiting the greatest degree of correlation with specific gravity was the width of the latewood band, not growth rate
Pseudotsuga menziesii	Littleford 1961	Ring width and latewood formation are not related. With rapid growth, some trees produce wood with a higher strength than others
Larix leptolepis	Pearson and Fielding 1961	Latewood specific gravity showed some relation to rate of growth, with narrow rings having the densest material. The relationship of juvenile wood to growth was not clear
Abies magnifica	Schniewind 1961	Ring width and latewood percentage have a significant negative effect on specific gravity
Pseudotsuga menziesii	Smith et al. 1961	Specific gravity was negatively correlated with growth rate ($r = -0.402$)
Pseudotsuga menziesii	Knigge 1962	Growth rate had a considerable negative influence on specific gravity only during the first decade
Pseudotsuga menziesii *Tsuga heterophylla*	Wellwood and Smith 1962	There was a negative relationship of specific gravity with growth rate in hemlock but no relationship in Douglas-fir
Picea abies	Mergen et al. 1964	Trees with narrow growth rings had a higher percentage of latewood
Picea abies	Klem 1965a	Density decreased with increased ring width; variation in ring width accounted for 18–42% of the density, depending on where the trees were grown
Pinus strobus	Thor 1965	Growth rate was not important as a determinant of specific gravity
Tsuga heterophylla	Krahmer 1966	Rate of growth as measured by width of annual rings accounted for 23% of the variation in specific gravity of the mature wood
Picea abies	Stairs et al. 1966	There was no difference in specific gravity between fast- and slow-grown trees

Table 5.3 (continued)

Species	Reference	Comments
Picea sitchensis	Brazier 1967	Selection for vigor favors trees with reduced wood density and often causes an increase in spiral grain
Abies amabalis	Kennedy and Swann 1968	Rate of growth is the single most important parameter (negatively) related to specific gravity. Growth rate can be used to identify high or low density wood
Picea abies *Larix decidua*	Petrik 1968	No difference exists in specific gravity with growth rate. Fast growth rate does not mean inferior wood or pulp quality
Abies grandis	Polge 1968	Wood density and growth of young trees are negatively correlated
Picea spp.	MacDonald and Franklin 1969	A definite negative relationship exists between fast growth (ring width) and basic density
Conifers (General) *Pseudotsuga menziesii* *Pinus strobus*	Polge 1969a	Wide-spaced plantations in Europe produced low density wood because of fast growth except plantations of Douglas-fir and white pine, whose wood was not affected by growth rate
Picea abies	Stairs 1969	The effect of growth rate on specific gravity is strong, with a correlation of −0.867 between diameter and specific gravity
Pinus strobus	Thor and Bates 1973	Specific gravity was not strongly affected by growth rate
Picea abies	Olesen 1973	A negative correlation exists between ring width and basic density, especially within trees when the ten internal rings are omitted; 80% of the variation in wood density can be ascribed to variation in ring width
Pinus strobus	Lee 1974	Fast-growing sources had a high specific gravity (0.271) compared to slow-growing sources (0.258)
Larix leptolepis	Lee 1975	These was no specific gravity to diameter growth correlation in juvenile wood
Picea abies	Nanson et al. 1975b	A slight but inconsistent negative relation of specific gravity with growth was found
Picea abies	Olesen 1976	About 45% of the variance in specific gravity can be attributed to variation in ring width
Pseudotsuga menziesii	Keller and Thoby 1977	Relationship between torque and width of rings is non-existent or very weak
Picea abies	Lewark 1979	Ring width is always closely correlated to average density and in most cases to minimum density but not with density contrast
Pseudotsuga menziesii	McKimmy and Campbell 1982	Specific gravity of 64-year-old progenies varied considerably among sources with a given ring width. The rate of increase of specific gravity with increasing age and decreasing ring width along the radius varied among families
Picea glauca	Taylor et al. 1982	A slight negative, but significant, correlation was found for some growth zones in some stands
Cryptomeria japonica	Mio and Matsumoto 1983	The faster growing cultivars have lower basic density and longer tracheids than the slower growing ones
Picea abies	Mothe 1983	Vigor and density were negatively correlated at all genetic and environmental levels
Pseudotsuga menziesii	Smith and Kennedy 1983	More than 10000 trees were evaluated as to whether rapid growth diminishes specific gravity. The impact of fast growth on specific gravity is slight. The claim that a high growth rate affects log quality was not confirmed
Larix leptolepis	Tsutsumi et al. 1983	Volume and ring width are essentially not related to wood basic density
Picea mariana	Hall 1984	In 12 natural stands in Newfoundland, growth rate and specific gravity were not closely related

Fig. 5.5. With many exotics, such as this *Pinus oocarpa* growing in Colombia, South America, growth rate does not affect the specific gravity of the wood (see Ladrach 1986b for details)

Tsuga, and others do not have severe juvenile wood, so the age of ring is not as critical as it is in the hard pines, where ring width usually is large closer to the pith. Although some studies are difficult to interpret because of the many unknowns, it seems safe to generalize that increased growth rate in the conifers generally lowers wood specific gravity with the exception of the hard pines, *Pseudotsuga* and *Larix.* As stated by Ladrach (1986b) "faster growth may result in lower wood density for species of *Abies, Picea* and *Cupressus,* but it does not affect the wood density of many hard pines and hardwoods." This statement was based upon work done with exotic species in Colombia, South America (Fig. 5.5).

5.3 Growth Rate and Wood Specific Gravity in Hardwoods

It is not appropriate to discuss the growth rate-wood relationship in hardwoods as an entity. As discussed in Chapter 1, there are two distinctly different groups of hardwoods, namely, ring-porous and diffuse-porous. Wood of these groups is affected differently by growth rate. As outlined by Fukazawa (1983) "in the ring-porous hardwoods changes of basic density are influenced by the ring width, while those of diffuse-porous hardwoods are almost independent of the ring width." Because of this, we have separated discussions of the two groups of hardwoods in the two following sections.

5.3.1 Specific Gravity and Growth Rate in the Ring-Porous Hardwoods

One area related to wood quality where many, but not all, are in agreement is that fast growth in ring-porous hardwoods results in dense wood, while in diffuse-porous woods, growth rate differences have little effect on specific gravity (Khurshudyan 1958; MacDonald and Franklin 1969; Kazumi 1983). It was stated plainly in the textbook by Panshin et al. (1964) that in ring-porous hardwoods the ring width is directly related to the average specific gravity of the wood. As an example, Kazumi (1983) reported that this effect from fast growth is evident in ring-porous hardwoods; similarly, Wheeler (1987) cites a positive correlation of growth rate and specific gravity in both the juvenile and mature wood zone of ring-porous hardwoods. The reason she gives is that the earlywood zone is nearly constant from year to year and the wider rings therefore contain more latewood with fewer vessels. Guiher (1965) and Baas (1982b) also report that high specific gravity is associated with fast growth in red oak.

However, a few other workers do not concur; for example, Taylor (1977) found no consistent relationship between ring width and specific gravity for southern red oak (*Quercus falcata*), water oak (*Quercus nigra*), post oak (*Quercus stellata*), shagbark hickory (*Carya ovata*), or mockernut hickory (*Carya tomentosa*). In another study on *Carya illinoensis* (pecan) Taylor and Wooten (1973) also failed to find any relationship between growth rate and specific gravity. They concluded that "growth rate is not a reliable indicator of specific gravity." Similarly, van Eck and Woessner (1964) found no relation between growth rate and specific gravity in red oak. In Europe, Polge and Keller (1973) also working with oak, state that within plots of the same age this relationship of high density to wide rings does not always hold, although within a given tree wood density increases with increased growth rate. They reported that it is not unusual to find individual trees with rather wide rings that have low specific gravity.

In general, the earlywood volume of ring-porous hardwoods has a relatively high proportion of vessels and the width of earlywood remains fairly constant from year to year; the latewood percent, with fewer vessels, increases as ring width increases. However, the juvenile wood formed near the pith does not necessarily follow this pattern.

The effect of the proportion of different types of cells which influence specific gravity has been determined by several researchers. These were summarized by Knigge and Koltzenburg (1965) who reported for *Quercus* and *Fagus* that an increase in the proportion of fibers and a decrease in that of vessels occurred with increasing annual ring width. For a group of 11 North American oaks, Maeglin and Quirk (1983) found substantial amounts of parenchyma tissue by weight in slow and medium growth trees. Since large volumes of small (but old) oak trees are now being harvested, the excess of parenchyma causes "fines" which result in poor water drainage and screen clogging on paper making. The heavy proportion of vessel elements results in "picking" with resultant poor printing.

5.3.2 Specific Gravity and Growth Rate in Diffuse-Porous Hardwoods

According to Knigge and Koltzenburg (1965), only limited studies on the effect of growth rate on specific gravity had been made with diffuse-porous hardwoods prior to 1965. A number of reports are available since 1965; some of these have been summarized in Table 5.4. As for all wood studies, there is no unanimity, even though authors such as Panshin et al. (1964) state that there is little relationship between specific gravity and ring width in diffuse-porous hardwoods; this was also claimed by Kazumi (1983). Similarly, in papers by Johnson (1942), Brown et al. (1949), and Wangaard (1950) it was concluded that there was no definite relationship between growth rate and specific gravity in diffuse-porous hardwoods.

Table 5.4. Relationship of growth rate to wood properties in diffuse-porous hardwoods

Species	Reference	Comments
Tectona grandis	Scott and MacGregor 1952	Faster growing trees had a slightly higher specific gravity
Acer saccharum	Radcliffe 1953	There was no correlation between rate of growth and specific gravity nor with modulus of elasticity, modulus of rupture, or the proportional limit
Populus spp.	Kennedy and Smith 1959	There is an increase in specific gravity with faster growth
Populus spp.	Göhre 1960	Density of the wood was not related to ring width
Swietenia macrophylla	Briscoe et al. 1963	Specific gravity increased with growth rate
Populus deltoides	Kaeiser and Boyce 1964	Relationships between growth and specific gravity were upset by lean and crook which produces layers of high density and gelatinous fibers
Liriodendron tulipifera	van Eck and Woessner 1964	Wood density in yellow poplar was independent of ring width or growth rate
Liquidambar styraciflua	Webb 1964	No growth rate to specific gravity relationship exists
Salix alba	Herpka 1965	The specific gravity of wood decreases with an increase in annual ring width, but this is caused by the fact that juvenile wood has the widest rings and lowest specific gravity
Liriodendron tulipifera	Taylor 1965	Faster growing trees had the same specific gravity as slower growing ones
Populus deltoides	Farmer and Wilcox 1966a	Specific gravity was slightly less in faster growing trees
Fagus sylvatica	Koltzenburg 1966	There is little relationship of ring width with specific gravity, although rings narrower than 2 mm tended to be denser
Nyssa sylvatica	McElwee and Faircloth 1966	There was no relationship between specific gravity and growth rate

Table 5.4 (continued)

Species	Reference	Comments
Populus tremuloides	Einspahr and Benson 1967	There was no relationship between growth rate and specific gravity
Liriodendron tulipifera	Kellison 1967	There is only a slight reduction in specific gravity associated with an increase in growth rate
Prunus serotina	Koch 1967	When variables of age and genetic differences were eliminated by measuring sprouts of different sizes from the same stump, the diameter of the stem (growth rate) had no effect on specific gravity
Fagus sylvatica	Koltzenburg 1967	There was no correlation of wood specific gravity with increased ring width
Populus spp.	Mutibarik 1967	A slight negative correlation exists between ring width and specific gravity in Euramerican poplar hybrids in Yugoslavia
Acer saccharum	Sajdak 1967	Growth rate (rings/inch) explained only a very small part of the specific gravity variation
Eucalyptus alba	Ferreira 1968	Tree basic density does not bear any relationship to rate of growth
Populus spp.	Guiher 1968	Rings per inch is not a good indication of specific gravity
Gmelina arborea	Lamb 1968	Fast growth does not change the density of *Gmelina* wood, making it ideal for rapid production of uniform timber
Liriodendron tulipifera	Taylor 1968	Specific gravity is not affected by changes in growth rate either within or between trees
Platanus occidentalis	Taylor 1969a	Normal variations in growth rate have little or no effect on specific gravity
Liriodendron tulipifera	Sluder 1970	Basal area growth was not related to specific gravity
Celtis laevigata	Taylor 1971	Growth rate does not affect specific gravity
Eucalyptus robusta	Skolman 1972	There is no relationship between growth rate and specific gravity among stands
Swietenia macrophylla	Chudnoff and Geary 1973	No relationship was found between tree size and wood density
Eucalyptus grandis	Taylor 1973a	Radial growth rate in South Africa does not have a significant effect on specific gravity; however, trees with fast height growth have a lower specific gravity
Numerous hardwoods	Taylor and Wooten 1973	There were no differences in specific gravity related to growth rate
Nyssa sylvatica *Liriodendron tulipifera* *Liquidambar styraciflua*	Taylor 1977	In all three species, no meaningful growth rate to specific gravity correlations were found
Eucalyptus grandis	Brasil et al. 1979	In Brazil, there was no relationship between specific gravity and diameter class
Eucalyptus spp.	Ferreira et al. 1979	There is no correlation between wood volume increment and wood specific gravity
Eucalyptus saligna	King 1980	In coppice, the largest shoot had the highest specific gravity
Platanus occidentalis	Land and Lee 1981	There was no correlation between growth rate and specific gravity in sycamore
Eucalyptus tereticornis	Bhat and Bhat 1983	Faster growth is slightly related to lower basic density and higher moisture content. Height and diameter growth rate explain only 9% of the variation in basic density and 31% in fiber length

Table 5.4 (continued)

Species	Reference	Comments
Terminalia ivorensis	Durand 1983	There was no statistical significance between basic density and ring width
Various hardwoods	Kazumi 1983	Basic density of diffuse-porous hardwoods is independent of ring width; examples are *Betula maximowicziana* and *Tilia japonica*
Eucalyptus spp.	Kininmonth et al. 1983	Internal growth stresses in the eucalypts are usually less serious in fast-grown than in slow-grown sawlogs of the same age
Platanus occidentalis	Land et al. 1983	Genetic correlations of volume with specific gravity were near zero
Triplochiton scleroxylon	Oteng-Amoako et al. 1983	Fast growth produced thin-walled cells and lowered specific gravity but had no relationship with fiber length
Betula pendula	Nepveu and Velling 1983b	There was no correlation between growth rate and wood density; fast-growing trees had less volumetric shrinkage
Eucalyptus spp.	Wilkes and Abbott 1983	Fast-growing trees have a larger vessel volume
Tectona grandis *Gmelina arborea*	Keiding et al. 1984	Evaluation of a series of provenance trials showed a strong effect of the diameter on wood density
Eucalyptus spp.	Delwaulle 1985	In the Congo, some clones that are the fastest growing have the densest wood, while others have less dense wood
Populus tremuloides	Jefferson and Yanchuk 1985	The majority of the studies show that there is a negative genetic relationship between relative density and volume

A lack of relationship was definitely indicated for the eucalypts by Hillis and Brown (1978), who state "it appears that the mean wood density for a tree is not influenced by, or is insignificantly correlated with, growth rate." They mention there are a few exceptions such as described by Hans et al. (1972; Fig. 5.6). Early work and many of the papers cited in Table 5.4 make it clear that in diffuse-porous hardwoods, growth rate usually has little effect on specific gravity.

There are, of course, some exceptions to this generalization, although it appears to hold well for the eucalypts, for many of the temperate climate species, and for some tropical hardwoods. In fact, Bhat and Bhat (1983) recommend "any treatment for faster growth would be desirable." For example, all 19 of the references to *Liriodendron, Liquidambar, Platanus,* and *Populus* in Table 5.4 report little or no effect of growth rate on specific gravity. For eight important hardwood species in the southern United States (Taylor 1977) no consistent relationship between specific gravity and ring width was noted.

The exceptions are few, such as reported by Schönau (1982) that fast growth increases specific gravity in black wattle (*Acacia mearnsii*) and in *Swietenia* (Briscoe et al. 1963), and by Kennedy and Smith (1959) who found higher density in faster growing *Populus*. For *Tectona* and *Gmelina*, Keiding et al. (1984) showed a strong effect of diameter size on wood density, but Lamb (1968), also working with *Gmelina*, says that growth rate does not affect density, and that rapid growth of this species is desired. A negative relationship between growth rate and density was found for *Triplochiton* by Oteng-Amoako (1983).

Fig. 5.6. Specific gravity of most diffuse-porous, fast-growing hardwoods, such as *Eucalyptus*, is not much affected by growth rate. This 5-year-old stand at Aracruz, Brazil produces wood with very desirable properties

5.4 Summary of Growth Rate Effects on Specific Gravity

One of the most difficult tasks in developing the chapter on growth rate and its relationship to specific gravity was to assure that the trees being compared were of the same age. The error of comparing wide-ringed, low specific gravity wood of young trees with narrow-ringed, high specific gravity wood of older trees has been repeated many times, and the relationships found were interpreted as the effect of growth rate rather than of juvenile wood.

Whether the forester can have fast growing, high density conifer forests is determined by what the silviculturist can do, and what success the geneticist has. Stairs (1969) mentions that when there is a negative relationship between growth

Fig. 5.7. Fast-grown plantations become merchantable at an early age. Wood density is low and tracheids are short. The reason is the high proportion of wide-ringed juvenile wood, as evidenced on the cross-section of a tree taken from a young pine plantation

rate and specific gravity, the initial selection should be made for fast growth, and specific gravity should be secondary. For most species in which specific gravity and growth rate appear to be negatively correlated, wood of the desired type can be produced regardless of growth rate achieved because of the relative genetic independence of these two characteristics. The potential for combining fast growth and high specific gravity is discussed in Chapter 8 on genetics.

In a general summary in their book, Panshin et al. (1964) state that the specific gravity to ring width relationship in conifers only applies to a limited range of ring widths. Correctly and importantly, they point out that ring width alone has little meaning as an indicator of the properties of wood, and that the age of the cambium at the time of the formation of the growth increment is of key importance. As referred to earlier, juvenile wood of conifers usually has wide rings and low wood specific gravity, but the wide rings are not the cause: narrow-ringed juvenile wood also has low specific gravity (Fig. 5.7). In overmature timber, for example, the very narrow-ringed wood at the outside of the tree may have nearly as low specific gravity as the wide-ringed wood in the center of the tree.

The effect of a growth rate increase on wood usually depends on its cause. For example, when the growth rate of poplars was stimulated by the use of gibberellins, the trees that had been induced to grow more rapidly by this chemical treatment had greatly reduced specific gravity (Swan 1961).

Despite the conflicting results and the confusion relative to growth rate and wood specific gravity, it is evident that this relationship has enough potential importance that it must always be considered and cannot be ignored for any species or group of trees.

5.5 Cell Characteristics and Growth Rate

Cell size and relative cell dimensions have a major influence on the quality of pulp and paper products (Clark 1962) and on solid wood products. Fiber length and width, wall thickness, and lumen size have an effect on the bulk, burst, tear, fold, and tensile strengths of paper. Fiber characteristics are very important, and coarse fibers have the same effect on paper as short fibers. Cell length is often thought to be of only academic importance for solid wood products; however, definite correlations between tracheid length and certain lumber mechanical properties exist (Bisset et al. 1951). For example, wood with long tracheids has greater tensile strength than wood with short ones.

In recent years a more general interest has developed in cell size characteristics. Numerous studies have been made on the effect of cell length, diameter, wall thickness, and various ratios (such as cell length to cell width) on the final product. Because of this a parallel interest is evident on the effect of growth rate on cell dimensions (Fig. 5.8).

In addition to the work of Bailey (1920), Bannan (1967), and others, numerous studies have been devoted to the question why cell characteristics vary with growth rate. As an example, Richardson (1959, 1964), in studies on the physiology of xylem development, found that the major determinant of wall thickness was the availability of carbohydrates and hormonal activity. He also found for Sitka spruce (*Picea sitchensis*) that an increase in temperature increased cell length but not wall thickness. Richardson concluded that increases in cell length are directly related to the temperature but only indirectly to increased carbohydrate formation caused by increased light intensity. The main determinant of cell diameter is hormonal, as influenced by day length.

Some authors such as Richardson (1964) and Nanson et al. (1975b) report that the tracheid length of mature trees is correlated with height growth so that the length of cells of older trees will be predictable from the rate of height growth of seedlings. Although such a relationship is complex and sometimes does not hold, Denne (1971) felt that the shoot length to tracheid length correlation was strong enough to use in planning for large-scale planting. Similarly, although not universally accepted, Bisset et al. (1951) and Dadswell (1957) report that trees which have initially short tracheids will have short tracheids throughout the life of the tree. Somewhat opposite to this, the present authors find that it is not until

Fig. 5.8. Growth rate affects both cell morphology and wood density. Number and size of vessels, wall thickness, and amount of parenchyma are all affected in hardwoods. However, changes in cell morphology are minor in diffuse-porous hardwoods such as that shown in this cross-section of *Gmelina* compared to changes in the conifers (courtesy of Companhia Florestal, Monte Dourado, Brazil)

one is outside the zone of juvenile wood of loblolly pine that there is a good relationship in tracheid length between young and older woods in the same tree.

In his paper relating anatomical characteristics of wood to growth rate in pine, Taylor (1982) points out that no matter what the cause of increased growth rate (genotype, environment, hormonal substances) the increased growth is caused by an increased rate of cell division in the cambium. He states that many researchers feel that radial cell diameter and cell wall thickness are independent. A manipulation of photoperiod, light intensity, and water stress will result in the production of tracheids with any combination of radial diameter and wall thickness possible. Results related to hormonal substances and their control of wood properties were reported by Wodzicki (1964) in larch and Wodzicki and Wodzicki (1973) in *Pinus sylvestris*. For hardwoods, it is generally recognized that, especially in the ring-porous species, growth rate can play a major role in determining paper properties or strength and utility of lumber by altering the proportion of fibers, vessels, and rays.

5.5.1 Summary of the Relation of Cell Length to Growth Rate

A confusing picture seems to emerge with respect to the effects of growth rate changes on cell length in both conifers and hardwoods. To aid in clarification, Table 5.5 for conifer species and Table 5.6 for several hardwood genera were

Table 5.5. Some publications on the relation between growth rate and tracheid characteristics in conifers

Species	Reference	Comments
Pinus radiata	Bisset et al. 1951	Trees with faster growth had shorter tracheids
Picea abies	Nylinder and Hägglund 1954	As stem diameter increased, tracheid length was less
Pinus sylvestris	Echols 1958	Tracheid length increased with an increase in width of the annual ring
Picea abies	Hildebrandt 1960	Wide annual rings have short tracheids
Pseudotsuga menziesii *Tsuga heterophylla*	Smith et al. 1961	Growth rate had no important effect on tracheid length
Pinus echinata	Ralston and McGinnes 1964	Width and length of cells were unaffected by growth rate
Pinus virginiana	Thor 1964	There was little reduction in tracheid length with increased growth rate
Pinus strobus	Thor 1965	A relationship between growth rate and tracheid length exists only in juvenile wood
Picea abies *Larix decidua*	Petrik 1968	There is no relationship between tracheid length and growth rate
Pinus sylvestris	Dorn 1969	Growth rate did not affect tracheid length
Pinus ponderosa	Echols 1971	Faster grown trees had shorter tracheids
Pseudotsuga menziesii	Erickson and Arima 1974	Increased growth rate caused only small changes in fibril angle
Picea abies	Nanson et al. 1975b	Tracheid length is positively correlated with height growth
Southern pines	Taylor 1982	There were no significant differences in tracheid length between fast- and slow-grown stands
Picea rubens	Jagels 1983	Tracheid length increased with a reduction in growth rate resulting from spruce budworm attacks
Pinus taeda	Megraw 1985	Tracheid length is directly negatively correlated with growth rate

prepared. The generally accepted rule is that faster growth, whatever the cause, results in shorter cells, as was reported by Hildebrandt (1960). However, he went on to state that different results are being obtained and "the above-mentioned general rule must be verified." In fact, this "general rule" is not so general; for example, for conifers (Table 5.5) 7 of the 15 papers cited showed little effect of growth rate on tracheid length, and 2 reported longer tracheids with faster growth; for example, Echols (1958) observed longer tracheids for Scots pine with wider annual rings, and in *Pinus nigra* Nanson et al. (1975) found longer tracheids with increased height growth. Six papers (e.g., Thor 1964; Petrik 1968; Dorn 1969) found that growth rate differences did not affect tracheid length. One very obvious situation is that the kind of growth which is used in the comparisons must be specified, because it appears that radial growth may cause a different change in cell dimensions than height growth. In all instances, comparisons must be made between widths of rings of the same age. Otherwise, juvenile wood, usually with wide rings and short cells, will lead to an incorrect assessment of the effect of growth rate.

The effects of growth rate on cell properties have not been studied nearly as intensively in hardwoods as they have in conifers. Since the fibers are generally

Table 5.6. Relationship of fiber length and anatomical characteristics of some hardwood genera with growth rate or with factors that result in faster growth

Species	References	Comments
Juglans	Mell 1910	Better sites produced trees with longer fibers
Eucalyptus	Amos et al. 1950	The shortest fibers were produced during the fast growth period
Fagus	von Pechmann 1953	There was no relationship of cell length with site quality and thus with growth
Eucalyptus	Scaramuzzi 1957	A decrease of fiber length occurs on better sites
Populus	Scaramuzzi 1958	There was no change in the proportion of wood elements with growth rate
Populus	Kennedy and Smith 1959	There is greater fiber length on better sites. Fast-grown sprouts had fibers 0.540 mm, slow grown were 0.475 mm
Fagus	Aytug 1962	Site quality had no effect on cell size
Liquidambar	Webb 1964	Fiber length was not related to growth rate, but the interlocked grain index was positively correlated with growth rate
Fagus	Koltzenburg 1966	Wider rings had less fibers and less parenchyma, while the proportion of ray tissue and vessel diameter increased. Wide rings had a decrease in the number of vessels per mm^2
Nyssa	McElwee and Faircloth 1966	There was no relationship between fiber length and growth rate
Liquidambar	Hunter and Goggans 1969	Fiber length was only weakly related to growth rate
Salix, Celtis, Carya	Taylor and Wooten 1973	Wide growth rings contained proportionately fewer vessels and more fibers than narrow rings. Wide rings also had short fibers
Eucalyptus	Bhat and Bhat 1983	Faster growth is negatively correlated with fiber length ($r = -0.31$)
Triplochiton	Oteng-Amoako et al. 1983	Growth rate had no relationship with fiber length, but fast growth increased parenchyma volume
Populus	Yanchuk et al. 1983b	There was a slight positive correlation between fiber length and growth rates
Eucalyptus	Wilkes and Abbott 1983	Fast-grown trees had more vessel volume, although there were more vessels in slow-grown trees. The percentage of rays and fibers were not affected by growth rate. Tree growth rate did not influence fiber length

short in hardwoods, any changes in fiber length that might be achieved through silviculture or genetics have relatively little practical significance. Studies that had been made on the effect of growth rate on fiber length were summarized in 1954 by Spurr and Hyvärinen; there have been many later investigations. Most interest has centered on the fiber length differences in juvenile wood compared to mature wood, but some research has also been done on the relationship of growth rate to fiber length in similarly aged woods. Usually, the faster growing trees are found to have shorter fibers.

Most interest in cell properties in the conifers is related to tracheid length per se. Assessment of the true contribution of cell length variation is made difficult since cell wall thickness expressed as specific gravity plays such a major role in determining the properties of the final product. Many incorrect ideas result from a failure to consider the effects of specific gravity and cell length separately. For wood of the same age, specific gravity was found to have a much greater effect on paper quality than tracheid length (Barefoot et al. 1970). It is only when tracheids are short, such as in juvenile wood (or in short-fibered hardwoods), that cell length becomes of major importance for paper properties. Many feel that it is the ratio of cell length to width, or cell length to wall thickness, or similar ratios that are most important for product quality.

Although many reports indicate that as growth rate increases, cell length decreases, exceptions to this general rule occur (see Table 5.5). An example is spruce pine (*Pinus glabra*) for which Manwiller (1972) reports that fast-grown stems had tracheids 0.2 mm longer than slow-grown stems of the same age. Tangential diameters of the tracheids were larger in the fast-grown trees. Also in contrast to the general trend, other researchers, such as Stairs et al. (1966), found no relationship of cell length, cell diameter, or wall thickness to growth rate in spruce.

Results reported for hardwood genera in Table 5.6 are also confusing. Variations often do not represent growth rate differences per se, but result from growth on sites that are more conducive to rapid or slow growth. Although the general trend again was for shorter fibers with faster growth on the better sites, many exceptions occur. For example, Mell (1910) found longer fibers of walnut (*Juglans*) on good sites with the best growth, whereas von Pechmann (1958) observed no relationship between growth rate and cell length in beech (*Fagus*). Kennedy and Smith (1959) noted greater fiber lengths in *Populus* on the best sites.

As an opposing example, in the genus *Eucalyptus*, Amos et al. (1950), found that the shortest fibers were produced during the period of fast growth, while Scaramuzzi (1957) observed shorter cells on the better sites where *Eucalyptus* grows most rapidly. In South Africa for *Eucalyptus*, Taylor (1973a) noted that fibers were longer in trees growing rapidly in height but there was no difference in fiber length with differing radial growth rate. Depending on the species and conditions, longer, shorter, or equal fiber lengths in the hardwoods were associated with fast compared to slow growth.

No consistent pattern is evident for the genera listed in Table 5.6, which is contrary to the earlier literature and to the commonly accepted dictum as explained by Bannan (1967) of shorter fibers with faster diameter growth. Based upon our experience and the literature reviewed, it is safe to say that fast growth of hardwoods does not consistently result in meaningfully shorter fibers if wood of equal age is compared.

Growth rate in hardwoods does have an effect on the cellular composition and percentage of tissues in the hardwoods (see Table 5.6). No generalizations about vessel numbers, vessel volumes, or ray tissue as related to growth rate can be made, however.

If one must generalize, it seems best to say that the most common response is no change in fiber and tracheid lengths with increased radial growth and no effect, or somewhat longer tracheids and fibers, with increased height growth. Since height and radial growth are almost always closely correlated, the results shown in the literature sometimes appear to be contradictory.

It seems obvious that using current methods and products as a guide, changes in tracheid or fiber length of mature wood will be of only minor importance in their effect on product quality. It was stated by Taylor (1982) "the conclusion is that southern pine trees can be rapidly grown without affecting these properties" i.e., tracheid length and specific gravity. However, in juvenile wood of conifers, with its marginally short tracheids, changes in cell length can be of importance. Responses may also be greater in juvenile wood; Thor (1965) found that the only effect of growth on tracheid length in white pine was in the juvenile wood.

5.6 Growth Rate and Aesthetic Qualities of Wood

Although the literature pertaining to this subject is rather sparse, the effect of growth rate on the aesthetic value of wood for use for products such as furniture and trim can be very large, especially in some of the ring-porous hardwoods. There has even been considerable discussion on the effect of growth rate on the value of the highest quality diffuse-porous hardwoods, such as walnut (*Juglans nigra*) and cherry (*Prunus serotina*).

Certainly growth rate differences will have some effect on species that produce quality woods but it will not become evident until the young and fast-grown trees are harvested, rather than the more mature, slower grown trees from natural stands which now supply the bulk of the high-quality wood products (Fig. 5.9).

Growth rate has a major effect related to wood quality in the closeness of the annual rings. Ring width influences the desirability of solid wood products; for example, Lutz (1964) points out that wide-ringed conifers are not well suited for

Fig. 5.9. Growth rate has a major effect on wood properties as related to the closeness of the annual rings. Logs of these old, mature pine trees have very fine grain, producing products with characteristics not possible with fast-grown, wide-ringed wood

veneer or plywood. This was recently reported by MacPeak et al. (1987) who found that fast-grown loblolly pine gave lower yields and poorer grade veneer than that cut from slower grown trees of the same size. The fast-grown trees yielded less than 1% of the better grades of lumber (better than C), while the control trees exceeded 20%. He recommends that two or more annual rings should be in the thickness dimension of veneer to reduce shelling, depth of knife checks, and warping of the veneer.

For lumber, ring width (closeness of grain) can affect the appearance, weathering, and paint retention qualities of the boards. Thus, growth rate has an effect through ring size on product quality, even though it may not affect the wood properties per se.

Narrow-ringed wood is often referred to as dense wood by foresters. This causes considerable confusion because in this case there is no reference to specific gravity (wood density), only to how close together the annual rings are. Paul and Smith (1950) refer to the density rule of grading lumber and suggest that wide-ringed wood should be excluded from dense structural grades unless it possesses more than 50% latewood.

5.7 Practical Application of Wood Changes Caused by Growth Rate Differences

Knowledge of the effect that growth rate might have on wood is vital if good forest management is to be practiced. Wood formation is a physiological process, and any changes in the environment in which the tree is growing can affect the tree's physiology and thus its pattern of growth and wood properties. For example, any management practice that will lead to a high proportion of latewood cells or to thicker-walled earlywood will result in wood of greater specific gravity and usually greater strength (Crow 1962). Unfortunately, it is usually not possible to predict what effect a change in the environment will have on the wood produced. This was strongly emphasized in Chapter 2 in relation to growing exotics and moving provenances into different environments and is discussed in Chapter 7.

Some of the difficulties in using growth rate to assess the value of wood are related to an error in measurement techniques. Usually growth rate is expressed as ring width, such as rings per inch. This can be very misleading, as pointed out by Crow (1962), who emphasized that ring width will generally be reduced as the tree grows older because it cannot continue to put on the same width of ring around an ever-increasing tree circumference. The basal area of the tree, the real measure of growth rate, becomes larger as the square of the radius of the tree. Crow feels that growth rate should be expressed by basal area rather than ring width when assessing wood properties. This is rarely done; as seen in all the tables presented, growth rate is commonly expressed by ring width. As early as 1959, Jeffers pointed out that many studies on the effect of ring width are unconvincing because they have not taken into account the effect of the location in the tree (the age of the ring) where width was measured; hence, the confusion with specific gravity and ring width in juvenile and mature woods.

The authors have for many years noticed that the response of wood properties to growth rate differences may vary with the cause of the increased growth. In the southeastern coastal plain of the southern USA, trees that grow more rapidly as a response to thinning have essentially the same specific gravity as those from the unthinned stand. Yet, when the same group of trees on a similar site grow faster as a result of heavy nitrogen fertilization, the wood specific gravity of the fast-growing fertilized trees is much lower than that from the unfertilized forest or from the fast-grown trees released by thinning. Here, two groups of trees were stimulated to grow fast, yet one group responded by producing low specific gravity wood, while the other did not (see Fig. 5.2). Contrary to the above, Erickson and Harrison (1974) found that both thinning and fertilizing Douglas-fir caused a reduction in specific gravity. When trees of the same growth rate were assessed, the differences in specific gravity disappeared. Such anomalous results make any growth rate to wood quality relations difficult to interpret.

Frequently, therefore, it is not growth rate per se that causes a change in wood properties, but it is the physiological stimulus that resulted in the difference in growth. Expressed another way, growth rate and wood properties sometimes are not directly related (Echols and Dorman 1962; Isebrands and Hunt 1975; de Guth 1980; Franklin and Bridgwater 1981; Zobel 1984).

Sometimes in species where faster growth results in a different quality wood, it is advised that one should not breed for fast-growing strains of trees on the assumption that these trees will have poor wood. This does not necessarily occur since trees that grow more rapidly because of an improved environment may produce wood different from those that are inherently fast growers. When growth rate and wood properties are not strongly genetically related, as frequently happens, it becomes quite feasible to develop fast-growing trees that also have desired wood. It appears to be a common situation that there is enough genetic independence between growth and wood properties that a proper selection of parents can result in fast-growing, high density trees, even though there may be a mild, negative growth to specific gravity relationship (Stairs 1969). This subject is dealt with in more detail in Chapter 8.

As mentioned earlier, a serious and common cause of error in assessing the relation of growth rate to wood quality results when the age of the wood as indicated by the number of rings from the tree center is ignored (Nicholls and Fielding 1964). It has not been uncommon to select randomly boards or logs with wide annual rings and then to compare their wood properties with boards having narrow annual rings, thus confusing the location within the tree and the possible presence of juvenile wood.

Several major publications have compared wood specific gravities of two species and related them to growth rates. The conclusion was that one species had fast growth and low specific gravity while the other was slow-growing and had a high specific gravity. Upon examining the sampling method, it became evident that the wood samples from one of the two species actually were less than 20 years of age and had wide rings, consisting mostly of juvenile wood, while the trees of the other species were over 60 years of age and had narrow annual rings of mature wood. When grown on the same site and at the same age, the woods of these two species are nearly identical regardless of ring width (Cole et al. 1965). Yet the studies mentioned, which supposedly showed that low density was related to fast growth, were used by the forest industries for many years as a reason to prefer the second species over the first, despite the fact that the basic research approach had been biologically incorrect.

The confusion in assessing growth rate when juvenile wood is not recognized was well described in an early paper by Rendle (1959) who stated, "...it is believed that fast-growth coniferous timber – that is, timber with wide annual rings – is necessarily of inferior quality. For this reason foresters have tended to restrict the rate of growth by delaying thinning operations. It has now been established that it is only in the central core of the so-called juvenile wood that wide rings are associated with weakness in structural strength of the timber." Another quotation is relevant here. Discussing the effect of age, Larson (1969b) states, "when age is taken into consideration, the relation between ring width and latewood percentage is either weak or non-existent. As an example, it is frequently possible to find growth rings of the same age and of the same width possessing either a very low or very high percentage of latewood."

Several studies were reviewed for this book which compared wood following silvicultural treatment (thinning, fertilizing, or other) with wood formed prior to the treatment. Especially in young trees this will lead to erroneous results because the "before" and "after" wood differences are partly the result of age effects (juvenile wood) and would occur whether or not any treatment had been applied. Such a comparison was made by Posey (1965) who found that the wood 7 years after fertilization had lower specific gravity than that produced 7 years prior to treatment. The importance of wood age was recognized by Posey who adjusted the results based upon untreated checks and the properties of the normal untreated wood before and after the time when treatment was applied. After adjustment, it was still evident that high nitrogen fertilization resulted in lower density wood.

Chapter 6

Wood Properties Affected by Environmental, Biological, and Other External Agencies

6.1 Introduction

As has already been mentioned several times in this book, anything that changes the growth pattern of a tree can affect wood properties. This was emphasized by Paul (1959) and many others, who showed that differing environmental conditions can result in the production of wood with varying properties. In 1962 (Anon 1962) TAPPI published Monograph No. 24 with 883 references, related to the effect of environmental and genetic factors that would affect pulpwood quality. Only a few of these references are repeated in this chapter.

Environmental factors include climate, soil and overall site, as well as the influences of the trees on one another within forest stands. Sometimes ecological influences or other external agencies only affect wood to a minor degree but sometimes the effect is major.

6.2 Effect of Site, Soil, and Climate on Wood Properties

The most difficult environmental "characteristic" of all to relate to wood qualities is a measure of the overall soil and climate which is often referred to as site index or site quality. Site actually combines many things that can each, on its own, influence wood. Thus, site differences are due almost totally to interactions, and as such it is not a very suitable measure for predicting the type of wood that will be produced. Since the factors that jointly determine site may differ from area to area, it is of limited value to use site as a general assessment of wood quality, although it is commonly done, such as by de Villiers (1973) who states that "in *Pinus patula, P. caribaea* and *P. elliottii* overall density also appears to be markedly affected by site factors."

Site classification is frequently used locally to predict wood characteristics, where foresters know that certain wood properties are associated with growing conditions of certain sites. As Jayne (1958) states, "the many conflicting reports on the effect of site on the specific gravity of wood indicates that variation in wood density needs to be studied in relation to specific environmental factors which make up site ... it is highly probably that ... seasonal fluctuations of soil moisture and the texture of the soil are closely associated with the specific gravity of the wood."

There are many miscellaneous ecological factors that can have an effect on wood. As just one example, Lawton (1984) reported for a tropical montane forest that: (a) "Shade-intolerant species tend to have less dense wood than shade-tolerant species from the same habitat". (b) "Species found on windy sites tend to have denser wood than species characteristic of sheltered habitats."

Although the relationship of the environment to wood properties will not be extensively discussed in this chapter, a few examples will illustrate its complexity. For instance, Crow (1962) presented a paper on "Site, Soil, and Silviculture" but the contents of the paper dealt more with desired wood density than the cause and effect relationship with soil and site. Gilmore et al. (1966) reported that the characteristic most closely related to specific gravity was site index, but they go on to explain that moisture availability was probably the controlling factor.

Polge stated in 1973 that the effect of ecological factors on wood quality is so great that the utility and value of timber can change markedly from one site to another; he used *Quercus petraea* and *Q. robur* as his examples. Polge emphasized that sometimes the effect of site factors is more important than species variation; this was illustrated with differences within Norway spruce and Scots pine as they occur in temperate and more northern climates. In his article on wood quality of radiata pine, Harris (1981) emphasized that sites with differing growing conditions can be a source of variability in wood. He found that pulp yields may vary by as much as 20% for trees of the same species, age, and growth rate when grown on differing sites. As summarized by Bamber and Burley (1983), "it would be surprising if density were not site related." They cite several studies on radiata pine that show a small to considerable effect of site conditions on wood properties.

For the convenience of the reader, a few papers dealing with site effects on wood properties have been summarized in Table 6.1, divided into sections on hardwoods and conifers. Papers were included in the table to cover different species growing under varied site conditions.

Table 6.1. The effect of site and soil on wood properties

Species	Reference	Results
	Hardwoods	
Populus trichocarpa	Kennedy and Smith 1959	As site changed from good to medium to poor the specific gravity increased from 0.331 to 0.383, respectively
Populus tremuloides	Wilde and Paul 1959	Only 10% of the variation in specific gravity and chemical composition of wood could be related to differing soils on which the trees were grown
Conystylus bancanus	Murthy 1960	There was a higher proportion of fibers in wood from the coastal swamp compared to the deltic swamp and Podamy forest
Eucalyptus camaldulensis	Chudnoff 1961	In Israel, wood from all sites was similar; the wood was almost the same as that of Australian timber
Liriodendron tulipifera	Thorbjornsen 1961	No relation was found between wood properties and site index
Liriodendron tulipifera	Barefoot 1963b	The best sites produced the lower density woods; toughness of wood follows the same pattern

Table 6.1 (continued)

Species	Reference	Results
Liriodendron tulipifera *Quercus rubra*	van Eck and Woessner 1964	There was a limited trend for better sites to produce higher density wood in yellow poplar. There was essentially no relationship between forest site quality and wood specific gravity in oak
Fagus sylvatica	Koltzenburg 1966	Vessels were larger on the better sites, but fiber length showed no clear relationship to site quality. Specific gravity of seedlings was different on contrasting soils
Acer rubrum	Saucier and Taras 1966b	Specific gravities were highly different between sample locations or sites
Fagus sylvatica	Koltzenburg 1967	Cell distribution and wood density showed a closer relationship to site than to light intensity
Liriodendron tulipifera	Gilmore 1971	The specific gravity of plantation-grown trees is not related to site
Fagus sylvatica	Keller et al. 1970	In France, the ecological factors that reduce site fertility also reduce wood density
Populus spp.	Provin and Lassen 1970	No significant differences were found in specific gravities and site quality on the 23 sites sampled
Fagus sylvatica	Polge 1973	Site differences have an effect on the presence of red heart
Cordia alliodora *Virola koschyni* *Cedrela mexicana* *Brosimum* spp.	Howe 1974	In Costa Rica the specific gravities differed when trees were grown in different environments except for *Brosimum* spp. which were not affected by site
Quercus rubra	Maeglin 1974	Site changed the proportion of cell types. Fiber proportions increased and ray proportions decreased with better site quality. There was an 11% increase in fibers and a 4% decrease in vessels when trees were grown on site indices varying from 25 to 85 ft
Tropical Hardwoods	Chudnoff 1976	In a gross assessment of site, the density of tropical timbers was related to several tropical life zones. The tropical moist zone had the highest density species and the tropical dry zones contained species with a large variation in specific gravity
Quercus rubra	Hamilton et al. 1978	Northern red oak had no significant differences in wood properties related to site quality
Populus x *euramericana*	Ferrari and Scaramuzzi 1980	There was considerable influence of site on wood basic density with higher quality sites giving lower density wood. To a lesser extent, fiber length was also affected. Chemical composition of the wood was not affected by site

Table 6.1 (continued)

Species	Reference	Results
Populus grandidentata	Richardson and Koerper 1981	There was no difference in stemwood specific gravity of old trees growing on three different sites
Eucalyptus camaldulensis	Sesbou 1978	Trees grown on soil with a high lime content in Calabrica have a higher density and less shrinkage than those growing in Sicily
Platanus occidentalis	Land et al. 1983	Trees on sites producing the fastest growth had the lowest specific gravity and the highest moisture content
Quercus robur	Nepveu 1984b	Environment (climate) strongly affects width of the latewood zone and fiber percentage, while width of the earlywood zone and vessel percentage have strong genotypic control
	Conifers	
Pinus resinosa	Jayne 1958	The best site produced lower specific gravity wood
Cryptomeria japonica	Yazawa and Fukazawa 1959	Variations in earlywood are closely related to the environment but latewood is more stable
Pinus sylvestris *Picea abies*	Ericson 1960	No site to wood relationships were found
Cryptomeria japonica	Fukazawa and Aoki 1960	There was no relationship between specific gravity and site quality
Picea mariana	Risi and Zeller 1960	There is no close relationship between site conditions and wood properties
Pseudotsuga menziesii	Knigge 1961	The better site produced more volume of lower density wood
Abies magnifica	Schniewind 1961	Site had no overall effect on specific gravity, modulus of rupture, or modulus of elasticity. The radial distribution of specific gravity values differed from site to site
Pinus radiata	Harris 1963	Specific gravity of outer wood varies significantly between sites
Pinus radiata	Reid 1963	The effects of site on wood are greater than those of silvicultural treatments
Pinus radiata	Warren and Mirams 1963	There was as much as 17% difference in basic density among differing sites
Pinus elliottii *P. taeda*	Hamilton and Harris 1965	Loblolly pine is more responsive to site differences than slash pine. The better site produced wood with large diameter cells and thicker cell walls. In slash pine, longer tracheids were produced on the better sites. Overall, wood differences related to site are small
Pinus radiata	Harris 1965	Specific gravity varies significantly among 37 sites. Temperature and phosphate deficiency had major effects upon the wood

Table 6.1 (continued)

Species	Reference	Results
Pinus taeda	Zobel et al. 1965	Site indices from 65 to 95 (at 50 years) had very little effect on tree specific gravity and percentage of juvenile wood
Pinus taeda	Lantz and Hofmann 1969	Differences in specific gravity between sites were greater than differences between seed sources
Pinus radiata	Wright 1970	Lower density wood was produced on the better sites; this resulted from a better moisture availability
Pinus caribaea	Andrews and Hughes 1973	Tracheid length in Jamaica varied (3.5 to 4.4 mm) between sites, with longer tracheids and higher densities on the better sites
Soft pines *Pseudotsuga menziesii*	Polge 1973	Site differences have no effect on the wood of the five-needle pines. In Douglas-fir, permeability of the wood varied with sites
Pinus strobus	Thor and Bates 1973	Twenty-four percent of the variation in specific gravity and 58% of the wood extractives were accounted for by 21 soil-site characteristics as independent variables
Pinus radiata	Nicholls and Wright 1976	There is no relationship between environmental factors and maximum and minimum wood density
Pinus patula	Kromhout and Roon 1978	Tracheid length is little affected by environment; wood density is greater and spiral grain more severe in trees from warm, low-altitude sites
Pinus nigra	Tsoumis and Panagiotidis 1980	Tracheid lengths were the greatest on the intermediate sites
Pinus caribaea	Cown 1981	In Fiji, high elevation sites produced a basic density of 540, low elevation a density of 475 kg/m^3 at 25 years of age
Pinus caribaea	Plumptre 1983	Higher quality sites gave lower density wood. Differences in lumber strength between sites were small
Pinus taeda	Megraw 1985	Major soil differences and moisture deficiencies cause specific gravity variation
Pinus oocarpa	Lima 1987	All families produce reasonably uniform wood quality on the sites (environments) tested
Cupressus lusitanica	Ladrach 1987	Wood density did not vary notably among sites

6.2.1 Soil Properties and Wood

A major component of site is soil. Wood properties often change with soil characteristics, although variations of soil with wood specific gravity are usually small within a given geographic area. Usually, wood responses to soil occur when extreme soil differences are present, such as between sand and clay. More subtle differences, such as site index variation in similar soils, usually are not accompanied by major wood differences. The soil differences are so closely related to moisture and to the nutrient supply that it is not always possible to separate them from a general site category.

Only a few papers refer directly to soil properties as they affect wood. An example is Wilde and Paul (1959) who reported that aspen wood of the highest specific gravity is produced on well-drained, fine-textured soils, while lower specific gravity wood is produced on poorly drained and nutrient-deficient soils. In the western mountains of North America, Carlson and Nimlos (1966) found that specific gravity of larch and Douglas-fir differed significantly among some soils, but the wood of ponderosa and lodgepole pines did not. Even where differences did exist, Carlson and Nimlos feel that using soil series to predict wood specific gravity is of questionable value.

The grain of high quality hardwoods has often been related to soil properties, but published information is scarce. One of the most cited examples is the color of black walnut wood (*Juglans nigra*). Nelson et al. (1969) reported that Indiana-grown walnut had a higher luminance than did Missouri-grown trees. They state "the relationship of heartwood color to soil properties was greater than it was either to tree age or to diameter growth rate." In contrast, McGinnes and Phelps (1980) found that there was only a small effect of site differences on the color of the wood of walnut. They reported that most color differences were from tree to tree within a site.

The effect of site on the color of wood of black walnut is widely accepted. Whether the trees are growing on soils that give a "good" or "poor" color partially determines the price paid for walnut. Soil effects on wood color were also found for yellow poplar and black cherry by Sullivan (1967). He theorized that the cause of color differences is the concentration of color precursors in trees from different areas, resulting from differences in physiological processes. Polyphenols and other organic extractives, which are produced in the leaves and cambium, are transported to the heartwood by the rays where they undergo chemical change to produce heartwood color. For tanoak (*Lithocarpus densiflorus*) Prestemon (1967) could find no relation of heart stain with site quality, geographic location, or old versus second-growth timber.

In a detailed study of the effect of dark clays, chalk soils, and rich bottomlands on the wood of ash (*Fraxinus*), Krasnitskii (1959) concluded that soils had little effect on wood in general, but that latewood percentage, vessel diameter or porosity, and the volume of parenchyma cells were different on the differing soils. Soil parent material affects the specific gravity of northern red oak. Wood from trees on limestone soils had a specific gravity of 0.597 and on sandstone soils, 0.581 (Hamilton et al. 1978). Site quality was found to affect the anatomy of northern red oak (*Quercus rubra*). An increase of site index from 25 to 35 ft resulted in an increase of 11% in the amount of fibers and a decrease of about 4% in rays and vessels (Maeglin 1974).

6.2.2 Summary – Effect of Site, Soil, and Climate on Wood Properties

Very few positive statements can be made concerning site and soil differences as related to wood properties. For both conifers and hardwoods, reports vary from no effect to an important effect (see Table 6.1). The reasons for so much conflicting evidence is that site is really not definable biologically and is dependent on an interaction of factors such as soil, nutrients, moisture (amount and pattern), day length, etc. Each site is a unique combination of the above factors, each of which may influence the kind of wood produced.

Based on Table 6.1, about one-third of the hardwood species listed had wood properties unaffected by site. A few had a higher density on the best sites, and a few others had a lower density on the best sites. For yellow poplar (*Liriodendron tulipifera*), there were reports of denser wood produced on good sites, lighter wood on good sites, and no wood response at all to site differences.

For conifers, about half the studies in Table 6.1 show no site relationship to wood properties. Results were not clear; for example, in radiata pine reports are available showing a large effect of site on wood, and others showing no relationship of site to wood properties.

In the hardwoods, differences in growing conditions sometimes change the cellular components and proportions of fibers, vessels, and rays. In a special review on the theoretical considerations of wood density with respect to site quality in upland hardwoods, Zahner (1968) discussed the ring-porous and diffuse-porous hardwoods separately. He concluded that wood density of the ring-porous species (such as oaks and hickories) is strongly influenced by site because these species grow in a wide range of environments which affect latewood production. Since the size and proportion of large vessels in the earlywood are relatively constant at all sites, the highly variable production of high density latewood among sites can increase wood density up to 20% on poor to good sites. Zahner believed that better sites not only yield larger volumes of ring-porous wood, but also wood of greater density. This theoretical result is not fully borne out by the papers summarized in Table 6.1, although it is often true (see Chap. 5).

In the high density, diffuse-porous hardwoods with a specific gravity greater than 0.45 [examples given by Zahner were sweetgum (*Liquidambar styraciflua*) and maple (*Acer*)], some environmental influence on wood density is found because of the production of larger and more frequent vessels in a matrix of relatively thin-walled fibers on good sites. Thus, wood density should decrease from poor to good sites. The low density (less than a specific gravity of 0.40) diffuse-porous hardwoods, such as *Populus* and *Alnus*, show little difference in wood density on differing sites. In fact, there are many exceptions to the theory expressed. Richardson and Koerper (1981) concluded that the forest manager can concentrate on volume production of quaking aspen (*Populus tremuloides*) or bightooth aspen (*P. grandidentata*) without concern that there will a change in yield of fiber per unit volume of wood produced on different sites.

Fig. 6.1. Site differences within a given geographic area generally have little effect on wood properties. The excellent *Pinus taeda* stand shown, growing on a good site in coastal Georgia, has wood properties nearly the same as those on soils with more clay or more sand, growing in the same region (Zobel and Talbert 1984)

The following is a summary of our evaluation of the relationship of wood properties to site and soil:

1. Site or soil differences within a given geographic area usually are not associated with major wood differences, unless the sites or soils are very diverse. Thus, wood of a given species tends to be relatively constant from site to site within a given region when site differences are moderate (Fig. 6.1).
2. The large individual differences among trees within a site make it difficult to detect differences caused by site or soil variability.
3. Site differences related to large geographic variations, such as in different provenances, usually have an important effect on wood properties. These are sometimes predictable, as outlined in Chapter 2 on geographic race descriptions. For example, trees grown at high latitude or high elevation sites usually produce lower specific gravity wood and shorter cells within a given species than wood of that species when grown closer to the equator or at lower elevations.
4. We have observed that site differences seem to be more important in affecting wood at the extremes of a species range where environments become marginal for growth than in the center of this range.
5. Under some conditions, site differences seem to be reflected in the kind and amount of earlywood formed but usually it is the latewood that is affected

most (see Chap. 1). This is controlled by moisture availability, temperature, and day length, each of which determines the length of time a tree produces latewood cells.

6. Unusual extremes of weather conditions, such as cold or dry seasons, can change wood properties drastically despite the normal site conditions in the area.
7. Especially the low specific gravity, diffuse-porous hardwoods tend to have wood that varies less over differing sites than that of the conifers. High specific gravity, ring-porous hardwoods tend to be most sensitive to site with respect to wood quality.
8. Soil properties are commonly related to site and thus to wood variation. However, they are so closely tied to moisture and nutrient availability that a clear definition of soil effects is difficult to make. In localized environments, soil characteristics can sometimes be used to predict wood qualities, but this is rare.
9. The site and soil relationship to wood is strong enough that potential changes must always be considered when growing a species on different sites or different soils. The only way to know with certainty what kind of wood will be produced is to grow the trees of interest on the site of interest.

6.3 Moisture Availability and Wood Properties

Changing the moisture regime results in a change in environment that will alter tree growth patterns, which in turn will affect the site and the kind of wood produced. Moisture available to the tree varies naturally for many reasons, however, it can be controlled by several kinds of silvicultural actions.

In a study of the effect of varied environments on wood properties of *Pinus taeda*, Buckingham (1967) showed that the ratio of latewood to earlywood within a given year was not fixed but varied with the environmental factors, mostly moisture availability, during the growing season. He found that water deficits in the fall determined the latewood percent, but that it was temperature in the spring that controlled initiation of wood formation. In their studies on red pine, Kraus and Spurr (1961) concluded that the transition in xylem structure within an annual ring may be triggered by soil moisture decline. Individual trees appear to have separate and characteristic transition periods. As Polge (1965) states, "we investigated what could be the climatic factor most satisfactorily accounting for variation in maximum density – it is already apparent that the rainfall at the end of the growing period plays a predominant role – with a highly significant coefficient of correlation of −0.89 between maximum annual density and the precipitation for the months of August, September and October." These results showed that rainfall at the end of the growing period plays a predominant role. Polge emphasized the complexity and interaction, using the water reserve in the soil as an example.

Many studies and comments have been made on the effect of soil moisture on wood formation, e.g., the results shown in the section dealing with soil moisture by Bamber and Burley (1983). This is somewhat surprising since soil moisture is difficult for the silviculturist to control. Because of the minimal control of moisture by the forester, this section on moisture and wood includes both the effects of natural moisture differences and those brought about by the silviculturist. A few studies of both have been summarized in Table 6.2. As is evident, moisture effects are interrelated with all other factors affecting growth such as site, soil properties, climate, etc.

The most discussed possibilities for controlling the moisture available to the tree by the silviculturist are irrigation and drainage (Zahner et al. 1964; Howe 1970, Nicholls and Waring 1972). Although not a standard silvicultural tool, irrigation has been studied even though it has been rarely used on an operational scale. One example is the effect of irrigation on white and red oak by Szopa et al. 1976). Currently, however, there exists several large forest-tree irrigation programs.

There are other workable methods applicable by the forester which will change the amount of moisture available to the tree, such as the control of competition, good site preparation, proper use of bedding, drainage, and thinning (see Chap.

Table 6.2. A partial list of publications relating to soil moisture and wood properties

Species	Reference	Results
Pinus caribaea	McWilliam 1955	Inadequate moisture in poor soil on exposed pine ridges in Queensland is responsible for the high percentage of latewood found there
Pinus taeda	van Buijtenen 1955	High soil moisture levels were associated with large tracheid diameters but did not affect wall thickness. In natural forests, irrigation did not affect the type of wood formed
Pinus elliottii	Larson 1957	All significant factors of the site influencing percentage latewood were related to soil moisture availability
General conifers	Hildebrandt 1960	Influence of soil moisture was evident on cell wall thickness and amount of earlywood
Pinus sylvestris *Picea abies* *Betula verrucosa*	Ollinmaa 1960	There are only slight differences in wood between slow-growing trees before and fast-growing trees following drainage. There is a slight lowering of specific gravity in conifers but no change in birch
Pseudotsuga menziesii	Kennedy 1961	Differences in latewood were related to precipitation; highest percentages were found in trees with the longest period favorable for latewood production
Pinus elliottii	Foil 1961	The width of the latewood band was most strongly associated with moisture level in late fall (October)
Pinus caribaea	Schmidt and Smith 1961	Average annual rainfall is not a major factor in the determination of the percent latewood
Pinus elliottii	Jackson 1963	Latewood percent is related to seasonal rainfall and depth of the water table

Table 6.2 (continued)

Species	Reference	Results
Several conifers	Farrar 1964	Tracheids resemble latewood during drought by having smaller diameters and thicker, more lignified walls. With severe drough a band of thin-walled cells form, and the tracheids collapse. Upon rewatering, normal latewood cells are again formed
Pinus elliottii	Goddard and Strickland 1964b	Early rains caused a reduction in specific gravity, while summer rains caused an increase
Pinus resinosa	Zahner et al. 1964	About 100% more latewood was produced in irrigated trees, but the percentage of latewood was equal to that of the trees subjected to drought
Pinus radiata	Shepherd 1964	Extended drought produced cells much reduced in radial diameter; normal cell production was delayed following resumption of normal growth
Pinus banksiana	Grigal and Sucoff 1966	The wood variation among plots was poorly related to the moisture retention capacity of the soil
Pinus taeda	Hiller and Brown 1967	Soil moisture conditions directly affected tracheid diameter and wall thickness but not the fibril angle of the tracheids
Pinus taeda	Saucier and Taras 1967	Trees from high rainfall seed sources had a low percent latewood and low specific gravity when grown in Alabama
Pinus ponderosa	Howe 1968	The number of latewood tracheids that were not flat were more than doubled with irrigation, producing a more gradual transition from earlywood to latewood and an increase in the amount of latewood and specific gravity
Pinus taeda	Gilmore 1969	The longest tracheids were in well-watered plots, but this was not statistically significant. It was concluded that heredity plays the primary role in tracheid length
Liquidambar styraciflua	Hunter and Goggans 1969	Longer fibers were produced in regions with longer growing seasons and more summer rainfall
Pinus ponderosa	Howe 1970	Trees irrigated for 7 years had their percent latewood and specific gravity increased. The effect was greater at the base of the tree than at the base of the crown
Pinus resinosa	Murphy and Brisbin 1970	Specific gravity of stem wood was higher in the controls than in the trees irrigated with sewage plant effluent; specific gravity was 11% less for the heavy irrigation

Table 6.2 (continued)

Species	Reference	Results
Liriodendron tulipifera	Sluder 1970	Environmental variability affects specific gravity by its effect on moisture availability
Populus tremuloides	Einspahr et al. 1971	Specific gravity was only slightly reduced as a result of irrigation plus fertilization
Pinus radiata	Nicholls 1971 a	The proportion of latewood was increased 20–30% and mean density 19–28% by supplementary watering
Pinus elliottii	Smith et al. 1971	Specific gravity was not affected by irrigation
Pinus strobus	Thor and Bates 1973	Specific gravity increased with greater soil moisture holding capacity
Pinus taeda	Hsu and Walters 1975	Maintaining soil moisture in the range of 30–60% of field capacity plus use of nitrate fertilizer increased bending strength and stiffness. There was 57% latewood in nonirrigated plots compared to 36% on the irrigated plots
Ouercus rubra	Szopa et al. 1976	Irrigation studies were made to determine the effect on wood
Pinus elliottii	van der Slooten et al. 1976	Rainfall and its distribution are a major determinant of the quality of wood of slash pine
Pinus radiata	Nicholls and Waring 1977	Specific gravity was greatest in irrigated trees due to an increase in maximum density. In trees subjected to drought, specific gravity also increased due to an increase in minimum wood density
Picea abies	Madsen et al. 1978	The basic wood density level is negatively correlated with the moisture content of the soil
Pinus caribaea	Plumptre 1983	Trees on poor, dry iron latosols had a spiral grain angle of 4.3% compared to 3.2% in trees on humic latosols

7). Although these are practices used widely by the silviculturist, relatively few studies have been made of their effect on wood properties (Fig. 6.2). In a general summary, Gilmore (1968) reported that factors which extended the production of earlywood, such as growing trees at the bottom of a slope, or growing trees on northern slopes or where more rainfall was preserved, produced a lower wood specific gravity.

It has been reported by several investigators that irrigation in summer and early fall results in higher wood specific gravity because of greater latewood production (Paul and Marts 1931). After a tree is triggered to switch from the production of earlywood to latewood, generally at the time of cessation of height growth, moisture in some climates becomes limiting, and the growth rate is rapidly re-

Fig. 6.2. Intensive forest management practices, such as the bedding shown in the coastal area of Colombia, change the moisture conditions, and thus growth patterns, of the trees grown there. However, very few studies have been made to show the effects of such practices on wood properties

duced (Zahner 1955). If irrigation is carried out at this time, radial growth is continued longer than normal. The extra wood produced is latewood, and the specific gravity is thus increased.

A few studies have been made in which drought stresses have been imposed upon the tree. For example, Zahner et al. (1964) found that trees under moisture strees initiated flattened cell formation common to late formed latewood earlier than those that had sufficient moisture from irrigation. The cambium in irrigated trees was nearly as active in mid- and late summer as it was early in the growing season and this produced more latewood. A number of researchers indicate that the highly flattened latewood cells, commonly found at the border of the growth ring, are accentuated by a deficiency of soil moisture. Paul (1959) stated that cessation of growth during a dry summer will result in curtailment of the latewood portion of the annual growth ring and thus will reduce wood density. In a study related to the effect of moisture on cell formation, van Buijtenen (1955) found that cell dimensions were more related to soil moisture content than was wall thickness.

It is expected that the rainfall distribution pattern will have an effect on wood properties. This result has been reported in many forms, for many species, and by many authors. One example is for slash pine planted as an exotic in southern Brazil, where van der Slooten et al. (1976) concluded that differences in wood quality were related to rainfall and its distribution. The same was found for Douglas-fir by Kennedy (1961) who reported large variations in wood properties among years related to precipitation. Highest latewood percentages were found in trees having the longest periods available for latewood formation. However, in the mahoganies, Boone and Chudnoff (1970) report that there does not appear to be a meaningful relationship between rainfall and wood density. They found that a specific gravity of 0.62 on the driest sites with 1250 mm annual rainfall was almost identical with that from the wettest sites with over 3000 mm rainfall. In Denmark,

Madsen et al. (1978) make the statement that their results support the hypothesis that the basic density of wood is negatively correlated with the amount of water accessible to the tree. In *Eucalyptus*, Ferraz (1983) showed that there was a considerable effect of moisture deficiency on wood properties, and that different species and sources did not have their wood affected in the same way.

Like all causes of wood variability, responses to moisture differences are very complicated, and interactions with other factors (such as nutrients) always occur. Thus, much of the later discussions of silvicultural actions, such as stocking control or site preparation, are partially related to moisture availability. Although widely recognized, the interactions cannot be easily determined. Zahner and Oliver (1962) and others report that the greater moisture availability resulting from thinning plays an important role in the response of the trees to wood production. It often keeps the trees producing latewood for a longer period of time (Kennedy 1961).

Several investigators have analyzed the wood produced in years with climatic extremes and have deduced some moisture effects on wood qualities. For example, Knigge and Schultz (1961) studied wood of oak, beech, spruce, and pine following the extremely warm and dry 1959 season. As expected, rings widths were reduced, but there were also other wood changes. In oak there was a much greater percentage of earlywood and many fewer latewood fibers and vessels; oak had clearly contracted vessels both in the middle and at the end of the annual ring. Beech showed no difference in wood structure in the dry year except that the width of the vessels was smaller.

The effect of moisture availability on wood properties can be illustrated by several examples:

1. The southern pines, such as *Pinus taeda*, produce quite dense wood. Yet, when this species is grown in climates similar to parts of California, where spring rains are common but where it does not rain for 5 or 6 months during the summer and early fall, loblolly pine produces a very low specific gravity wood. The result is wood with a preponderance of earlywood in the annual ring but essentially no latewood.

2. In parts of the tropics, rainfall (in addition to other environmental factors) is so ideal that species such as *Pinus caribaea* apparently never cease height growth and thus the formation of latewood, which is produced only when height growth has ceased, is very restricted. Trees growing in this kind of environment with plentiful moisture and suitable temperatures have a very low density wood.

3. The southern pines in Florida start growth early in the spring when moisture is limiting. Rainfall then becomes heavy throughout the summer, and the wood specific gravity is high, because the period when latewood is formed is extended through the summer. The same species, when grown at more northern latitudes or inland, where extensive periods of moisture deficiency occur in the summer, have much lighter wood because growth ceases after a shorter period of latewood formation.

Annual rainfall patterns can have a major effect on the specific gravity of the wood produced in any given year. Annual variations are often large. Lassen and Okkonen (1969) report that in coastal Douglas-fir the specific gravity of wood formed during a single growing season was 0.52 for dry summers and 0.45 for wet summers. This seems contrary to what is stated above, but the authors categorically state that there was a negative, linear trend in wood density with increased rainfall during the summer, independent of elevation.

In Australia, Nicholls and Wright (1976) tested six clones of radiata pine on five different sites. The number of growing days, determined by favorable water deficits, had a definite effect on the formation of latewood. No relationship was found between environmental factors and the minimum and maximum densities. Related to these results were those for eastern white pine (*Pinus strobus*) reported by Thor and Bates (1973), who found that greater specific gravity was related to soils with the greatest soil moisture capacity. In another paper, Thor and Core (1977) concluded that site factors (mainly moisture) accounted for only a small percentage of the variation in specific gravity, wood extractives content, or fiber length of natural stands of yellow poplar. This is in contrast to the specific gravity increase with greater soil moisture storage capacity that Thor and Bates (1973) reported for white pine.

6.3.1 Moisture Effects on Wood – A Summary

The major trends relating soil moisture to wood properties are:

1. Many site and environmental differences that are credited with having an effect on wood properties can be related to the moisture available to the trees. Moisture is recognized as being a major factor in controlling wood properties.

2. Soil moisture is generally determined by natural conditions of the environment. Foresters can only change it by such methods as drainage, bedding, competition control and, in a few instances, irrigation. Thinning is an important method for controlling soil moisture. This was shown for red pine (*Pinus resinosa*) in Minnesota (Bay and Boelter 1963) where thinned stands had their moisture recharged earlier, and the moisture deficits in the fall were less. This aspect of moisture availability is especially important in droughty, sandy soils in the tropics. Other studies indicate the importance of weed control in maintaining moisture for tree growth. Wilde et al. (1968) showed that weed cover used about 100 kg water per kg oven dry tissue. Loss of moisture from weeds in red pine plantations can reduce growth by as much as 50% and will affect wood properties. Unfortunately, the long-term effects of silvicultural moisture control on wood properties are not well known.

3. The control of wood properties by moisture is through its effects on growth patterns of the tree. Numerous studies report higher specific gravities by extending the growth period longer into the fall after latewood formation has started. Although this effect appears to be the most important, other studies indicate that lower specific gravity results when plentiful moisture is present throughout the season, enhancing the production of earlywood formation and delaying the initiation of latewood. There have been a number of studies involving both conifers and hardwoods, where irrigation was not found to have any effect on wood properties. If used to change wood, the timing of irrigation will be very important.

4. Drought stresses trigger the formation of latewood, specifically the production of the greatly flattened cells formed at the end of the year.

5. Several researchers report that plentiful moisture leads to the production of larger-diameter cells; drought produces smaller cells. Similar results were occasionally obtained for tracheid length.

6. The amount of moisture available has a limited influence on the specific gravity of the low density, diffuse-porous hardwoods, but has a major effect on the high density, ring-porous hardwoods. The effect is primarily through a change in the relative amount of vessels produced in the earlywood and latewood; latewood continues to be produced when moisture is available.

7. Differences in rainfall are sometimes related to the amount of latewood. This is often confounded with other site factors.

8. Annual rainfall patterns can have a major effect on the kind of wood produced during a given year.

6.4 Miscellaneous Biological Factors that Affect Wood Properties

There are many biological factors that can alter wood quality, but do not fit neatly into any category. Therefore, this section will contain short descriptions of various biological phenomena that can alter wood.

6.4.1 Forest Pests and Wood

Insects and diseases that do not actually kill trees often cause deformation or changes in wood properties that render the wood of limited utility. Others that kill trees can cause a rapid deterioration of wood quality, reducing its usefulness for some products and making the wood useless for others. Even a modest coverage of this subject would be enough for a separate book. We therefore will choose a few examples, so that it is clear that pest attack is a major source of wood variation and wood degradation.

6.4.1.1 Disease and Wood Properties

Especially in older forests, the main reason for changes in wood properties is the heart rots or root rots that also affect the bole. These frequently are caused by fungi (Fig. 6.3). There are two main types; one is the group of fungi that cause wood rot by attacking the cell walls. These occur as either white or brown rots; the former attack the lignin, resulting in lower yields and difficult bleaching of pulps, while the latter leave the lignin and attack the carbohydrates in the cell walls. The other type is represented by the sap stains that attack the stored food in the wood; they do little damage to the wood as such except to discolor it (Rydholm 1965). Vast volumes of timber are rendered useless or of low quality by these fungi. In a few instances, such as pecky cypress, a specialty, high-value product has been developed from the diseased wood, but this usage is rare.

The most common infections that cause variation in wood are the blue-stain fungi. These almost always occur as a secondary response to the death or severe

Fig. 6.3. Much damage to wood is caused by the heart rots, as shown for a number of logs from the jungle in the Amazon. In this region, only about 20% of the logs are sound, the rest being destroyed or damaged by the rot

decline of the tree by harvesting, insects, weather, or other fungi that in themselves have very little direct effect on wood properties. Although not a primary cause of death, the blue-stain fungi help weaken the tree by disrupting its water-conducting capacity in the sapwood (Levi 1981). The blue-stain fungi often enter the tree only hours after initial attack or stress caused by other agents and subsist on the sugars and proteins in the wood. At first they form blue-gray, wedge-shaped areas, but as staining progresses, all sapwood may become discolored. Although many are reluctant to use blue-stained wood, the stain has little effect on the physical properties of boards other than decreasing wood toughness and increasing permeability.

Blue-stained wood cannot be used for the best-grade lumber, although it is perfectly useful for framing lumber (except for poles and posts). It makes gluing difficult, and pulps made from it are darker and more difficult to bleach than pulps from normal wood, even though the strength qualities are about the same (Sinclair and Ifju 1977, 1979). Special products, such as blue-flecked paneling (Levi 1978) can be made from blue-stained wood.

The degrade and loss of wood quality of the final product due to the stains and heart rots are so large and important that they deserve special, full coverage in a separate publication and will be dealt with no further here, other than to recognize that severe wood variability is caused by the rot and stain organisms.

Fig. 6.4. Diseases such as fusiform rust can cause degradation of the most valuable basal log, reducing them in value for solid wood or fiber products. Examples of the effect of this disease are shown in the photograph of serial cross-sections from below to above the gall

There are diseases which do not kill the tree, but do a great amount of damage to the wood. One of these is fusiform rust (*Cronartium quercum* f. sp. *fusiforme*) which causes its most serious damage on the most valuable basal log. This disease often makes the basal log unsuitable for solid wood products and reduces its value for fiber products (Fig. 6.4). For example, Kinloch (1963) concluded that fiber length in the diseased wood is greatly reduced. Reduction in specific gravity of diseased wood is significant but of a lesser magnitude. The cells have thin walls and are deformed. The resin content is increased, sometimes dramatically. In the fusiform gall itself, no latewood is formed according to Jackson and Parker (1958), who also reported that tracheids are very variable in shape and less than 2 mm long and that there is an increase in the number of vertical resin ducts.

An intensive study of the wood properties of southern pine trees infected by fusiform rust was made by Zobel (1971) and Veal et al. (1974). Badly infected wood yielded 23% less pulp and used 20% more chemicals than nondiseased wood. Paper made from diseased trees had a low tearing strength and was difficult to bleach. Tall oil yields were increased considerably. Veal et al. pulped various mixtures of diseased and normal wood. In a 30% diseased–70% normal wood mix, 2–4% more alkali was required for comparable delignification, and the paper had a 5% lower burst strength. When trees from a heavily infected stand were pulped, 10% more alkali was needed, 1% less pulp was obtained, burst strength was 3% lower, and 25% more crude tall oil was obtained compared to a rust-free stand of the same species. Holley and Veal (1977) have reported the information shown in Table 6.3. Note where the gains and losses occur.

Table 6.3. Compensating quality factors in pulping a loblolly pine stand with 29% stem infection of fusiform rust for kraft linerboard[a]

Changes caused by rust	(%)	Gain ($/1000 lbs oven dry wood)	Loss
Pulp yield reduction	0.9	–	0.51
Increased alkali use	0.8	–	0.04
Pulp strength reduction	5 to 10	–	–
Tall oil increase	25	0.23	–
Turpentine increase	40	0.25	–
		0.48	0.55
Net loss (approx.)			0.07

[a] From Holley and Veal (1977).

Fig. 6.5. Wind often causes cracks or ring-shake as shown for this *Pinus elliottii* from South Africa. Disease infection and resin deposition follow the damage within the tree bole (courtesy of G. van Wyk, South Africa)

Diseases commonly attack wood of trees destroyed by fire, blown down from wind, or other causes. The wood can be severely affected and one of the most puzzling problems is what uses can be made of such wood. Aho and Cahill (1984) discuss this for a number of species such as *Thuja plicata, Pseudotsuga menziesii, Tsuga heterophylla, Abies amabilis, Pinus monticola, Abies grandis,* and *Picea engelmannii.* Product degrade was heavy, depending on species, age, and other factors. Certainly the wood is so changed that it involved much variation in wood properties.

Although not a disease as such, the wood formed following wounding can be changed significantly (Bauch 1980). Often fungi or bacteria are involved;

mechanized harvesting is a major cause of "wetwood", shake, and other wood problems related to wound effects (Fig. 6.5).

6.4.1.2 The Influence of Insects on Wood

There are both direct and indirect reasons for changes in wood properties following insect attack. When trees are killed by bark beetles, spruce budworm, defoliators, or similar insects, the wood is usually degraded by a subsequent fungal attack or from attack by secondary insects, rather than directly from the insect that killed the tree. This was shown for *Prunus serotina* by Rexrode and Baumgras (1984) where gum spots were caused by bark beetles. The gum spots cause grade defect for both veneer and lumber. This gives rise to several questions related to insect damage: how severe will the secondary degradation be; how long will wood be useful; for what kind of products will it be suitable? Levi (1981) states that bark beetle attack can severely reduce the value of the trees for solid-wood products because of both secondary fungal and insect problems.

One of the most extensive and widespread type of tree destruction by insects has been the huge kill by spruce budworm (*Choristoneura fumiferana*) in eastern Canada and the northeastern and western United States. Many studies have been made on the rate of degradation of the wood, and what the wood can be used for. It is clear, even in the cooler climates, that insect-killed wood is not suitable for solid-wood products for more than a year or two after death of the tree.

A study of the value of spruce budworm-killed balsam fir for flakeboard and waferboard by Hughes and Gertjejansen (1984) showed that balsam fir, dead for up to 2 years, makes good quality board, superior to that from aspen. According to Govette et al. (1983), who studied trees killed by spruce budworm for 6 months or less, for 12 months, or for more than 18 months, rapid decay eroded the potential for recovery of large percentages of No. 2 or standard and better grades of lumber. The lumber from budworm-killed trees had reasonable quality up to 1 year after death. After that time, 63% of all lumber was only suitable for studs, utility, or No. 3. An increase in the time after death caused increasing degrade of lumber into the economy grades. In a similar study, Barnes and Sinclair (1983) determined changes in wood of trees dead for 6 months, 12 months, and 22 months. The moisture content of the wood from trees killed by spruce budworm decreased rapidly following insect attack; this was evident in the outer portion of the tree after 6 months, and in the whole tree after 12 months. In still another study of spruce budworm-killed balsam fir, Sinclair and Barnes (1983) reported that wooden beams from budworm-killed balsam fir have a reduction in toughness where shock resistance is important. For pulpwood, it is reported that the wood is usable, although with some loss in yield and strength, for up to 5 years.

In a study of kraft and sulfite pulping of budworm-killed balsam fir (*Abies balsamea*) and white spruce (*Picea glauca*) in New Brunswick (Hatton 1984) yields and tearing strength of unbleached kraft decreased with length of time since tree death up to 4 years, but burst did not differ from normal wood. Loss of tear was severe in trees dead for longer than 4 years. Yields and brightness of bleached sulfite pulps from sound and dead woods were similar unless the wood had been dead for more than 4 years. In practice, however, most companies do not use wood that has been dead for more than 2 years. For pulp, this usually constitutes only a small percentage of the total furnish.

A major defoliator of hardwoods that often kills is the gypsy moth. Huge numbers of oaks and other species have died and there is concern about the use of their wood. In 1984, Garges et al. made a study of the wood of white (*Quercus alba*) and red oak (*Q. rubra*) killed by the gypsy moth. The trees studied had been dead, for 1 year, 2–4 years, and over 5 years. Lumber recovery decreased as time after death increased. For red oak, the percentage of high grades was reduced for trees dead for 2 years or more; trees dead for over 5 years produced half the No. 1 common grades compared to live controls. It was recommended that trees be salvaged within 1 year to reduce losses in lumber yield and grade.

Similar studies have been made on trees killed by bark beetles. Although reported recommendations vary greatly, for most species, excluding those such as redwood (*Sequoia*), cedar (*Thuja*), cypress (*Cupressus*), and a few others, the utility of the wood is rather short-lived after the trees have died. In the United States Forest Service's Southern Pine Beetle Handbook (Levi 1981) the utility of wood from beetle-killed trees is assessed. It is emphasized that the grades of lumber from beetle-killed trees are lower than from healthy trees. However, the quality of particleboard, hardboard, and pulp from beetle-killed wood is acceptable, if the trees have not been dead too long. Levi (1981) states that timber killed in the mountains by the southern pine beetle (*Dendroctonus frontalis*) is usable for some products for 4 years, but in the southern coastal area the wood cannot be used after about 1 year. Gajda et al. (1982) did not recommend the exclusive use of beetle-killed southern pine for paper production but suggested that it be used in a blend with normal wood for paper as well as for particleboard and hardboard. The lower moisture content of the beetle-killed wood requires the use of additional amounts of pulping chemicals or different pulping techniques. For solid wood products, they report that the trees with a few small limbs remaining cannot be economically used for boards. The blue stain causes further degrade to already weakened boards. The dead trees lose weight rapidly, the moisture content being only about one-half that of freshly cut trees (McNab 1983). This thoroughly disrupts any weight scaling method.

Snellgrove and Ernst (1983a) studied the wood properties of lodgepole pine (*Pinus contorta*) killed by the mountain pine beetle (*Dendroctonus ponderosae*). Volume recovery of veneer was not much different from that of live trees nor from those that had been dead for 1 year, but recovery from trees dead for 3 years was about 30% reduced. Grade was not affected by time of death, but a higher proportion of random strip was produced from the older, dead logs. In the same species, Lemastre et al. (1983) found that dead lodgepole pine was useful as boards but that most of the loss was caused by drying, with a high incidence of warp and wane. Dead trees with tight bark produced good products, but those with loose bark were less suitable for some products. A recent article on insect-killed lodgepole pine (Plank 1984) summarized that the value per unit volume recovered was less from the dead than from the live trees. Stud grade recovery was 86% of the volume in live trees and 73% in the dead trees.

In a study of the pulping properties of southern pines Ifju et al. (1977) assessed the wood of trees killed by beetles. Trees that had been dead for 2 years gave about the same pulp yield as live trees, but tear strength was reduced after as little as 6 months following the death of the tree. They concluded that one can use trees dead up to 2 years with little effect on yield and only a mild decrease in paper quality.

White spruce in Alaska that had been killed by *Dendroctonus rufipennes*, a spruce beetle, was pulped to determine the wood properties of trees that had been dead for 1 year compared to 50 years. In the Alaskan climate the wood of the dead trees produced paper with good strength properties; the authors conclude that dead trees can be used to produce high quality kraft pulps (Werner et al. 1983).

Insects also cause a secondary degrade. For example, some of the bark beetles and insects, such as *Sirex*, carry fungal spores. Often, the actual damage to the tree bole is slight but the associated fungi plug up the vascular system of the tree, causing its death. Usually, the fungi causing these problems are blue-stain fungi, and perhaps brown and white rots (Levi 1981). Also associated with the dead wood are ambrosia beetles, termites, and other insects that cause degrade and facilitate access of fungi.

In Chapter 4 there were several references relative to aphid infestation causing formation of compression wood (Smith 1967; Van der Sijde et al. 1985; Timell 1986). Foulger (1986) found that wood of grand fir (*Abies grandis*) formed during an aphid (*Adelges piceae*) infestation in the West had shorter tracheids, greater fibril angle, a greater percentage of latewood, higher specific gravity, a reduced modulus of elasticity, and greater longitudinal shrinkage when compared to noninfected trees. These wood properties are similar to those induced by *Adelges piceae* in *Abies balsamea* in the east. Foulger concluded that a widespread aphid infestation would be detrimental to the quality of wood produced, particularly in stems approaching merchantable size.

In 1955, Balch stated with regards to the balsam woolly aphid, "the salivary injections of *Adelges picea* into the bark of balsam fir cause a stimulation of cambial activity and the production of an abnormal type of wood very similar to compression wood ... Trees so affected are unsuitable for lumber owing to uneven shrinkage causing warping and splitting. ... pulp ground from this wood was short fibered and brittle; burst and tear were low and the density of the wood was high."

6.4.1.3 Effects of Other Pests on Wood Properties

Numerous other organisms that attack a tree can affect wood quality. For example, Pirito et al. (1974) studied the effects of mistletoe (*Phoradendron* sp. and *Arceuthobium* spp.), common pests of both conifers and hardwoods, on lodgepole pine. They found that the infected wood had a higher specific gravity, more alcohol-benzene extractives, shorter tracheids, and an increased fibril angle, resulting in an increase in longitudinal shrinkage of boards made from it. An important finding was that both infected and noninfected wood in the same tree was affected by mistletoe.

Defoliation is a common result of various pests attacking growing forest trees, but unfortunately, few studies have been made on the wood properties produced following the loss of foliage. In addition to spruce budworm, beetles, worms and sawflies, various diseases cause defoliation in total or in part. Considerable change would be expected because of the change in carbohydrate production and altering of the growth hormones produced because of the loss of foliage and/or the growing points or buds. Such a loss causes a hormone imbalance which in turn affects wood quality. One published result, for defoliation of sugar maple (*Acer saccharum*) by Gregory and Wong (1983), showed that wood production either declined drastically or ceased following defoliation. Cell enlargement, wall thickness, and lignification were all markedly reduced.

In order to assess the effect of defoliation on wood properties and to obtain more basic information on the physiology of wood formation, a number of defoliation experiments have been carried out. Richardson (1964) found that partially defoliating *Picea sitchensis* seedlings suppressed xylem formation and destroyed apical dominance. Greater defoliation resulted in reduced cell wall thickness and lumen diameter. The effect on the wood was not restricted to the defoliated zone. Using maritime pine (*P. pinaster*) Polge and Garros (1971) studied the effects of artificial defoliation on the wood structure of the defoliated tree. Trees that were defoliated, but whose buds were left intact, had normal cell size, but the wall thickness was much reduced.

6.4.2 Compass Direction in the Tree Stem

Especially in the higher latitudes, it is sometimes found that wood formed on the side of the stem that receives more sunlight or is subject to prevailing winds is different from wood on the shaded or protected side of the tree. Nylinder (1935) demonstrated a difference in basic density between the north and south side of the tree. He found basic density to be 2% higher on the south side of the 53 Norway spruces that he sampled. However, Göhre and Götze (1956) could not find differences with compass direction in Douglas-fir. In the more tropical areas, the effect of the compass direction is less (Bissett and Dadswell 1949).

Knigge and Koltzenburg (1965) have discussed the effect of compass direction on wood properties of hardwoods. They cite a number of papers which show that fiber length is always the shortest on the sunny side of the tree. In his review, Hildebrandt (1960) quotes several authors who showed that cells on the sunny side of the stem were shorter than on the shady side. Cambial activity started earlier on the sunny side.

Some foresters consider the compass differences within trees to be minor (Yanchuk et al. 1983a) for poplar or nonexistent for yellow poplar (Taylor 1968), but others, such as Olesen (1973), emphasize the importance of a knowledge of this within-tree variation for sampling of wood and assessment for genetic evaluation of the tree (Fig. 6.6). Olesen made his study with 56-year-old Norway spruce trees. He found significant differences in basic density within the tree, the density being the highest on the south side of the tree and lowest on the north. Ring width was

Fig. 6.6. Although compass, or cardinal, direction is important in some areas, especially at high latitudes, and in some species, in others such as the yellow poplar (*Liriodendron tulpifera*) shown, compass direction has been reported to be of no importance relative to wood quality

less on the south and greater on the north side, while fiber length was less on the north side.

For *Pinus contorta* grown in Europe, Schütt (1962) found that specific gravity values of the outer wood were different when taken from different cardinal directions. Samples from the east side were always heavier than those from the north or south side. In *Pinus pinaster* in France, Polge and Illy (1968) found wider rings and higher specific gravities on the east compared to the west side of the bole. The wood on the east side of the tree had some characteristics similar to compression wood but no excessive longitudinal shrinkage, although it produced low pulp yield with short fibers. In *Picea abies*, only half the 500 stems studied showed dependence on cardinal direction and it was not considered to be of importance by Stern (1961).

Cardinal direction can also have an effect on wood properties in tropical species, as reported by Briscoe et al. (1963) for mahogany. At higher levels in the tree the wood was heaviest on the western stem side.

In diffuse-porous hardwoods, cardinal direction seems to have little effect, even in the far North. For example, Yanchuk et al. (1983b) found no significant differences with cardinal direction in trembling aspen (*Populus tremuloides*) in Alberta, Canada. There was also no difference between the north and south part of the tree in *Populus deltoides*, according to Walters and Bruckman (1965). Knigge and Koltzenburg (1965) cited several authors who concluded that fiber length in the poplars was greatest on the sunny side of the tree. They then cite several authors (for *Eucalyptus* and *Populus*) who found no differences. Meyer-Uhlenreid (1957) reported a smaller proportion of fibers and more rays and vessels on the sunny side of poplar trees. Several of the authors cited remarked that the position of an individual tree within the forest stand can modify the effect of compass direction.

The direction of prevailing winds can also have a considerable effect on wood. One of the summary points of Olesen (1973) on Norway spruce is "... wind is thus indirectly the main cause of the systematic variation in basic density between directions within a stand."

6.4.3 Temperature, Light, and Wood Production

Although temperature is one of the integral parts of the climate, only a few studies have been made directly on temperature effects, other than those cited above related to compass direction. In one made under controlled conditions, van Buijtenen (1955) found thinner cells in loblolly pine as temperature was increased. When young red pine trees from widely separated natural stands were grown at four different temperatures for 23 weeks, differences were found for most wood characteristics although only minor differences were observed among source populations (Larson 1967).

In *Picea sitchensis*, Richardson (1964) found that higher temperatures increased tracheid length and radial lumen diameter, but cell wall thickness was not affected. In an earlier, general paper, Richardson (1959) reported that wood density and cell wall thickness increased with night temperature but showed no relationship with day temperature. Tracheid length increased with increased temperature, both day and night. From this he hypothesized that cell elongation is a direct function of temperature, while cell wall thickness is determined by net assimilation rate.

In his large study on *P. radiata*, Harris (1963b, 1965, 1973b) found a correlation of 0.939 between temperature of a given site and specific gravity of the wood of the trees growing there; the correlation was better for mature than for juvenile wood. For tracheid length, the correlation with temperature was $r = 0.718$. In Douglas-fir, Knigge (1962) reported that the influence of both temperature and rainfall on wood properties were of minor importance.

It is often considered that the total sum of light hours is important for growth patterns and thus for wood properties. To determine this for tracheid wall formation in *Larix decidua* var. *polonica* grown in a greenhouse, Wodzicki (1961) treated the plants with differing combinations of day length. He found that when daylight was supplemented by low light intensity during the night, thick-walled tracheids were formed. Similar results were obtained when 2- and 6-h periods of light and darkness were used. Contrary to most reports he did not find a relationship between cessation of shoot growth and tracheid wall thickening.

Wodzicki's experiments (1961) made it clear that the photoperiod controlling growth process is a major controlling factor in cell types formed in young trees. For example, 42% of the variation in wood specific gravity was associated with phenological variables, with growth initiation of the leader and cambium used as an index of crown activity.

6.4.4 Position of the Tree in the Stand

A surprising number of studies have been made relative to the position of the tree in the stand, i.e., whether the tree is dominant, codominant, intermediate, or suppressed. Much of the European literature refers to this as the "sociological position" of the tree. Most researchers state that the crown location does have an effect on wood properties (Knigge 1962) but many are somewhat vague as to just what the position effect is. Certainly sociological position affects crown function through light availability and crown size so it would be expected to have some effect on wood properties (Larson 1963a). Hildebrandt (1960) generalizes that in even-aged stands dominant trees have a lower wood density than the codominant ones.

In a study of beech (*Fagus sylvatica*) Koltzenburg (1967) found that the larger amount of light available to the more dominant crown classes produced larger vessels in the wood along with wider annual rings. On the other hand, fiber length of beech showed no relationship to crown class or light availability. Koltzenburg found that the tracheid length of the different tree classes changed with age; up to 20–30 years the tracheids were longer in the subdominant and suppressed trees, but at later ages they were shorter than in the codominant trees. Hildebrandt (1960) concluded that the effects of growth conditions on the properties of beech wood are not important enough to establish rules for silvicultural treatment of beech stands.

For Norway spruce (*Picea abies*) von Pechmann (1958) found that the dominant trees had the lightest, weakest wood and suppressed stems the densest and heaviest wood. He explains this as the result of the fact that trees with a larger crown have more conductive tissue. Similar results were reported by Hildebrandt (1954), who generally found the highest density wood in the suppressed and intermediate spruce trees. Similarly, Holubcik (1969) reported that the dominant spruce had 1–4% less latewood (for the same annual ring width) than the codominants and intermediate trees. But he warns that one can find complete irregularity of stem density related to stem class. Although it is implied by other authors, von Pechmann (1958) stresses that the dominant spruce trees with large crowns also have larger diameter knots which affect wood quality. Schultz-Dewitz (1960) found that intermediate and suppressed spruce trees have shorter fibers and smaller rays as well as fewer rays than the dominant trees. The height of the rays in spruce was also increased, with the result that dominant stems have more ray cells than intermediate ones (Schultz-Dewitz 1959). For the intolerant pines and larches he found that the dominants had larger diameter earlywood tracheids and thicker walled and larger diameter latewood tracheids. The proportion of latewood was less.

In *Pinus sylvestris* and *P. mugo* the dominant trees have larger diameter tracheids and greater wall thickness than the intermediate class trees (Schultz-Dewitz 1958). For Douglas-fir, Knigge (1962) also mentions the importance of the effect of the "sociological position" of the tree in the stand. In the dominant tree class, ring width had little relation to wood weight, but suppressed trees had lower specific gravity when ring width was larger.

A position effect was also reported for Japanese red pine (*Pinus densiflora*) with the subdominant (intermediates and suppressed?) trees having a higher specific gravity, but crown class did not affect tracheid length (Kano and Saito 1970). In contrast, Hughes and Mackney (1949) state that tree class has no effect on the wood moisture content in radiata pine.

In summary, it is evident that the position of the tree in the canopy can have an effect on wood properties. It is of little practical importance in well-managed forests where the intermediate and suppressed trees are removed. Little wood variation occurs between the dominant and codominant trees within a stand.

6.5 Unusual Causes for Wood Variability

One type of defect in wood that causes extensive degrade, especially for solid wood products, are cracks and ring shakes caused by frost damage, wind, and internal tree stresses (see Fig. 6.5). These are usually found in severe climates and only in some species, but the overall degrade is much larger than commonly realized.

Cold temperatures and unseasonable frost sometimes have a direct effect on wood properties. Saucier and Taras (1966a), for example, discuss the possibility that frost resistance was associated with specific gravity differences among six sources of longleaf pine, especially the Florida source whose wood differed from the others. Kubler (1983) summarized the mechanism of frost crack formation in trees, a frequent cause of degrade in wood properties. The cracks result from tension stress in the tangential direction of the annual ring. The defect can be termed "frost shrinkage" from freezing out wall moisture into the lumens of the wood cells. Healed wounds in stems enhance the stress level, which causes cracking. Special efforts must be made to avoid wounding trees in cold climates.

Dietrichson (1963) studied the relation between wood maturation and climate in *Picea abies*. After a warm summer there is late maturation, which results in hidden frost damage. This occurs most in the southwest continental European provenances. Dietrichson et al. (1985) feel that the triggering mechanism for stem cracks is a late summer drought. They also state that a growth rhythm out of phase with the climate makes the trees more susceptible to stem cracks.

In what must be the most unusual cause of wood variability, Resch et al. (1983) made a study of changes in wood properties resulting from the eruption of the Mount St. Helens volcano in the northwestern USA. Huge areas of good forest were damaged. A large salvage operation was launched, but there was considerable concern that the wood subjected to the heat and force of the blast would be substandard. Contrary to expectations, Resch et al. (1983) found that there was little effect on wood properties from the blast, even on wood located close to the surface of the tree. Considerable ash and dirt in crevices and checks in the blown-down trees caused some troubles in sawing, but the fear that rocks and ash would penetrate the wood did not materialize (Snellgrove et al. 1983b).

An unusual degrade of wood quality is caused by sapsuckers, a bird that pecks holes in the cambium, which results in formation of abnormal wood and sometimes in the death of the tree. Trees released by thinning are more severely damaged by this bird than unreleased trees (Erdman and Peterson 1972). For some species, abnormalities in the wood caused by pecking are used for decorative purposes.

Although there is little positive to report, there is great concern about the effects of air pollution (acid rain) on wood properties. A special working party has been organized by the International Academy of Wood Antomists and IUFRO to study the problem and to coordinate current work (Baas 1984). The warning was given not to confound pollution effects with growth rate effects on wood properties. The effect of air pollution on wood properties can be an important source of wood variation, which will need to be clarified. In one current study in Germany, no significant difference in wood quality was found between the wood of healthy and damaged trees after long storage; however, air pollution in Germany results in extended wetwood in the bole of fir, caused by bacteria, while in spruce wood-degrading fungi impair the wood quality, as reported by Bauch (1983).

6.6 Growth Stresses and Strains in Tree Stems

As trees grow and develop, internal stresses are produced in the stem, as has recently been reviewed in detail by Archer (1986) and by Kubler (1987). These stresses may be minor or, as frequently occurs in tropical hardwoods and some eucalypts, they are of such magnitude that they can render the wood useless for some quality products. The internal stresses can be so strong that they constitute a danger; upon cutting, the released stresses can cause the cut tree to spring apart rapidly and injure and even kill the logger. For some trees, such as in *Eucalyptus*, ring barking of the tree and later cutting it after the tree has dried prevents much of the damage from growth stresses.

In 1965, Dinwoodie reviewed growth stresses in European species. He emphasized that growth stresses had been of little importance in Europe until fast growing exotic conifers were planted. For some of these the external growth rings are in a high state of tension relative to the compressive stresses of the wood near the pith. Dinwoodie mentions that fast growth rate is often said to be the cause of growth stresses but that it is not really clear what part growth rate plays in the formation of growth stresses.

Turnbull (1965) reviewed the attitudes relative to growth stresses. He states that they are much more general than realized, and many foresters do not seem to be aware of their occurrence. They are found in many species, both in hardwoods and in conifers. In fact, growth stresses appear to be present in all trees. According to Turnbull, stresses arise because of a shortening of the stem as it grows in diameter which imposes compression on the inner wood. He feels that stresses are so little understood that improving on this defect in timber is most difficult. Watanabe (1965) discussed the origin of longitudinal growth stresses in trees; these are the result of radial growth. He describes the cellular changes that result in the growth stresses, especially lignification. Wood properties are greatly affected by the stresses.

Many of the ideas about growth stresses have been expressed by M. Jacobs in Australia. In one publication, Jacobs (1965) summarizes the rules for the strain gradients induced by the stresses. He reports a shortening in longitudinal length between two points as the tree grows in diameter, which imposes a longitudinal compression on the inner wood. The stresses result in changes in shape and dimensions of pieces cut from erect main stems of trees. The stresses and strains are more evident in hardwoods than in conifers but they exist in both.

The growth stresses and strains can have a significant effect on the utilization and properties of wood. They are predictable to some extent; the degrade caused in solid wood products can be considerable. In certain tropical hardwoods, or species such as *Eucalyptus cloeziana*, the release of the stresses cause so much trouble that special methods must be taken to use the wood, if it can be used for solid wood products at all.

Chapter 7

The Effect of Silvicultural Practices on Wood Properties

7.1 Introduction

The major activity of the silviculturist is geared toward changing growth patterns of the tree in the desired direction; for example, Schmidtling (1973) obtained an eightfold increase in yield in the southern pines after 9 years by fertilization and cultivation. As a result of the growth differences, wood properties often are affected. The changes that will occur in wood are not always predictable, and the silviculturist cannot be certain whether one will even occur. Most foresters assume that a varied growth pattern resulting from forest management will not affect the wood produced. This attitude has developed largely because there have been few good studies that have been underway long enough to determine the true effect of silvicultural activities on wood.

Silvicultural manipulation can either alter wood directly through physiological changes within the tree or it can change the tree form, which in turn may have an effect on wood properties (Larson 1969b). For the southern hardwoods, Kellison et al. (1983) make a broad statement that silvicultural practices applied to natural stands and plantations improve tree growth and form without adversely affecting wood properties. However, some investigators do find that wood properties are changed as a result of silvicultural manipulations.

The situation regarding silvicultural manipulation and wood properties is very complex. Growth rate can have an effect on wood, and the actions of the silviculturist usually affect both the pattern and rate of growth. As described in Chapter 5, the interaction between growth rate and wood characteristics is not clear-cut and certainly is not always predictable. The assessment of what a silvicultural treatment can do to influence wood cannot be based upon a simple determination of its effect on growth, since so many other factors and interactions are involved.

The objective of this chapter is to clarify what changes in wood properties will result from various silvicultural treatments. However, the only sure way to determine the kind of wood that will result from silvicultural treatment is to establish definitive silvicultural tests and then to assess the wood produced. This takes many years.

We have reviewed some of the many studies related to the silvicultural effects on wood and from these will broadly summarize how various silvicultural manipulations can cause variation in wood. Many studies are incomplete and some are not reliable. As much as possible, we will emphasize results from the best investigations. Two detailed summaries were published in 1962 and 1966 that dealt with silvicultural effects of wood: "The Influence of Environment and Genetics on Pulpwood Quality," in the TAPPI Monograph Series No. 24 in 1962 and a publication by Weiner and Roth (1966) at the Institute of Paper Chemistry. To our knowledge, no summaries of these types have been done in recent years except those by Megraw (1985, 1986) which include a number of references related to silvicultural effects on wood properties. Timell (1986) recently reviewed the effect of various silvicultural practices on the formation of compression wood in conifers.

Silvicultural manipulations can be of many kinds. For simplicity and ease of assessment we have grouped them into: (1) Change of the nutrient status including fertilization, (2) stocking control, through initial planting or thinning, (3) pruning, and (4) miscellaneous activities such as planting techniques, resin tapping, and coppicing. The silviculturist also can manipulate age of harvest, a subject so important to wood properties that it is discussed separately in Chapter 3; similarly, growth rate and growth patterns can be altered, as described in Chapter 5.

All of the above actions by the silviculturist are interrelated to some extent. Larson (1963 a) explains these interactions by showing how the silvicultural treatments result in trees being grown under widely differing environmental conditions, which change the crown of the tree. According to Larson, the crown is the source of control of growth and wood development and the changed environmental factors that alter the growth of the foliage or the developing meristems will be reflected in a different cell size, wall thickness, and perhaps chemistry of the wood produced.

Silvicultural operations such as fertilization are often combined with stocking control by thinning, which affects the crown development as well as the nutrient and moisture regime in the forest. Therefore, although we will discuss the changes in wood caused by specific silvicultural manipulation and refer to them as specific treatments, the complexity of the effects on the tree growth pattern and thus on wood properties always makes clear-cut cause and effect determinations difficult. This complexity was described by Choong et al. (1970) who found differences in wood properties resulting from four different silvicultural treatments applied to *Pinus taeda.* They concluded that when a rapid growth rate is the result of silvicultural treatment, specific gravity and moisture content of the wood may be independent of growth rate or other effects of the silvicultural manipulation. Similar complexities were demonstrated by Cown and McConchie (1981), who found that wood response was much greater when thinning and fertilization were combined than when either was applied alone.

There are numerous papers dealing with the effect of silvicultural manipulation on tree growth and tree form, but only rarely have these also included the direct effects of the manipulations on wood properties. Even though Kellison et al. (1983) emphasize the improvement in wood that results when natural stands are managed for better tree quality, the question is raised as to whether the economic benefits resulting from the changes in wood caused by silvicultural manipulations will be enough to justify the costs involved.

In a paper dealing with the silvicultural management of *Eucalyptus,* Hillis (1972) mentions the major changes in wood that will result from improved silviculture in both plantation forestry and natural stand manipulation. He points out that eucalypts from plantations will have a wood with lower density, shorter and thinner-walled fibers, a lower proportion of heartwood, and less extractives than trees grown in natural stands with longer rotations. In a study on 27- to

37-year-old *Eucalyptus regnans*, Higgs and Rudman (1973) indicated that important wood quality and yield effects were related to silvicultural treatment. One result was the relatively small reduction in wood quality in relation to the added growth from improved silviculture; the change in wood is minor compared to the increased volume production.

In his article on silvicultural control of wood, Larson (1967b) brings out the concept that no matter how much one changes the environment, the inherent large within-tree variation is usually present and tends to mask other causes for wood variation. (This is covered in Chaps. 3 and 8.) Knigge (1961) also emphasizes the pronounced variation within and between stems. He states that the major problem is to produce timber of more uniform and higher quality. This will only be possible when both the genetic and environmental causes of variability are better known and can be manipulated by the silviculturist.

7.2 Effect of Nutrient Differences and Fertilization on Wood

Fertilization is a silvicultural tool just coming into broad operational use in forestry. Widespread fertilization of conifer forests is rapidly being accepted and on some soils is essential for reasonable tree growth and survival. Although there are a few exceptions, fertilization of hardwoods, especially hardwood plantations, is widely practiced and accepted. Some studies have been made of the effects of fertilization on wood properties but these are sometimes hard to interpret because fertilizer response depends on the age of the trees, kind of fertilizer, or time of test, which are not always clearly defined. As stated by Elliot (1970), "...generalizations on the role of a nutrient or group of nutrients are so confounded by site and species effects as to be at best empirical if not dangerous." Nevertheless, studies have been made on enough species so we can now see some trends in the effects of nutrient differences on wood properties.

Assessment of the value of fertilization must include both volume improvement and changes in wood properties. Frequently, the added growth more than offsets the loss in specific gravity when nitrogen fertilization is used (Walker 1960). The same concept applies to Swedish forests as outlined by Tamm et al. (1960) who showed that the improved growth measured in dry weight of wood will be somewhat less than if the fertilizer effect is measured only in volume per unit area; this is due to the lowered wood weight per unit volume resulting from the fertilization.

For slash pine, Gooding and Smith (1971) found that fertilizing increased volume production by five or six times, while weight loss from reduced specific gravity was only about 4%. Also for southern pines, Schmidtling (1973) emphasizes that an increase in dry wood production can be more readily obtained by an increase in volume than by improving wood specific gravity. Similarly for yellow poplar, Ross et al. (1979) stated that if gains are proportional to growth increases and losses are proportional to density losses, there would be a considerable net gain from fertilization.

A further consideration must be the quality of the wood, especially for solid wood products. Sometimes following high nitrogen fertilization the wood formed

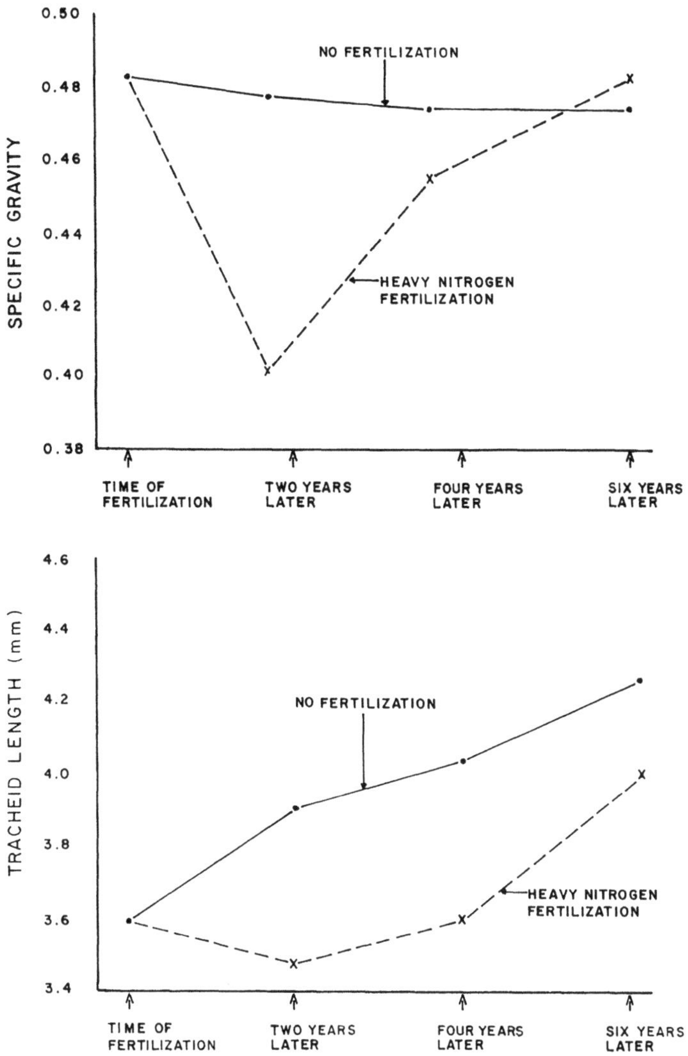

Fig. 7.1. Large amounts of fertilizer, especially nitrogen, can have a great effect on wood properties. This is shown for wood specific gravity (*top*) and tracheid length (*bottom*) for loblolly pine fertilized at 16 years of age (Posey 1965). Note that the specific gravity returned to normal after 6 years, but that tracheid length remained below normal

has characteristics much like juvenile wood with its inherent instability and low fiber yields (Posey 1965; Fig. 7.1). As Larson (1967b) mentions, heavy fertilization in young stands increases crown and branch size and delays natural pruning. He advises fertilization in pole-sized stands where crowns are already formed, and the effect on wood quality is therefore small.

It is not correct to talk in general terms about the effect of fertilizers on wood because one must always qualify any results with "what kind of fertilizer and how much was used, and when was it applied?" For example, it appears that wood characteristics in general are more responsive to nitrogen than to phosphorus fertilizers. The amounts are critical, and as Posey (1965) and Nicholls (1971b) showed, fertilizing in several small doses may not greatly affect wood characteristics, while the same amount of fertilizer applied in a single dose can have a major effect from changed specific gravity and tracheid length to formation of juvenile like wood.

Although covered in detail in a previous chapter, the effects of growth rate and fertilization cannot be separated since increased growth is usually a result of fertilization (von Pechman and Wutz 1960). Especially interrelated are the effects of fertilization combined with thinning, usually done at the same time and both resulting in an increased growth rate (Cown and McConchie 1981). For example, Schönau (1982) found that fertilized black wattle (*Acacia mearnsii*) trees had a wood of lower density than those that were unfertilized. This was also true for trees growing at a lower stocking density where they also grew more rapidly. For this species and under the conditions described, Schönau came to the conclusion that timber quality of black wattle was not affected by fertilizers or by stocking density per se; instead he credits the wood variation to the faster growth stimulated by the fertilizers and by the lower stand density.

The above is contrary to the senior author's observations that loblolly pine trees that grow faster when fertilized with nitrogen have a low wood specific gravity, while similar trees that grow more rapidly as a result of thinning have an unchanged wood specific gravity. A similar pattern was noted by Higgs and Rudman (1973), who reported that fertilizing 27- to 37-year-old *Eucalyptus regnans* trees caused a lowering of specific gravity and formation of shorter fibers, whereas thinning had an opposite effect. They state that the specific gravity differences are not the result of a change in growth rate, but of the thinning treatment. Parker et al. (1976) reported that wood density and pulp yield with Douglas-fir were adversely affected by fertilization, but not by thinning, even though each increased growth rate. A third and different pattern was obtained by Erickson and Harrison (1974) with Douglas-fir, where both fertilization and thinning produced lower density wood and somewhat shorter tracheids; the effect was mainly in the first 3–4 years after treatment.

Although numerous other studies could be cited relative to the effect of fertilizer compared to thinning for other species, it is evident that complex interactions exist and no single pattern will always hold. Whether the wood change is caused by the treatment per se, or only be resultant growth changes, is strongly debated (Cown and McConchie 1981).

Not all fertilization is with standard inorganic fertilizers. One system much talked about, sometimes tried, but rarely used operationally, is to fertilize with sewage or pulp plant effluent, which is an automatic combination of fertilizer application and irrigation. Increased growth sometimes results from such applications. A special study of the effects of the use of sewage effluent on the wood produced was made by Murphey and Brisbin (1970) with 24-year-old red pine. The application of 2 in. of effluent per week was equivalent to heavy fertilization. The wood specific gravity of the faster growing trees that were treated with effluent was reduced, although the moduli of elasticity and rupture of the wood produced were higher. They concluded that a modest use of effluent would not unduly influence the properties of the wood produced.

Another type of fertilization now being extensively studied is the use of certain nitrogen-fixing plants to supply nutrients (Davey and Wollum 1979). In a recent study (de Jesus et al. 1988) *Eucalyptus-Leucaena* mixtures were grown with the hopes that *Leucaena* would supply nitrogen to the *Eucalyptus*. After 7 years, there were no measurable fertilizer effects from the *Leucaena*. There were no significant differences in the wood density or holocellulose yields of the eucalypts whether grown in pure stands or in admixture with the nitrogen-fixing *Leucaena*. Few workers have investigated how nitrogen-fixing plants affect wood. Working with spruce and pine, von Pechman and Wutz (1960) studied wood formation following planting of lupines. They found an immediate response in growth rate. In older trees, more earlywood was formed but there was also more latewood produced. Despite the much wider annual rings, overall wood density was nearly the same as before planting of the lupines. In spruce and younger pine, the growth response from planting lupines resulted in a reduction of wood specific gravity. At the present time, it appears that the use of biological nitrogen fixers in forestry has little effect on the wood produced.

An important concept relative to fertilization, especially when nitrogen is used, is that the effect on wood usually disappears after a few years and that the wood produced after that time returns to nearly normal specific gravity (Posey 1965; Parker et al. 1976). The time period before wood normalizes can vary from 5 to 10 years for conifers and is reported to be somewhat longer in hardwoods. One example has been reported by Cown and McConchie (1981), who found that the wood of fertilized radiata pine trees returned to the pretreatment level after 5 years.

7.2.1 Fertilizing Conifers

Fertilization of conifers is currently being done on a large scale on several continents. Costs are considerable, and numerous economic analyses have been made. Usually the assessment deals primarily with growth response although some investigations have dealt with the effect of fertilizers on wood propeties; a number of these have been summarized in Table 7.1. This table in no way contains all available references on conifer fertilization and wood properties. Those listed were chosen to show the response of a number of species that have been fertilized under a range of differing conditions. An annotated review of the effects of fertilizers on wood was prepared by Cleaveland et al. in 1974. Studies on fertilizer effects and wood were also reviewed by Cown (1977).

One early fertilizer study on loblolly pine by Posey (1965) gave interesting results for plantations fertilized at 12 and 16 years of age. Posey found a 15% drop in specific gravity (from 0.48 to 0.39) immediately after fertilizing with nitrogen (Fig. 7.1). All fertilizer treatments caused a decrease of at least 0.04 in specific gravity, with reductions being proportional to the amount of nitrogen used. After the 6th year, the specific gravity returned to normal, or a little above normal. A reduction in tracheid length was also evident, with 12% shorter tracheids immedi-

Table 7.1. Wood quality of fertilized conifers

Species	Reference	Comments
Pseudotsuga menziesii	Erickson and Lambert 1958	Fertilizers caused a significant decrease in wood specific gravity
Picea abies *Pinus sylvestris*	von Pechman 1958	Fertilizing slow-growing pine increased both ring width and latewood production, while N fertilization of spruce stimulated earlywood growth. Heavy nitrogen fertilizer resulted in increased earlywood but little change in latewood
Picea abies *Pinus sylvestris*	Hildebrandt 1960	Repeated fertilization with N after fertilization with Ca and P increases the earlywood of spruce and decreases the wall thickness of latewood. No change in the wood of pine was evident after fertilization
Picea abies	Tamm et al. 1960	Growth increase from fertilizing with N resulted in no, or only moderate, decrease in wood density
Picea abies *Pinus sylvestris*	von Pechman and Wutz 1960	Fertilization with N had little adverse effect on specific gravity of old pine but reduced specific gravity for young pine and young spruce. Wood of fertilized trees on poor sites was similar to that of trees on fertile sites
Pinus elliottii	Williams and Hamilton 1961	Following fertilization, specific gravity was lowered from 0.513 to 0.479; wall thickness was reduced by 5.8%
Pinus taeda	Zobel et al. 1961	The application of high N results in a decrease in specific gravity and tracheid length
Pinus sylvestris *Larix leptolepis* *Picea abies*	Seibt 1963	The mean basic wood density of 30-year-old trees on poor heath sands was reduced after treatment with Ca, N, P, and K simultaneously
Pinus taeda	Moehring 1964	Wood specific gravity was reduced only slightly by irrigation combined with fertilization but significantly by application of fertilizer alone
Pinus taeda	Posey 1965	High nitrogen fertilization resulted in a drastic reduction of wood specific gravity which returned to normal 5 years later
Pinus densiflora *Cryptomeria japonica* *Chamaecyparis obtusa*	Tajima 1967	The specific gravity of the wood of *Chamaecyparis obtusa* increased after fertilization but that of *Pinus densiflora* and *Cryptomeria japonica* did not. In pine, fertilization caused no change in anatomical features
Pinus radiata	Gentle et al. 1968	There was no reduction of wood density when phosphate fertilizers were used
Picea abies *Pinus sylvestris*	Klem 1968	No changes in wood of practical importance occurred as a result of fertilization when done independent of thinning
Pinus radiata	Rudman and McKinnell 1968	N and P alone, or with other chemicals, caused a marked drop in wood density of 7-year-old trees
Picea abies	Seibt et al. 1968	Fertilizer application increased latewood widths. Wood density decreased with increased ring width after fertilization. Fiber length and spirality were not affected by fertilizers
Pinus spp.	Larson 1969	Fertilizer can be most deleterious to wood quality because it increases crown and branch size and delays natural pruning. Fertilization extends the period of juvenile wood formation. In pole-sized stands fertilization causes increased growth of earlywood and latewood with relatively little change in wood quality
Pinus pinaster	Polge 1969b	Use of P increases growth rate and decreases wood density. Fiber length increases slightly but not significantly
Pinus taeda	Linnartz et al. 1970	In young trees specific gravity was lowered by N alone. In combination with P specific gravity was increased

Table 7.1 (continued)

Species	Reference	Comments
Larix occidentalis	Murphey et al. 1970	Cell development was not affected by nutrient deficiencies, but the wall characteristics were altered. The specific gravity of the N-treated trees was the lowest
Pinus radiata	Rudman and McKinnell 1970	There was a reduction in wood density by 20% with phosphate fertilizers, but only 10% with potassium
Pinus ponderosa	Echols 1971	Fertilization increased specific gravity from 0.342 to 0.352
Pinus elliottii	Gooding and Smith 1971	A slight reduction in specific gravity (120 – 225 lbs/cord) was found at the basal portion of the tree following fertilization
Pinus pinaster	Nicholls 1971b	Tracheid length is abruptly reduced, but otherwise wood quality is not affected by fertilization
Pinus resinosa	Siddiqui 1971	Specific gravity and fiber length were not affected by fertilization but the percent latewood was lower and the cell walls in the earlywood were thicker
Pinus elliottii	Smith et al. 1971	Specific gravity was relatively unaffected by fertilization; there was no effect on spiral grain in 6-year-old trees
Pseudotsuga menziesii	Megraw and Nearn 1972	Trees genetically superior in density maintain their superiority regardless of subsequent fertilization or thinning, even though fertilizers affected the within-ring wood density
Pinus palustris *P. elliottii* *P. taeda*	Schmidtling 1973	Growth was greatly increased, but specific gravity and fiber length of juvenile wood was not decreased by fertilization
Pseudotsuga menziesii	Erickson and Harrison 1974	Fertilization caused immediate production of low-density wood
Pinus resinosa	Gray and Kyanka 1974	Bending strength was a little less and stiffness of boards was the same, but overall, wood from fertilized trees had about the same characteristics as that from unfertilized ones
Pseudotsuga menziesii	Megraw and Munk 1974	Specific gravity reduction after N fertilization was small on average sites but the specific gravity was 6.5% less on the best sites where high N was used
Pinus resinosa	Gray and de Zeeuw 1975	Fertilizer caused a decrease in tracheid length; density of both earlywood and latewood decreased after fertilization
Larix leptolepis	Isebrands and Hunt 1975	Fast growth, stimulated by fertilizers, caused a loss in specific gravity for several years, after which specific gravity increased
Pseudotsuga menziesii	Resler et al. 1975	Yield and properties of pulp from wood of fertilized trees are the same as those of non-fertilized trees
Pseudotsuga menziesii	Parker et al. 1976	Fertilization caused a decrease of up to 10% in density for the first few years; the density then returned to normal
Pinus radiata	Cown 1977	Nitrogen fertilizers reduced latewood percent 10 – 15%. Phosphate fertilization restored wood to normal density
Pinus banksiana	Hunt 1977	Specific gravity was slightly reduced but in light of the 46% volume increase from fertilization, there was no adverse effect
Pinus radiata	Nelson 1977	Density of fertilized wood was 6% lower than that of the control; pulp yield was higher; wood consumption per ton of pulp was similar to that of nonfertilized trees
Pseudotsuga menziesii	Brix and Mitchell 1980	Fertilization delayed earlywood-latewood transition by 1 – 2 weeks. Latewood percentage was reduced until the fourth year after fertilization. Latewood cell wall thickness was less in both fertilized and thinned plots

Table 7.1 (continued)

Species	Reference	Comments
Pinus radiata	Cown and McConchie 1981	Maximum decrease in density was 20% at butt level and 12% at 12 m height; whole tree wood density differences were 5%. Tracheid length was not affected. Fertilizer treatment will have little technological significance on wood utilization
Pinus radiata	Cromer et al. 1985	Phosphate fertilization had a large effect on diameter growth but had very little influence on wood density
Pinus elliottii	Hodges 1985	Wood specific gravity at age 10 was 4% higher on plots fertilized with phosphorus and 2% higher where phosphorus and nitrogen were used together
Pinus taeda	Megraw 1985	Nitrogen fertilization reduced specific gravity 5–10%. More earlywood and less latewood were formed
Pinus caribaea	Brito et al. 1986	There was a decrease in earlywood and latewood densities on fertilization. The greatest decrease was in the latewood
Pseudotsuga menziesii	Megraw 1986	In Douglas-fir there is almost always a lowering of wood specific gravity following fertilization with nitrogen

ately after fertilizing and 6% after 6 years (Fig. 7.1). Posey found that the younger trees responded more rapidly to fertilization than the older ones, but the effect of fertilizers lasted longer in the older trees. He also found that fertilizer did not affect all trees to the same degree; trees with higher initial specific gravity or longer tracheids were more affected by fertilizer treatments than those with initially lower wood density or shorter tracheids (Fig. 7.2). Some individual trees were not affected by fertilizer or even had a high wood density following fertilization with nitrogen. Posey concluded that trees receiving nitrogen without phosphorus, or nitrogen and phosphorus combined, responded with an immediate increase in growth, but a decrease in specific gravity, tracheid length, and radial wall thickness. The results of Posey were not all confirmed by Megraw (1985) (who also cites several others), who stated "...trees have been shown to maintain their same relative specific gravity ranking after ferilization as they had before... Fertilization does not appear to alter preferentially either high specific gravity trees or low specific gravity trees...specific gravity gains made through genetic improvement will not be reduced by fertilization."

There is a general belief that fertilization will change the chemistry of the wood formed following treatment. Few studies about this subject seem to exist, but one by Tajima (1967) reported that fertilization of Akamatsu (*Pinus densiflora*) did not affect holocellulose yield but resulted in a slightly higher alphacellulose yield. For loblolly pine, Posey (1965) found some reduction in holocellulose content, as shown in Fig. 7.3.

Based primarily on the results shown in Table 7.1, a few generalizations can be made regarding the effect of fertilization on the properties of conifer wood. None of them hold for all species and all conditions, but those listed commonly exhibited responses.

1. Nitrogen fertilizers cause the greatest change in wood properties. Sometimes the combination of nitrogen plus phosphorus also causes variation. Most

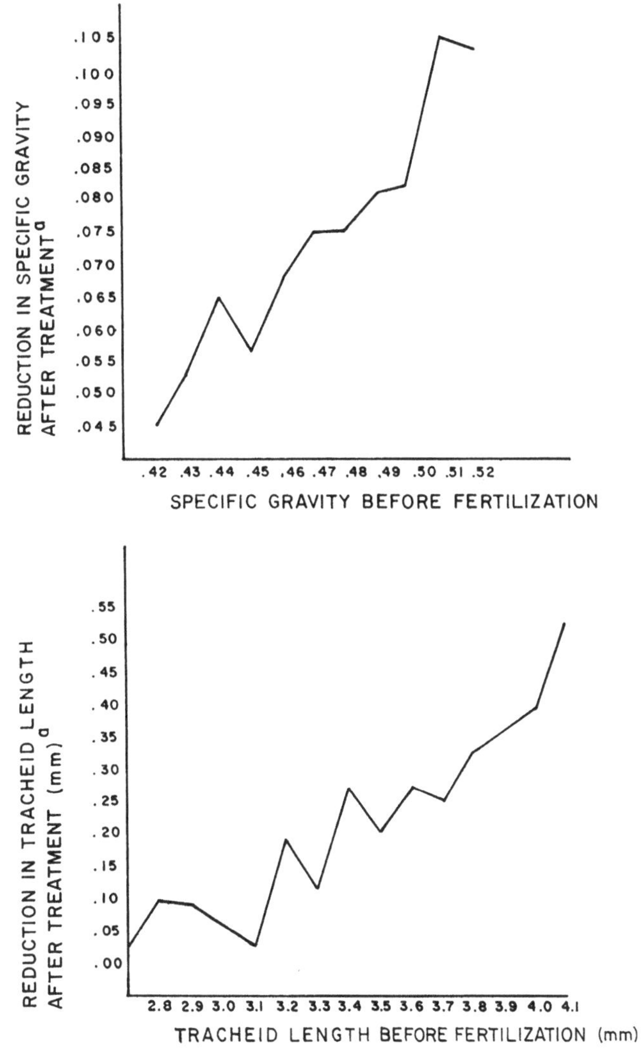

[a]Computed for 100 trees after treatment with 160 pounds of N±P±K

Fig. 7.2. Nitrogen fertilizer does not affect all loblolly pine trees in the same way. Trees with a high wood specific gravity are more affected than those with low specific gravity, and trees with initially long tracheids are more affected than those with short tracheids (Posey 1965). See text for discussion

other nutrients have only a relatively small effect on wood, although with some species phosphorus alone has a considerable effect.

2. Nitrogen fertilization often causes a severe to mild decrease in specific gravity. This lasts from 5 to 10 years, after which the specific gravity returns to near normal (see Fig. 7.1). Nitrogen fertilization usually causes the production of significantly shorter tracheids. There are reports of increased specific gravity and

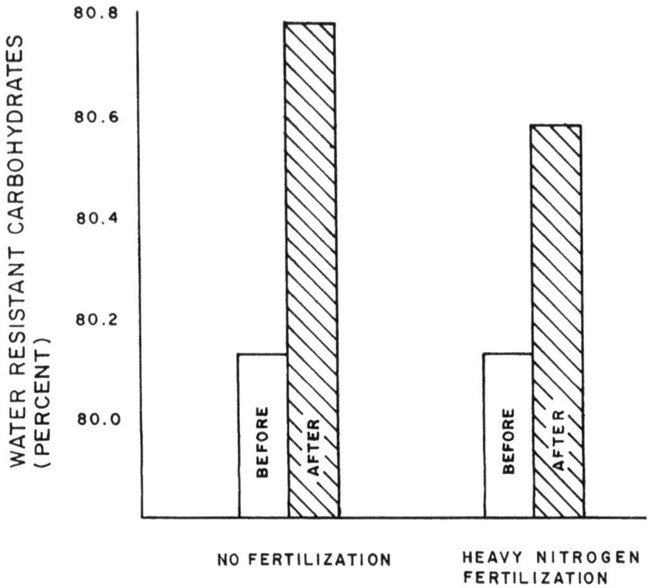

Fig. 7.3. Fertilizer application usually has only a small influence on the chemistry of the wood formed. Some effect is evident, as shown for holocellulose in *Pinus taeda* fertilized at 16 years of age, but it is not large. Results show values at time of fertilization and 6 years after fertilization (Posey 1965)

increased tracheid length following fertilization, but these are rare. In a surprising number of instances, wood properties were not affected by fertilization.

3. Frequently, fertilization increases the relative amount of earlywood production and usually has little effect on the amount of latewood. It was reported by several authors that fertilization increases the time when juvenile wood is formed, but this cause and effect is not yet clear.

4. Although combined fertilization and thinning increase growth rate and frequently lower specific gravity, this reduction is primarily associated with fertilization and does not occur following thinning alone.

5. For best quality wood products, nitrogen fertilization should be light and frequent. A heavy application causes formation of a few rings of juvenile like wood that causes difficulty in curing and machining of the boards.

6. In all studies reviewed, the added value of the greater volume of wood resulting from fertilization more than offset an adverse change in wood properties (McKinnell and Rudman 1973; Hunt 1977; Cown and McConchie 1981). Only for specialty woods or wood with marginally useful characteristics, such as juvenile wood, did the fertilization produce overall negative results.

As for nearly all relationships with fertilizers and wood density, the literature is contradictory. A good example is the use of phosphorus on *Pinus radiata*. Nicholls (1971b), McKinnell and Rudman (1973), Cown (1977), and Nelson et al. (1980) report reductions of 3 to more than 20% in wood specific gravity when phosphate fertilizers were used. However, Gentle et al. (1968) found no reduction following use of phosphate fertilizer alone but when nitrogen was used, wood density was reduced.

Even though there is variation with species, age, and site, fertilizing conifers with nitrogen frequently has the following general effect within rings:

1. Produces earlywood fibers (tracheids) with thicker cell walls.
2. Produces latewood fibers (tracheids) with thinner cell walls.
3. Increases wood uniformity within the annual ring.

7.2.2 Fertilizing Hardwoods

More than for the conifers, where the usual interest relates to tracheid characteristics and how these will affect yield and quality upon pulping, fertilizer studies on hardwoods have been related to quality of solid wood products; this holds except for species such as the eucalypts, aspens, sycamore, etc., which are predominantly used for pulping.

One study by Foulger and Hacskaylo (1968) dealt with the actual altering of wood anatomy when nutrient deficiencies exist in cottonwood (*Populus deltoides*). They report that nitrogen and sulfur deficiencies are most serious, resulting in reductions in fiber length and width, and vessel width. Lack of boron, potassium, and to some extent, phosphorus produced the same results. The nitrogen caused an increase in the ratio of vessels to fibers.

A number of studies related to wood and fertilizing hardwoods is shown in Table 7.2; this is not a complete list but presents some trends that have been observed with hardwoods.

Results of the effect of fertilizer on hardwoods are inconclusive and contradictory. For example, Thor and Core (1977) found that fertilizing *Liriodendron tulipifera* did not affect wood properties, while Ross et al. (1979), working with the same species in the same geographic area, reported that fertilizer caused a significant decrease in wood density.

Some generalizations related to the effect of fertilizing hardwoods are:

1. Fertilization can cause degrade in the quality of solid wood products by changing tree form, epicormic branching, or inherent wood properties, but usually this is not the case. Frequently, it is reported that fertilizer has little effect on wood properties of hardwoods and that the wood of fertilized trees is suitable for both solid wood and fiber products.
2. Some mineral deficiencies such as calcium or phosphorus cause a high wood specific gravity. It has been reported for one species that manganese deficiency causes a low specific gravity.
3. There are some reports that nitrogen reduces cell size, wall thickness, and specific gravity, just as it often does in conifers, but these results are not consistent, especially in the diffuse-porous hardwoods.
4. It is apparent that the better volume growth following fertilization more than offsets a loss in quality of wood properties, with the possible exception of wood used for specialty products.

Table 7.2. Effect of fertilizers on the wood of hardwoods

Species	Reference	Comments
Populus spp.	Borsdorf 1965	Different clones responded differentially to fertilization. The "Oxford" clone had the greatest reduction in wood density following fertilization
Robinia pseudoacacia	McAdoo and Murphey 1968	Ca deficiency produced high specific gravity and Mg deficiency the lowest gravity. All mineral deficiencies except B and N caused changes in specific gravity
Platanus occidentalis	Saucier and Ike 1969	Specific gravity, fiber length, and volume of fibers and vessels were not affected by fertilizers. Ray volume was slightly increased by N and N + P
Quercus rubra *Liriodendron tulipifera* *Fraxinus americana*	Mitchell 1971	Fast-grown wood produced by fertilizers was as suitable as the slower growing controls used for furniture, millwork, and paneling
Quercus rubra *Liriodendron tulipifera*	Auchmoody 1972	N, P, and K fertilizers did not stimulate epicormic branching but N increased vigor of already formed epicormic branches
Eucalyptus globulus	Cromer and Hansen 1972	Fertilization had no major effect on wood density, extractives, lignin, or sugar content of wood from young trees
Fraxinus americana	Foulger et al. 1972	Only N concentration affected wood anatomy characteristics; N had a negative effect on vessel length and fiber percentage and a positive effect on amount of ray tissue following fertilization
Eucalyptus regnans	Hans and Burley 1972	There was no significant effect of fertilizers on wood properties, although treatments that increased growth tended to decrease density
Eucalyptus regnans	Higgs and Rudman 1973	Fertilization significantly decreased fiber length and specific gravity of 27- to 37-year-old trees
Quercus robur	Polge 1975	Fertilization causes an increase in density and in radial shrinkage and thus a reduction in wood quality
Eucalyptus globulus	Farrington et al. 1977	Fertilization resulted in an increase in wood yield per hectare without detriment to pulp strength
Liriodendron tulipifera	Thor and Core 1977	A combination of fertilization and irrigation resulted in 30% increased growth but had no effect on specific gravity, extractive content, or fiber length
Populus spp.	Cutter and Murphey 1978	Vessel elements and fiber length and diameter were all affected by potassium but fiber wall thickness was not
Liriodendron tulipifera	Ross et al. 1979	Significant decreases in wood density were found following fertilizer treatment
Southern hardwoods	Kellison et al. 1983	Intensive silvicultural practices applied to both natural stands and hardwood plantations improve tree growth and form without adversely affecting wood properties

7.2.3 Wood Uniformity Following Fertilization

In addition to the effect fertilizers might have on changing the overall specific gravity and cell dimensions, it has been reported that fertilization often leads to the production of a more uniform wood. This results from a decrease in wood specific gravity differences within an annual ring due to an increase in wall thickness in thin-walled earlywood cells and to a concurrent decrease in the thickness of latewood cells. This combination produces a more uniform wood in ponderosa pine and loblolly pine (Echols 1971; Gladstone and Gray 1972). A similar pattern was reported for spruce by von Pechmann (1958) and by von Pechmann and Wutz (1960), who found that following fertilization the cell walls in the latewood are thinner, while cell structure is the same in the earlywood. In young Douglas-fir, the trend toward wood uniformity after fertilization was apparent when different rates of nitrogen fertilizers were used (300, 150, and 100 kg/ha) (Megraw and Nearn 1972). Older Douglas-fir fertilized with 400 lbs/acre with a nitrogen fertilizer also produced a more uniform wood (Siddiqui et al. 1972).

When red pine (*Pinus resinosa*) was fertilized with 200 lbs of potassium per acre, on potassium-deficient sites, it was found that earlywood wall thickness of the fertilized trees was much more homogeneous than in the unfertilized ones (Siddiqui 1971). Similar results were reported by Polge (1969b) for *Pinus pinaster*, who found that the maximum density areas are reduced more than the areas of minimum density so that the wood becomes more uniform. Working with Japanese larch (*Larix leptolepis*), Isebrands and Hunt (1975) found that fertilizers produced a definite increase in uniformity of specific gravity across the growth ring and lower overall specific gravity. They stress the value of the more uniform wood produced by fertilization, resulting in more efficient wood processing since the cell walls are more uniform. In his work with spruce and pine, Klein (1968) emphasized that controlled and repeated fertilization reduces the variation in wood properties among trees in a stand. This uniformity is very important from a utilization standpoint.

It is evident that essentially all studies have shown that fertilization of the wood of conifers results in a more uniform wood by reducing the differences in cell anatomy between the earlywood and latewood. No information could be obtained relative to cell uniformity and fertilization of hardwoods. Uniformity is considered to be so important that Olson and Arganbright (1977) developed a uniformity factor to express uniformity in specific gravity.

7.3 Tree Spacing and Wood Properties

Stocking is one of the tools most used by the silviculturist to control growth and yield. As Larson (1969b) states, "...stand density has a tremendous influence on the quality of the wood formed." Spacing influences branching characteristics and rate of growth, both of which affect wood properties (Bamber and Burley 1983). Stand density can be controlled in two ways: (1) by initial spacing in plantations and (2) by thinning already established stands to a desired stocking. These two methods of stocking control may result in the production of different kinds of wood.

Stocking differences affect wood properties not only through their effect on crown development and growth rate but also through their effect on the utilization of nutrients and water. In addition, the temperature of the soil, moisture availability from the soil, penetration of radiant energy, and illumination of the crown are affected and result in a changed growth pattern which, in turn, can affect wood properties. For example, Savina (1956) states that more moisture remains in the soil in thinned plantations. Stocking control influences wood by its effect on tree form, with closer spacing producing trees with smaller and thinner branches. This is reflected in the quality of the products made from the tree.

As for all silvicultural actions that are commonly employed, the effects of stocking control on wood properties are varied and controversial. There are so many interactions involved that it is not always possible to predict how the wood will be affected, but there are some patterns that appear to be quite general. For example, in Section 7.2.1 on nutrients and fertilizers, several references were given, indicating that faster growth resulting from thinning usually has a minor effect on wood properties; this is often contrary to the effect of faster growth caused by the use of nitrogen fertilizers. There are, however, exceptions: for example, Echols (1971) reports a lowering of specific gravity following thinning and a specific gravity increase following fertilization with nitrogen of ponderosa pine. Because of the occasional difference in the response of wood to thinning and to initial planting as a means to achieve a given stand density, each will be treated separately, below.

7.3.1 Stocking Control by Thinning as Related to Wood Properties

Most studies in conifers relating stocking differences to wood properties involve some type of thinning operation. Thinning results in the tree suddenly being in an environment quite different from the one in which it had been growing (Fig. 7.4). There is nearly unanimous agreement that when done correctly, thinning results in improved growth (Bertoloti et al. 1983) and usually in improved tree quality and often a more desired species composition. Each of these has an effect on the properties of the wood.

It is of importance to recognize that thinning can have a major effect on wood quality by increasing the number of quality logs that are available for sawtimber or plywood. Thinning will also have other effects related to the age of the wood, as discussed in Chapter 3. For example, Cown and McConchie (1982b) point out that the change in raw material supply from untended old-crop trees to the thinnings and final harvest from intensively managed forests of radiata pine will be accompanied by age-related differences in wood properties, to which industry will have to adjust. Logs from the managed forest will be heavier because of a higher average moisture content, but will have a lower basic density. Since most trees are harvested when they are of a given size, rather than at a given age, the thinned stands are finally harvested at a younger age with a higher juvenile wood content; the thinnings themselves are primarily juvenile wood. The improved quality of the thinned stands, and the added value from an earlier harvest, are somewhat offset

Fig. 7.4. Whether done mechanically (see Fig. 7.5), or with horses as shown in South Africa, thinning changes the growth conditions of the residual stand and this affects the wood produced. Also, since the poor and defective trees are removed in the thinning operation, the wood quality of the final harvest is much improved (Courtesy of G. van Wyk, South Africa)

by a shift toward a higher proportion of juvenile wood from younger, thinned stands (Fig. 7.5).

Hundreds of references could be cited related to wood quality and thinning. Bamber and Burley (1983) report that the effect of thinning on radiata pine has been studied in detail. They list, for example, Sutton and Harris (1973) and Cown (1973a) who believed that the practical effects of thinning on wood properties were small. However, Siemon (1973) and Cown (1974) reported that wood density of radiata pine was reduced by 8–10% following thinning. However, the effect of reduced wood density was temporary and the density rose to normal values after thinning. A number of these publications for conifers are shown in Table 7.3. Below are a few publications on general concepts that will be discussed, which will supplement Table 7.3.

The work of Savina (1956, translated from Russian), on the physiological justification for thinning, provides a good review of the subject of wood response to thinning. He states that thinnings alter the overall pattern of environmental conditions necessary for satisfactory growth and development for the remaining trees. Thinning brings about an intensification of the physiological and growth processes, which in turn increases photosynthesis and causes intensified cambial activity. Savina reported that the cambium becomes active 5 to 10 days earlier in thinned than in unthinned stands of pine. He also emphasized that the greatest effect of thinning was evident the first 4 years. Savina (1956) found that of the three species he studied, pine, oak, and aspen, the most extensive changes in wood structure following thinning took place in oak and the least in pine.

Among several other studies on the fundamental response in tree development following thinning, Zahner and Oliver (1962) reported that the transition from earlywood to latewood was almost uniform from tree to tree in red pine (*Pinus*

Fig. 7.5. Thinning is usually done at relatively young ages. Thus, most of the wood removed is juvenile as shown by the load of thinned loblolly pine. Until recent technology had shown how thinnings can be better used, the excess of juvenile wood was considered a major drawback in a thinning program

Table 7.3. Effect of thinning on wood properties of conifers

Species	Reference	Comments
Pinus palustris	Paul 1941	Release of trees does not have a detrimental effect on the lumber produced
Seven *Pinus* species	Orman 1956	No examples were cited in which thinning caused any reduction in wood density
Pinus sylvestris	Savina 1956	Thinned pine had tracheids up to 25% longer than trees on the unthinned plots
Pinus palustris	Paul 1958	Sometimes release results in an increase in specific gravity, sometimes a decrease. Accelerated growth following release often results in wood with a high specific gravity, which decreases several years later
Pinus taeda	Echols 1960	There was little effect of thinning on wood properties
Picea abies	Hildebrandt 1960	Wood density is greater after heavy thinning and is most affected on good sites
Pinus ponderosa	Myers 1960	There was no difference in wood density between thinned and unthinned stands
Pinus taeda	Zahner and Whitmore 1960	Specific gravity and latewood content were about the same for radically thinned and unthinned young trees, with a specific gravity of 0.45
Pseudotsuga menziesii	Wellwood and Smith 1962	Thinning results in shorter tracheids
Pinus resinosa *P. banksiana*	Zahner and Oliver 1962	No significant difference in percentage of latewood was found related to thinning, although inception of latewood was delayed up to 2 weeks
Pinus radiata	Reid 1963	Thinning sometimes stimulates latewood production, leading to increased wood density. Usually no change is observed and thinning has less effect on wood than is predicted from changes in growth rate

Table 7.3 (continued)

Species	Reference	Comments
Picea abies *Pinus sylvestris*	Ericson 1966	In pine, thinning did not greatly affect basic wood density, even though the trees had a high latewood content. In spruce, the density was 7% less on the thinned plots
Larix occidentalis	Lowery and Schmidt 1967	Thinned 50-year-old trees increased in wood specific gravity for 15 years after thinning
Pinus taeda	Smith 1968	Specific gravity and latewood content increased for 4 years after thinning a 9-year-old stand
Pinus ponderosa	Echols 1971	Thinned trees had a specific gravity of 0.350 compared to unthinned trees which had a specific gravity of 0.366
Pinus pinaster	Nicholls 1971b	There was only a very minor effect of thinning on wood specific gravity and on tracheid length
Pinus radiata	Cown 1973a	Thinning produced shorter tracheids and more juvenile wood, which results in lower density
Pinus radiata	Sutton and Harris 1973	Thinning had only a minor effect on wood properties
Pinus radiata	Cown 1974	Wood density and tracheid length were reduced by thinning. Compared to older unthinned stands of the same diameter, wood density was 8–10% lower and tracheid length was shorter by 10%
Pseudotsuga menziesii	Erickson and Arima 1974	Accelerated growth from fertilization and thinning caused a slight increase in fibril angle but only for a few years
Pseudotsuga taxifolia (menziesii)	Erickson and Harrison 1974	Thinning and fertilization caused immediate production of low density wood and a small decrease in tracheid length. Effects were maintained for 3–4 years after treatment
Pseudotsuga menziesii	Parker et al. 1975	Wood density and pulp yield of 25-year-old Douglas-fir were not affected by thinning, although they were reduced by fertilization
Picea abies	Madsen et al. 1978	Basic density is not affected by thinning
Pinus radiata	Inglis and Cleland 1982	Equations were developed to predict final branch size in stands of thinned radiata pine
Pinus radiata	Bamber and Burley 1983	Normal thinning has no effect on wood but heavy thinning reduces wood density
Pinus caribaea	Bertoloti et al. 1983	A thinned 7-year-old stand had two to three times the number of 25-cm-diameter trees as the unthinned stand
Pinus ponderosa	Markstrom et al. 1983	Although ring width was increased by thinning, there were no differences in specific gravity, latewood percentage, tracheid length, or microfibril angle compared to unthinned trees
Southern pines	Martin 1984	Tight initial spacing combined with thinning will result in less juvenile wood in the final cuts. If moisture is limiting, thinning increases specific gravity
Pinus taeda	Megraw 1985	Thinning allowed latewood growth to increase, producing an overall increase in specific gravity. Tracheid length was less. There was an increase in the incidence of compression wood
Pseudosuga menziesii *Pinus taeda*	Megraw 1986	Thinning may have either a positive or negative effect on specific gravity, but it is of minimal practical importance

Fig. 7.6. For most hardwoods, except the very high-value decorative species, thinning of commercial age stands is usually not practiced. It is sometimes done, however, such as in this *Eucalyptus grandis* plantation in South Africa. The 30-year-old stand has very high quality logs useful for veneer or lumber. Wood properties are little affected by thinning

resinosa) but the change in individual trees varied over 3 weeks in jack pine (*P. banksiana*). It was found that in both species thinning delayed the date of inception of latewood formation about 2.5 weeks at breast height and 1 week at the base of the crown. However, there was no change in latewood content, showing that added moisture from the thinning operation kept the trees growing longer during the period when latewood cells are produced.

In addition to the references listed in Table 7.3, there are some publications with detailed analyses such as the one by Ericson (1966), which dealt not only with wood density, but also considered latewood and heartwood in Scots pine and Norway spruce following thinning.

Very little has been published relative to the effect of thinning on wood properties of hardwoods. It is, of course, evident that tree quality and form are greatly improved by removal of poor trees, thus leaving the higher quality trees (Fig. 7.6). One study by Savina (1956) indicated that released oak (*Quercus*) trees had wood that became less porous, and that the wood elements became longer after thinning; similar, but less extensive changes occurred in aspen. Most change was observed in 35- to 55-year-old oak trees; no changes occurred in trees 80 years or older.

In some species of hardwoods, the results of thinning are not positive in relation to wood quality. Opening up the stand for more light can cause bole (epicor-

Fig. 7.7. Sometimes thinning has adverse effects on the residual trees, if the species produces adventitious branches when light is available. Shown is a recently thinned stand of sweetgum (*Liquidambar styraciflua*); some of the trees will sprout vigorously, producing many small knots, while others will not

mic) sprouting which results in numerous knots (Fig. 7.7). Erdman and Peterson (1972) recorded the formation of epicormic branches but stated that in yellow-birch (*Betula allegheniensis*), most epicormic branching was in the upper part of the stem in the less valuable logs.

Sometimes thinning causes the formation of poorer quality trees, although this is rare. Depending on age when thinned, thinning density, and tree species, the thinned stands develop somewhat larger branches with bigger knots. This problem can largely be avoided by thinning at the proper age and in the proper way. Increased forking of the main stem has occurred in American elm (*Ulmus americana*) and sugar maple (*Acer saccharum*) following thinning (Conover 1958).

In summary, the results of thinning on wood properties in the conifers are remarkably uniform, although occasionally researchers report diametrically opposite results. The general trends are:

1. Thinning improves wood quality because it increases the number of logs in the final cut that are larger and of better quality for solid wood products.

2. The wood removed in thinning of a young stand will have a high proportion of juvenile wood, but the final cut will have less. As pointed out by Bond (1985), thinnings will have uniform fiber properties, enabling more efficiency during

manufacture. The proportion of juvenile wood in the final harvest will be substantially less in a thinned than in an unthinned stand.

3. In thinned stands the cambium becomes active earlier (5 to 10 days) than in unthinned stands but latewood production is delayed.

4. Thinning has the most pronounced effect on ring-porous hardwoods, followed by diffuse-porous hardwoods, then conifers.

5. Release of trees usually improves the quality of solid wood products and does not noticeably reduce the yields and quality of pulp.

6. Normally, specific gravity and tracheid length are not greatly affected by thinning. Some researchers show a decrease in both; one example is Hildebrandt (1960) who states "...when a suppressed tree is released either by silvicultural treatment or soil improvement the resulting rapid burst of growth is accompanied by a decrease in cell-length." However, an occasional increase in cell size has been reported as a result of thinning.

7. In hardwoods, thinning can cause degrade for quality products through establishment of epicormic branches with associated knots (see Fig. 7.7). Larger branches sometimes result but if thinning is done at the correct time, this is not of great importance.

7.3.2 Plantation Spacing and Wood Properties

A series of studies have been made on the effect of initial spacing of plantations on wood properties. This is somewhat different from thinning in that the spacing exists from the beginning, and there is no sudden change in light intensity, moisture availability, or competition. Differences in spacing do affect the environment such as in the availability of moisture and nutrients, but this occurs slowly, as the trees develop. The long-term effect of spacing differences can be considerable as explained by Geyer and Gilmore (1965) who showed that in areas where moisture is limiting it may be possible to increase wood specific gravity of loblolly pine by planting at a wider spacing.

Plantation spacing can also have a major effect on wood properties by altering tree form, especially limb size and its associated defects and reaction wood. For example, Polge (1969a) found that wide-spaced but unpruned plantations have larger knots, more juvenile wood, and more taper than trees grown at denser stocking. Risk of infection in the pruning stubs is less in wide-spaced plantations because the pruning scars heal more rapidly. Polge summarizes that "...one can predict that wide-spaced plantations will not produce products of equal quality with those of denser plantations." Although it makes common sense and most foresters talk about the relationship of spacing to tree quality, few actual studies have been reported. In those done, e.g. Grah (1961) with young Douglas-fir, a close positive linear relationship was found between knot size and increased spacing from 4×4 to 26×26 ft. Tree quality reflected both log and knot size and frequency. In a study on Scots pine, Persson (1977) also found large diameter limbs related to wide spacing.

The general finding has been that a normal spacing variation (i.e., those stand densities commonly used by the silviculturist) has little effect on wood properties such as specific gravity or tracheid length. This was shown to be true for *Pinus banksiana* (Grigal and Sucoff 1966) and for *P. caribaea* in Brazil by Montagna et al. (1983), who reported that there was no effect of spacing on wood specific gravity. Little effect of spacing has been shown for a number of species, including *P. elliottii*. Smith (1977) reported that in slash pine, a range of stocking levels from 6×6 to 12×12 ft had no effect on wood density through the 14th year. Martin (1986) obtained the same results for loblolly pine. As a variation, Geyer and Gilmore (1965) reported that the specific gravity of juvenile wood of loblolly pine was greater in the 10×10 ft spacing compared to the 4×4 ft but in mature wood, there was no effect of spacing upon wood density. At the wider spacings, the cell length of both juvenile and mature woods was greater than at the closer spacing. Based on a large number of *Pinus taeda* trees in the southeastern USA, Zobel et al. (1965) found no effect of stand densities from 50 to 135 ft^2 basal area on wood specific gravity.

For species with indistinct growth rings, such as Sitka spruce and Norway spruce, latewood percentage and wood specific gravity may decrease with increased spacing between the trees, according to Larson (1972). Earlywood increases more rapidly than latewood. He states, however, that "contrary to prevailing opinion, wide spacing has relatively little effect on either the percentage of latewood or wood specific gravity of young pines..."

Contradictory results have been reported within some species. One good example is *Pinus caribaea* where in Fiji, Cown (1981) reported that a closer spacing produced lower wood density. With the same species in Puerto Rico, Whitmore and Liegel (1980) found that for merchantable trees the close 5-ft spacing produced trees with a wood density of 440 kg/m^3 compared to 520 kg/m^3 for the 14-ft spacing.

With red pine, Barse and Laidley (1980) found that stand density had little effect on specific gravity, and they concluded that the use of narrow spacing is not a practical method to increase specific gravity. Similar results were reported for red pine by Hayne (1958), who found wood to be little affected by spacings of 4×4 to 8×8 ft. The same lack of spacing effect was observed for ponderosa pine by Echols (1960). After 10 years, differences were found in specific gravity of the wood associated with varied spacings but in comparison, 20 years after thinning, changes in specific gravity were not related to spacing. In contrast, Baker and Shottafer (1967) reported that a dense stocking of red pine will result in higher wood specific gravity and a reduced stem taper.

In a comprehensive study on spruce (*Picea abies*) Klem (1942) investigated the effect of five different spacings on plantations over 30 years of age. He found that with greater spacing the specific gravity decreased and the percentage of knots increased. He noted that up to a spacing of 2.5×2.5 m, there was no difference in the paper properties or yield of pulp per unit weight of wood. At a spacing of 3.5×3.5 both pulp yield on a weight basis and pulp strength were reduced. The differences Klem observed were considerable; for example, it takes 4.7 m^3 of solid wood to make 1 ton of sulfite pulp from trees with the 1.25-m spacing compared to 6.2 m^3 of wood per ton required for the 3.5-m spacing.

In another large-scale study with *Picea abies*, Persson (1975) reported on wood quality related to five spacings, from 1×1 to 2×2 m. The trees had been planted in 1908. He found, as Klem, that latewood content and basic density decreased with increasing spacing, with great differences being evident. There was no relationship between fiber length and spacing; contrary to Klem (1942), however, Persson found no relationship between spacing and pulp yields either on a volume or weight basis, but Persson's spacing differences were much less than Klem's.

In a paired study with Scots pine (*Pinus sylvestris*) and Norway spruce (*Picea abies*), Persson (1975) worked with spacings of 0.75×0.75 to 3.00×3.00 m; the trees had been planted in 1906. As for spruce, it was found that latewood content and specific gravity of pine decreased with wider spacing, while knot content increased greatly. Also, like spruce, tracheid length of pine was not related to spacing. There was a significant increase in the content of extractives at the wider spacings. Persson found that pulp yields per unit weight and volume of pine decreased with increased spacing, while this was not the case with spruce, although Klem (1942) had reported lower yields with wider spacing for *Picea abies*. Persson concluded that the most important overall effect of differing plantation spacings is on the yield rather than on the quality of the pulp, since the latter can be somewhat compensated for during the manufacturing process. He feels that the decrease in yield of 5% in pine, resulting from a 1-m increase in spacing, has considerable significance. The economic impact of increased knot content depend upon the product produced, being particularly great for solid wood products.

Polge (1969a) studied wood quality of Douglas-fir as affected by close and wide spacings. He concluded that the wider spaced trees have larger knots, more juvenile wood, increased taper, and wider growth rings. He assumed that wider growth rings indicated lower specific gravity but recognized that this did not always hold. In contrast, Bennett (1969) stated that spacing did not affect latewood percentage, specific gravity, or the production of quality dimension lumber in slash pine. He advocated wide spacing and pruning when the goal is sawtimber and veneer stock. Marts (1950) found that in longleaf pine the 6×12 ft spaced trees had a somewhat higher specific gravity than trees with a 4×6 ft spacing.

Very few studies on the effect of spacing on wood quality have been reported for hardwoods. Of those made, e.g. Schönau (1973) on *Eucalyptus grandis*, nearly all found no direct relationship between wood specific gravity and spacing, but there was an effect on growth rate. In a spacing study of *Eucalyptus* in the Congo, Delwaulle (1985) noted a wood density of 0.52 at a spacing of 1×1 m which increased to 0.60 at a 4×4 m spacing. At wider spacings the wood was not affected. This is a surprisingly large response to spacing in the eucalypts where we generally find only a small or no effect on wood by stocking density.

A few generalizations can be made for the effect of plantation spacing on wood properties:

1. Most foresters find that for plantation spacings normally used operationally, the initial stocking has little effect on wood properties, such as specific gravity and cell length. Exceptions to this generalization are some studies of spruce in Europe in which wider spacing resulted in lower wood density. There was, however, a considerable influence on wood quality from limb size and number, which were directly related to spacing.

2. Wider spaced plantations have trees with larger juvenile cores. According to Saucier (1987) trees grown at wider spacings are strongly tapered and the stem

has a large core of low specific gravity juvenile wood. Some researchers find that close spacing does not reduce the proportion of juvenile wood greatly at older ages, although it is increased at younger ages.

3. In a few species such as spruce, some researchers found that the closer spaced trees produced greater yields of pulp for a given volume of wood and in some instances, stronger pulps.

4. In areas with limited rainfall, wider spacing produces higher specific gravities because more moisture is available during the period of latewood formation.

7.4 Effect of Pruning on Wood

The improvement in wood properties possible by pruning are several (Scott and Stevens 1947). Improvement can either be from direct tree form changes, especially in limbs, or secondarily where wood is affected by the changes in growth pattern caused by removal of parts of the crown. The latter usually does not occur when pruning is done correctly, i.e., when no more than one-third to one-half of the live crown is removed, but there are exceptions. Slabaugh (1957) stated that up to 50% of the crowns of red pine can be removed without severely reducing growth. Trees pruned more severely become suppressed by those with larger crowns. Polge (1969a) concluded that pruning will have less effect on wood properties when it is done to simulate natural pruning physiologically.

It is usually recommended that pruning should start when the tree is tall enough so that pruning to the desired height results in the removal of no more than one-third to one-half of the live crown. When too much crown is cut, growth can be adversely affected. In a test of crown removal, Zahner and Oliver (1962) found that pruning half the live crown delayed the transition from earlywood to latewood for about 1 week in red pine but did not affect the date in jack pine. For most species, the tree diameter should be about 5 in. (12.5 cm) when pruning is carried out (Fielding 1965). Adding to this for healing outside the knotty core plus the swirled wood and included bark, the tree diameter will normally be 6–7 in. in most species before clear wood is formed. A really high quality wood will only be found outside that size cylinder. Park (1982) measured the occlusion and the defect of cores in pruned logs, based on sawing-out 85 *Pinus radiata* logs. From this he developed a model for predicting the occlusion and defect core related to the pruned branch stubs based upon tree diameter.

7.4.1 Direct Effect of Pruning on Wood Properties

Pruning is done to promote a higher proportion of clear wood. Additionally, in some species, such as *Pinus radiata*, pruning apparently causes an early cessation of juvenile wood formation and thus induces a change in the characteristics of the wood. Polge (1969a) reported for radiata pine that pruning resulted in an increase in latewood from 14 to 26%, an increase in wood density of 0.48 to 0.59, a reduction in spiral grain, and an increase in tracheid length from 3 to 4 mm. The effect of reducing the proportion of earlywood in radiata pine by pruning was mentioned by Fielding (1965). Also in *P. radiata*, Cown (1973a) reported that pruning increased specific gravity by 7–10% for 2 or 3 years after which it returned to normal. However, Bamber and Burley (1983) state that pruning did not have a deleterious effect on radiata pine wood. Pruning increased wood density of *Pinus caribaea* about the same amount as for *P. radiata* (Plumptre 1978). Pruning of Douglas-fir and loblolly pine also resulted in somewhat higher specific gravity below the point of limb removal (Megraw 1986). Pruning had no effect on the specific gravity of *Eucalyptus grandis* (Schönau 1973); however, the timing and speed of wood formation were altered.

There is much misinformation about the overall value of pruning; almost everyone assumes that it will all be positive. But if done improperly, the adverse results can negate the potential benefits. For example, if the saws used to remove the limbs cut deeply into the bark or swellings at the branch nodes, or if machetes are used which chop into the wood, the delay in production of good, clear wood can be long or it may never occur, and the occluded area and defective core will be overly large. In the wounded area, callus will grow and ultimately produce wood that has deformed tracheids, curly grain, or other defects. This formation of abnormal wood can continue for several years, and sometimes the cambium folds in upon itself, and the wound never heals cleanly (Polge 1969a). Equally serious problems occur when large or ragged branch stubs are left (Brown and Pawsey 1969), as frequently occurs when machetes are used. It takes a number of years to heal around and over the stubs to produce good wood (Polge 1969a).

It is amazing how often poor pruning is practiced. Many tree species have a swelling at the base of the branch; this is frequently sawed into when the limbs are cut and healing and production of clear wood is greatly delayed (Polge 1969a). One of the most common defects resulting from pruning are resin pockets; Park and Parker (1982) report that in radiata pine, resin pockets are the major cause of degrade in the clear wood zone of pruned logs. The causes of resin pocket formation are complex but it is known that they occur more frequently in radiata pine on some sites than on others. Within stands, certain individual trees appear to be more susceptible to resin pocket formation than others (Clifton 1969; Cown 1973b). In his report on the effect of pruning upon resinification in radiata pine, Harris (1963a) showed that pruned knots had a higher resin content than unpruned knots, with up to 15% resin increase.

To be effective in obtaining the highest quality wood, pruning should be done in log lengths, or half-log lengths, to coincide with the size of the lathe that will produce the plies. Usually, the desired length is either in units of 8 ft (2.5 m) or 16 ft (5.0 m). To prune to 6 or 10 ft is not efficient because the high quality wood produced will not coincide with lengths desired for plywood, although it will be helpful for lumber production if random lengths are manufactured. In a study of radiata pine in Australia (Brown 1965b), trees were pruned in stages up to 45 ft and a diameter of 8–9 in. When the trees were felled at age 35, 16 years after pruning, the pruned trees yielded 65% lumber greater than box quality, while the unpruned trees yielded only 35%. Trees green-pruned and harvested at 14-in. diameter gave clear wood in low yield because of taper, eccentricity, and nodal swelling occlusions (Brown 1965b).

Considerable research has been done on the growth response and formation of clear wood following pruning. As demonstrated by Brown (1965a), clear boards will not be generally obtained in trees less than 14 in. diameter at breast height; pruning too late will also be ineffective. The proper time and size of branches for pruning were assessed for Scots pine by Andersson (1967) but his results were not clearly defined. Directions for proper pruning to obtain the best returns have been prepared for a number of species; one for *Pinus radiata* (Wylie 1976) goes into specific details of the type of trees to select or to reject. In his publication dealing with pruning, Polge (1969a) thoroughly describes the best methodologies to be used.

One common concern is that pruning will reduce growth rate so much that the improved value from better wood quality is more than offset by the volume growth loss (Fielding 1965). There can be no question about this adverse result when severe pruning is carried out. Some studies have shown a significant reduction in growth even with less severe pruning. Andersson (1967) found growth reduction in Scots pine even when only one-third of the live crown was removed. In moderate pruning, a common result is that ring width becomes less at the base of the tree and greater toward the crown (Young and Kramer 1952; Fielding 1965).

Some foresters do not favor pruning because they feel that it facilitates the access of rot-producing organisms into the wood. For most species, this fear does not seem to be justified when pruning is done correctly. For example, Andersson (1967) reported for Scots pine that none of the pruned knots had been attacked by rot, and only one was discolored. Fielding (1965) stated that infection of the pruning wound is almost unknown in *P. radiata*. However, Shigo et al. (1978) found bands of discolored heartwood in association with both pruned and unpruned branch stubs. They stated that the advantage of pruning is to compartmentalize the defective wood in the small nonmerchantable central core. Similarly, Lowerts and Kellison (1981) studied rot infection in yellow poplar following wounding. They found that discoloration and rot could be extensive, depending primarily on the individual tree involved. They did not find a relationship between wood specific gravity and susceptibility to decay. Wounds from pruning appear to be a common source for entry of decay organisms in yellow poplar.

7.4.2 Pruning to Improve Tree Form and Quality

The desired effect of pruning is to improve the wood quality of the tree by improving tree form. For example, in Douglas-fir, Staebler (1963) reported that removal of one-third of the crown had little effect on growth or branch form, and that heavier pruning reduced growth, but in turn produced a more cylindrical tree. In many instances early shaping (pruning) as done regularly in the tropics results in the conversion of a multistemmed to a single-stemmed tree. This method of producing single stems is generally practiced in species with the highest quality woods, such as black walnut (*Juglans nigra*) (Beineke 1977) and high quality tropical hardwoods such as red ceiba (*Bombacopsis*) (Fig. 7.8).

Pruning is frequently done in conifers, with the objective of improving the quality of logs for lumber or plywood by reducing limbs and knot size. Some investigators feel that the form class of the tree is directly improved by pruning (Staebler 1963), but Polge (1969a) stated that such results have not been conclu-

Fig. 7.8. Pruning is used to improve wood properties. As shown for the 4-year-old red ceiba (*Bombacopsis quinata*), quality wood will be obtained early when multiple tops and large branches are removed

sive. Grand fir (*Abies grandis*) trees with 50% of the crown removed did show some form improvement, while those pruned to 35% showed no change; trees pruned to 20% were more conical. For loblolly pine, Young and Kramer (1952) found no reduction in height growth from pruning, but diameter growth below the crown was related to size of crown, while that in the crown was not affected. They felt that form factor improvement will be greater, the faster the tree grows. Often, however, ring width in the base of the tree is reduced while it increases in the crown area, thus changing tree form. Polge (1969a) found this to be true in the poplars, which resulted in an unusually rapid improvement in form following pruning. According to Megraw (1985), pruning of live branches accelerates the transition from juvenile to mature woods below the point of pruning. For loblolly pine, increases of about 5% in specific gravity can be expected below the point of pruning.

Sometimes pruning has the secondary adverse effect of epicormic branch formation, which reduces wood quality (Fielding 1965). This problem is most common in hardwoods; the maples (*Acer*) are particularly susceptible but light is not essential to trigger the dormant buds (Books and Tubbs 1970) so epicormic branching will occur even when pruning has not been done. Black cherry (*Prunus serotina*) has excessive epicormic branching if pruning is done up to 75% of tree height (Grisez 1978). Epicormic branches also sometimes appear in conifers such as Sitka spruce (*P. sitchensis*), which forms such branches on the portion of the tree exposed to light after pruning.

It has been reported by Fielding (1965) that if one prunes a portion of the crown, the branches left will become larger. This result is apparently not too common since Polge (1969a) and others found no significant differences in diameter growth of branches for Douglas-fir after pruning.

7.4.3 Other Considerations About Pruning

Although not a major subject of this chapter, the question is raised as to how high on must prune to improve wood quality. The general concensus is that the first log is often economically feasible, the second marginal, and that higher pruning will not pay because wood improvement is not that pronounced. For example, Brown (1965a) stated that for radiata pine in Chile, pruning over 6.5 m (22 ft) in height is not justified for rotation ages of 25 to 30 years. This was confirmed (1965b) for radiata pine in New Zealand by Brown (1965b), who concluded that pruning the first log will pay and pruning the third will not; returns from pruning the second log is doubtful. In a summary of pruning tests in Australia, Lewis (1965) reported that to be economically profitable (by improving wood quality), pruning should be done only in plantations on the best sites; second log pruning produces less than acceptable quality lumber and improved milling methods may be a better way of obtaining better boards than pruning.

7.5 Planting Techniques and Wood Properties

Planting is a silvicultural practice that can have a major effect on tree form and thus on wood properties. This is rarely recognized. When trees are planted with trailing roots or with U or J roots, the results on wood properties can be very serious (Fig. 7.9). Trailing roots result when the foot of the planting machine is not deep enough, or the planting machine moves too rapidly. Trees that grow with the roots to one side (not under the stem) are rarely straight. Because of the imbalance the tree overcorrects yearly by formation of reaction wood, and such trees often are crooked throughout their lifetime. Trees with offset roots are particularly susceptible to displacement by wind or snow; the authors have observed whole plantations leaning, all trees in a given row leaning in the same direction. The trailing root or J and U root problem is particularly evident when machine planting is done in heavy soils, and when very deep planting is done in sandy soils, such as in some tropical areas.

Root deformation is not equally serious for all species. Some, such as the spruces, and some hardwoods, regenerate new roots in the deformed root system that correct the imbalance and form a relatively balanced root system, resulting in a well-formed tree. Others, such as most pines, will not regenerate roots to com-

Fig. 7.9. When trees are planted poorly, such as the J-rooted red pine shown, they are unstable and never grow straight. As a result they will contain an excess of compression wood. This results in poor solid wood products as well as poor pulp

pensate for root deformation, and the deformation will affect the tree and its wood properties throughout its lifetime. This silvicultural cause of wood variability is a rather widespread occurrence.

Silvicultural methods of site preparation or competition control have rarely been assessed as to their effect on wood. The studies must be well planned and long-term. There does seem to be an effect, however, as reported by Smith (1977). He found the control area (no site preparation) produced a specific gravity in slash pine of 0.415, while bedding gave 0.424, harrowed 0.422, and the specific gravity on both harrowed and bedded sites was 0.426. In this instance the treatments that caused faster growth resulted in higher specific gravities. We have found no references relative to the effect of competition control on wood properties but there is undoubtedly some effect.

7.6. Resin Tapping, Resin Induction, and Wood Properties

In some pine species tapping the trees for naval stores, turpentine, and resin, is a standard silvicultural process. It has long been recognized that the wood in the stem following tapping becomes resin-soaked and is difficult to use for certain products. Except for timbers to be used in contact with the soil, the resin-soaked wood is not desirable for solid wood products or for pulp products.

Pinus pinea trees used for producing naval stores for 2 years had the same specific gravity and only slightly different compression and bending strengths compared to untapped trees, according to Giordono (1959). When the resin content of the wood is increased by application of paraquat, the result is a lowering of the density of the extracted wood in the lower bole on both the treated and untreated sides of the stem (Squillace 1978). Studies that have been made on the effect of resin-tapping on wood (Encev and Kostov 1959; Giordono 1959) showed that the density of the wood directly under the face exceeded that at the side of the face by 8% in *Pinus sylvestris* and by 13% in *P. nigra*. Wood density of tapped trees exceeded untapped trees by 16% in *P. sylvestris* and 21% in *P. nigra*. When paraquat-treated trees were pulped (Hurley et al. 1976), no effect of treatment was evident in any kraft pulping characteristics or physical properties of the paper. Treatment increased turpentine yield by 70% and tall oil yield by 27% 12 months after treatment. Additionally, the treatment induced beetle attacks which created conditions for wood degradation and tree mortality (Goldman et al. 1979).

7.7 Wood Properties of Coppice and Root Sprouts

For some hardwood species, the standard forest management practice is to harvest the stand and let the trees sprout; this is called coppice or root sprout forestry. It is generally assumed by most foresters that the wood of the coppiced trees will be the same as that of the original tree if the two are of the same age. The authors are aware of only a few studies to assess the relationship between the wood of the original tree and the wood of its coppice sprouts. One such study on the wood of oak (*Quercus sessilifolia*) indicated that the specific gravities of high forest and coppice trees were similar (Todorovski 1969). A study on *Eucalyptus camaldulensis* (Sesbou 1981) showed that the wood properties of coppice were different from those of the original trees. There was a decrease in wood density, but fiber length increased. The latter was not enough to improve the quality of the paper, however.

Fig. 7.10. It is common practice with some trees, such as the eucalypts, to regenerate the stand by coppice (stump sprouts). Wood is often inferior when multiple sprouts remain, as for the two sprouts per stump illustrated. The trees on the stump tend to lean away from each other and to have most limbs on one side of the tree. This causes formation of much tension wood in the stem

In *Populus*, Phelps and Isebrands (1983) report that the specific gravity of the original trees (3-years-old) is greater than that of their coppice sprouts of the same age in the basal portion of the tree but not in the top.

The amount of reaction wood is often greater in coppice from stump sprouts, and the distribution of limbs is more irregular; this occurs in many genera such as *Eucalyptus, Quercus*, and *Fraxinus* (Fig. 7.10). This is especially true when two or three stems are left per stump. Such stems lean away from one another and have most limbs only on one side. The leaning stems have much more tension wood than straight, single stems and there is defect related to the heavier limbs on one side of the tree. This defect does not occur for species that sprout from the roots such as *Populus tremuloides* or *Liquidambar styraciflua.*

Consideration of wood properties of coppice is most important. This method of regeneration is being employed on an increasing scale throughout the world, especially with tropical genera such as *Eucalyptus* and *Gmelina.* Usually the volume of wood produced by the coppice is most important and little or no consideration is given to the properties of the wood produced. This is incorrect if the wood is inferior, as when multiple sprouts are used with their higher proportion of reaction wood which is less suited for the intended purposes.

Chapter 8

Control of Wood Properties by Breeding

8.1 Genetics of Wood Properties

As pointed out by Harris (1983), wood technologists spend much time overcoming defects in timber which could have been minimized while it was growing. One major method of reducing defects in wood properties is by using seedlings of good genetic quality obtained through the application of genetics to the large natural variability in wood properties. The opportunities for genetic control of wood are particularly attractive, since wood properties are generally strongly inherited. Many of the aspects of the genetics of wood are covered in Chapter 11 of Zobel and Talbert (1984) entitled "Wood and Tree Improvement."

The genetic method for improving wood from plantations of tropical pines was emphasized by Palmer (1977) who recommended selection of the better trees as seed sources for future generations to minimize the huge variation in wood properties now present.

Thousands of examples could be listed relative to inheritance of wood properties. In general, the patterns are similar to that quoted by Khalil for *Picea mariana* (1985) as "all characters except fiber wall thickness have a heritability greater than 0.30." There are, as expected, many exceptions but as a group wood properties are very responsive to genetic manipulation. It is important, however, to understand the relative effects of genetics and environment and their interactions on wood.

Genetic manipulation can change wood properties in several ways, including the alteration of tree form, growth rate, wood morphology, wood chemistry, and others; these are described for the southern pines by Zobel (1971). Another example is the book of Burley and Nikles (1973) where at least ten papers deal with aspects of genetic control of wood properties.

Breeding for improvement of wood has not always been considered of importance. As stated in 1971 by Zobel "breeding for wood is a secondary effort, within the primary effort of getting faster growing, better formed, well adapted, pest resistant trees. It can be super-imposed on the regular program with the major extra effort required being a greater number of parent trees to satisfy the wood requirements." Unfortunately, many major tree improvement programs do not include wood improvement, even though wood is usually the desired product of a forestry program.

Concern with genetics of wood is becoming greater as indicated by the hundreds of articles dealing with the subject. A publication such as that by Forintek Canada Corp. (Keith and Kellogg 1985), which deals with wood quality in tree improvement programs, is only one of many emphasizing control of wood by genetic manipulation.

Genetic improvement cannot be successfully done in isolation, but must be integrated with silvicultural methods which also improve wood properties. This dual approach was covered well by the TAPPI Monograph Series No. 24 (Anonymous 1962) which combined the effects of environment and genetics on wood properties in the 883 annotated citations which were included.

8.1.1 What Are Genetic Differences and How Are They Measured?

Before discussing inheritance in wood per se, it will be of value to review briefly a few concepts of genetics necessary for an understanding of the process. Details can be found in any of a number of publications, such as Zobel and Talbert (1984).

There are two methods to examine genetic differences and similarities within a species. One relies on the resemblance among relatives, while the other method involves the study of variation patterns in natural forest stands. Studies of the resemblance among relatives can take many forms. The most straightforward is the clonal test. Since a clone is a group of individuals derived from one original tree (referred to as the ortet) by means of vegetative propagation, all members of a clone (called ramets) share the same genetic makeup and are therefore genetically identical. Differences among ramets of the same clone are thus assumed to be the result of environmental differences.

An assessment of the resemblance between parents and offspring is another rather common way of studying genetic differences. One can also compare families with one another. If families are composed of individuals that have only one parent in common they are known as half-sib families; the progeny of a wind-pollinated tree is a typical example. If individuals in each family have both parents in common they are referred to as full-sib families. In all instances careful attention to experimental design and field layout is necessary, so that the effect of genetic and environmental factors can be separated.

The other major method, developed by Shrikhande (1957) and applied to forest trees by Sakai and Hatakeyama (1963), can be used to assess genetic differences in natural stands in which the family structure is unknown. By comparing the variation patterns of groups of different sizes with the total variation pattern, an estimate of the genetic variation present can be made, since individuals growing close together are expected to be more closely related to each other than individuals growing further apart. This method has rarely been employed and will not be discussed in further detail here because of its infrequent use.

8.1.1.1 Types of Genetic Variation

Genera and species can vary greatly in their variation patterns. This is sometimes referred to as their genetic architecture. Genetic variation can be assessed at various levels; obviously, variation among species is the highest level within a genus and is usually quite well known. Variation within the species is the level with which tree breeders are most concerned. The within-species differences can be subdivided further into geographic variation, stand to stand variation, variation among individual trees, and variation within individual trees.

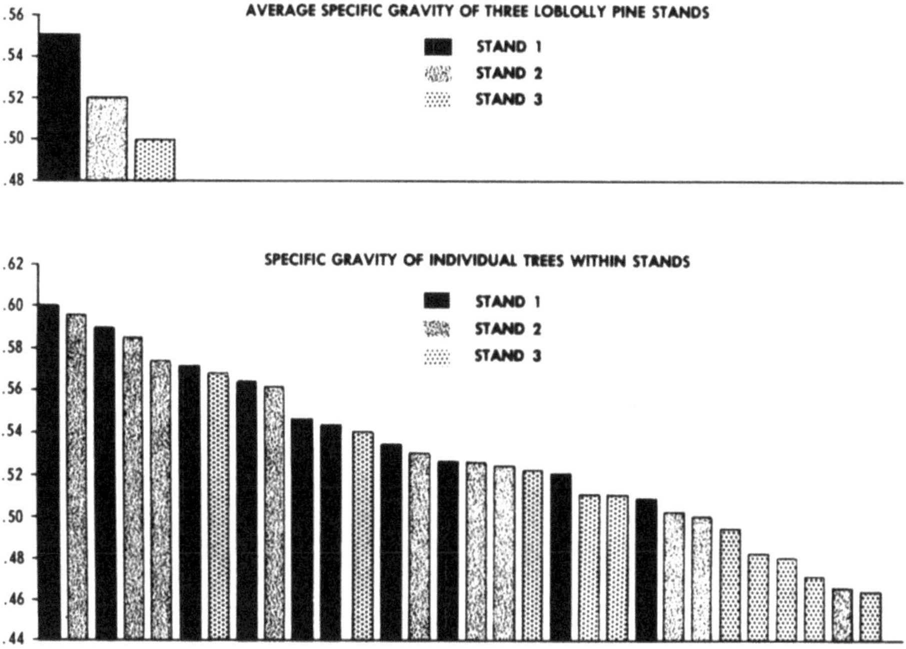

Fig. 8.1. Properties of wood from different geographic areas within a species range often vary considerably, as shown here for loblolly pine from Texas. Usually, however, the tree to tree differences, as indicated within each area, are very large, tending to mask the geographic differences and frequently making it impossible to detect them (Courtesy Texas Forest Service)

Geographic variation results from the action of natural selection, favoring individuals that are well adapted to the local environmental conditions, which include climate and soils. The local environment also modifies the expression of the genetic potential of these individuals. Geographic variation, as for all phenotypic variation, has two components: genetic and environmental. These must be clearly distinguished if one is to work with the genetics of wood successfully.

Within one physiographic region, considerable stand to stand differences may occur. The majority of these are due to differing site conditions, although genetic differences also occur. The most striking variability and the one most intensively studied by those who use wood is represented by the differences among individual trees within a stand. Every forester should be familiar with the tremendous differences among trees in the same stand (Fig. 8.1). Genetic causes can either be large or unimportant in causing the differences. For example, tree to tree competition can greatly exaggerate small initial differences resulting from genetics or from microsite variation. The combined effects of environment and competition can often mask genetic differences.

The amount and distribution of genetic and environmental variation among wood properties differs greatly from species to species. It is determined by the species range, the species distribution, i.e., whether the species is continuous or broken up into smaller populations, which often occurs in mountainous areas, and by the evolutionary history of the species. Some species such as red pine (*Pinus resinosa*) and yellow birch (*Betula alleghaniensis*) are extremely uniform, while other species such as loblolly pine, ponderosa pine, and Douglas-fir are highly variable.

8.1.1.2 What Kind of Tests Are Necessary to Determine Genetic Variation?

The determination of the variation present in natural stands or plantations is a first step in studying a species. Much has been done in this respect, as discussed in detail in previous chapters. Studies of plantations or natural stands are not sufficient to assess the genetics of wood, since natural variation is determined by both genetic variation and environmental causes.

Most traits of commercial importance are controlled by many genes with small effects, rather than by one or two major genes. Wood properties are no exception. As a consequence, genetic control of wood must be measured on a statistical basis. To do this, it is necessary to plant trees that have a known family structure in a normal environment, using a fairly sophisticated statistical design. Tests to determine the genetic variation are often referred to as progeny tests, but the term genetic test is more accurate. Progeny tests, by definition, assess the value of parents, while any tests of genetic variability are termed genetics tests. Trees to be assessed may be obtained by using sexual or vegetative propagation; tests are designed to study the genetic portion of geographic variation, as well as the variation among families or the worth of an individual parent.

In all genetic tests, related individuals are grown using a statistical design, such as a randomized, complete block design, in as uniform an environment as possible, with careful subsequent maintenance and measurements. Frequently nondestructive samples, such as those obtained from increment cores, are used to determine wood properties. Occasionally, such as when a thinning is done, entire trees may be harvested to study wood properties.

It is necessary to understand the most common genetic tests. They are briefly described below; detailed discussions of each can be found in the proceedings of a Workshop on Progeny Testing held in Charleston, South Carolina (Anonymous 1983).

Clonal Tests. The clonal test contains groups of individuals obtained by vegetative propagation, usually by grafts or rooted cuttings, or by means of tissue culture. Since all ramets of a clone share the identical genetic constitution, this is the simplest and most powerful way of determining genetic control. As shown by the pine grafts in Fig. 8.2, vegetative propagules within a clone (ramets of a clone) have wood that is very similar compared to wood from ramets from a different clone. Although the ramets of a clone are genetically identical, the trees produced do not always perform identically; such problems are evident with trees derived from tissue culture, since the tissue culture process itself may introduce variation that can be transmitted through sexual propagation.

Open-Pollinated Genetic Tests. This is the simplest and easiest kind of test to establish and is widely used. Seed is obtained from the female parent without

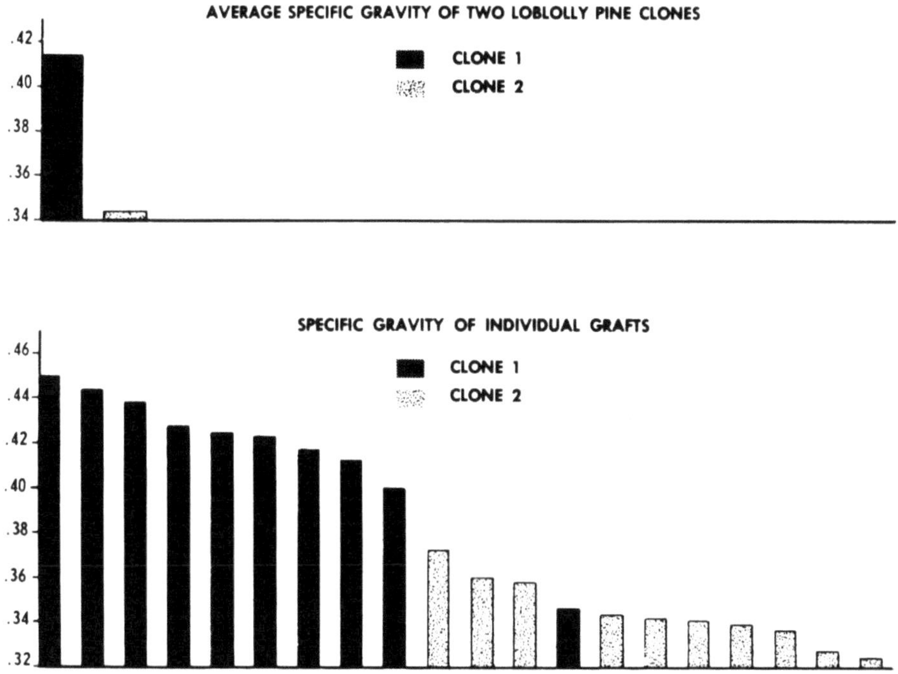

Fig. 8.2. Very uniform wood can be produced by use of vegetative propagation, as shown for these grafts of loblolly pine. The two clones had very different wood specific gravities; this was reflected in the ramets from those clones (Courtesy Texas Forest Service)

knowledge of the male parent. Normally, the offspring from the one parent are considered to be half-sibs, which implies that a large number of pollen parents contribute to each family, but this is not always a safe assumption. In many instances there are only a few male parents involved in an open-pollinated family which will lead to an overestimate of the actual genetic control. Despite their imperfections, open-pollinated progeny tests yield a great deal of useful information.

Control-Pollinated Genetic Tests. In this case families of which both parents are known are used to make the genetic tests. These are called full-sib tests and are by far the best means of obtaining information on the genetic control of traits of commercial importance.

8.1.1.3 How Is Genetic Control Measured?

Genetic control is measured by analyzing and subdividing the total variation into its components. The most convenient statistical means to quantify the variation is by computing the total variance and the individual components of which it consists. The total phenotypic variance (V_P) can be divided into two main components: the genetic variance (V_G) and the environmental variation (V_E). The genetic variance can, in turn, be separated into additive genetic variance (V_A) and dominance variance (V_D) (Zobel and Talbert 1984). When shown as a formula it is:

$$V_P = V_A + V_D + V_E.$$

Dominance variance is present when the progeny resemble one parent more than the other. In humans, for instance, dark eyes are dominant over blue eyes. Traits that are important for survival quite often show dominance under natural conditions. Genetic variation is additive when the progenies are intermediate between the two parents. This type of variation can be most easily captured in a breeding program and is most useful. Both types of variance are used in a vegetative propagation program.

The degree of genetic control is usually expressed by the term heritability (h^2 or H^2). Without further specification, heritability refers to the ratio between the additive genetic variance and the total phenotypic variance ($h^2 = V_A/V_P$). This form is referred to as narrow sense heritability. Broad sense heritability (or gross heritability) is the total genetic variance divided by the phenotypic variance ($H^2 = V_G/V_P$).

From clonal tests one can only determine the broad sense heritability, while half-sib tests can only give an estimate of the narrow sense heritability. Control-pollinated tests can be used to determine both the additive and dominance variance and consequently both the narrow and broad sense heritability. A brief summary of how these statistics are calculated from the various tests and their genetic interpretation is given in Table 8.1.

Table 8.1. Calculation of genetic parameters from progeny tests

Description	Symbol	Formula
Narrow sense heritability	h^2	$h^2 = V_A/V_P$
Broad sense heritability	H^2	$H^2 = (V_A + V_D)/V_P$
Coefficient of genetic prediction	CGP	$CGP = COV_{A_1A_2}/\sqrt{(V_{P_1}\ V_{P_2})}$
Genetic covariance between trait 1 and trait 2	$COV_{A_1A_2}$	Obtained from an analysis of an open- or control-pollinated progeny test
Additive genetic variance	V_A	$V_A = 4\sigma_f^2 = 4\sigma_m^2 = 2\sigma_f^2 + 2\sigma_m^2$
Phenotypic variance	V_P	$V_P = V_A + V_D + V_E$
Dominance variance	V_D	$V_D = 4\sigma_{mf}^2$
Environmental variance	V_E	Cannot be calculated
Clonal variance	σ_c^2	Obtained from ANOVA[a] of clonal test $\sigma_c^2 = 1/4\ V_A$
Female half-sib variance	σ_f^2	Obtained from ANOVA of open-pollinated or control-pollinated test $\sigma_f^2 = 1/4\ V_A$
Male half-sib variance	σ_m^2	Obtained from ANOVA of control-pollinated test $\sigma_m^2 = 1/4\ V_A$
Male × female interaction	σ_{mf}^2	Obtained from ANOVA of control-pollinated test $\sigma_m^2 = 1/4\ V_D$

[a] ANOVA: analysis of variance.

8.2 Genetic Control of Wood Specific Gravity

Wood specific gravity is chosen as the most important wood property to discuss because experience has shown it to be by far the most important in influencing the quality of solid wood products and both the quantity and quality of pulp and paper products (Barefoot et al. 1970; Pearson and Gilmore 1980). As stated by Zobel (1974) when referring to genetic control, "The most consistently responsive characteristic has been wood specific gravity."

8.2.1 What Are the Components of Wood Specific Gravity?

Specific gravity, its nature, development, and characteristics were covered in detail in Chapter 1. A few of the concepts are repeated here as a basis for understanding the following discussion on the genetics of wood. In one sense, it seems incorrect to consider wood specific gravity as a single wood property, since it is the result of several contributing factors.

A major component of wood specific gravity is the proportion of latewood. In both conifers and hardwoods the latewood is much denser than the earlywood and a high percentage of latewood therefore contributes to a high wood specific gravity (Saucier and Taras 1967). A second property which determines wood specific gravity is the cell wall packing density, which is the actual specific gravity of the cell wall material. It does not cause variation in wood specific gravity, however, because it is a nearly constant 1.54 g/cm^3 (Jayme and Kraus 1963; Yiannos 1964; Boyd 1967). A third factor is the cell wall thickness within the earlywood and latewood. This can be estimated by the light absorption in the latewood, and is the second most important factor related to the basic density of slash pine according to van Buijtenen (1963). The diameter of the earlywood tracheids also makes a small contribution to specific gravity variation, while the effect of the diameter of the latewood tracheids is often negligible. Final factors affecting specific gravity are the extractives and insoluble deposits present in the cells and cell walls. These factors were discussed and illustrated by Zobel and Talbert (1984).

8.2.2 Inheritance of Wood Specific Gravity in Conifers

8.2.2.1 Genetic Variation Among Provenances

Wood specific gravity differences among provenances are usually studied at an early stage in tree improvement programs, and much is known about provenance variation as outlined earlier in Chapter 2. Usually the trees are studied in their natural habitat where the effects of genetics and environment cannot be separated. Data are now becoming available from provenance tests which shed some light on provenance variation in wood density. Genetic variation among provenances usually is only slight or nonexistent, although there are exceptions. Some studies will be cited, in addition to those in the earlier chapter, to illustrate research results that have been obtained, with emphasis on those where genetically controlled differences in wood are related to geographic origin.

Barnes et al. (1977) reported on variation in diameter growth and wood specific gravity in 6-year-old *Pinus caribaea* provenance tests grown in Rhodesia (Zimbabwe), and showed specific gravities ranging from 0.339 to 0.356 among seven provenances. Provenance effects on wood can be very large when they also affect tree form. In loblolly pine some weak trends of specific gravity and seed origin are apparent. In one example (Thor and Brown 1962), it was found that there was very little difference in specific gravity among six provenances grown in Tennessee; in another Saucier and Taras (1967) observed a slight association between latitude of seed origin and specific gravity as well as percent latewood, both of which increased from southern to northern sources. Unpublished data from the Western Gulf Forest Tree Improvement Program indicate a similar trend with specific gravity decreasing from western to eastern and from northern to southern origins. This result from provenance movement is contrary to the variation pattern found in natural stands of loblolly pine (Zobel et al. 1960b) where the southern and eastern provenances have the highest wood density.

In lodgepole pine planted in Scotland the southern coastal provenance from the western USA has a 20% higher wood specific gravity than the central interior source (Henderson and Petty 1972).

In a study of 29-year-old ponderosa pine, the effect on wood of the elevation where the seeds were collected was apparent, according to Echols and Conkle (1970). They reported that wood specific gravity decreased as the altitude of origin of the parents increased, even though all genotypes produced wood of lower specific gravity when planted at a higher elevation.

In New Zealand, Cown (1971) found that the specific gravity of *Pinus nigra* in the North Island decreased 0.01 g/cm^3 for each 1° latitude and 150-ft elevation.

For the relatively uniform red pine (*P. resinosa*), Rees and Brown (1954) found that of the 19 seed sources tested in the Chippewa National Forest, only thy Bay City, Michigan, source had a significantly higher specific gravity than the other 18, namely 0.341 compared to a range of 0.310–0.326 for the others.

Wood specific gravity and tracheid length were examined in ten provenances of shortleaf pine by Posey et al. (1970). They sampled the species range along a north-south and an east-west transect, after which a separate test plantation was established for each transect. Variation in specific gravity, tracheid length, and total weight of the wood produced was greatest related to the north-south seed origins. In both tests the Ashley County, Arkansas, seed source had higher specific gravity, longer tracheids, and a greater weight of wood produced than trees from the local seed origin.

A 10-year study of geographic variation in Japanese larch (*Larix leptolepis*) in the Lake States (Lee 1975) indicated differences with seed origin but no definite geographic trends. In another test of the same species, Loo et al. (1982) found highly significant differences in specific gravity ranging from 0.385 to 0.417 in 20 seed sources of 17-year-old Japanese larch grown in Michigan. Lee (1979) found significant differences in tracheid length among 27 seed sources of *Pinus nigra* in Michigan, but there was no close association between specific gravity and geographic source for Norway spruce (*Picea abies*).

Contrary to most reports, Rink and Thor (1973) found that a gain of 14.4% in wood specific gravity could be obtained when the best 4 of 12 seed origins of Virginia pine were used. Surprisingly, they observed very small tree to tree differences.

In *Pinus pinaster* planted in Australia, marked differences occurred among the four provenances tested, with the Leiria race combining high density and good tracheid length with superior vigor and acceptable form (Nicholls et al. 1963; Nicholls 1967c).

One of the oldest provenance tests for which wood data are available is for Douglas-fir; in 1965, McKimmy reported results for a 46-year-old planting. There were significant differences among sources, but no pattern with seed origin. If a high specific gravity is desired, a careful match between seed source and site needs to be made, since a particular source will not always produce wood of the highest specific gravity. In another study of Douglas-fir, Cown and Parker (1979) used a subset of a cooperative provenance test consisting of five provenances at five locations but failed to demonstrate any provenance effects.

Although variation in wood specific gravity of Norway spruce (*Picea abies*) is not large, there are some weak relationships with latitude. For example, Ericson (1960b) and Worrall (1975) observed that northern sources produced less dense wood. In addition, Ericson found that specific gravity decreased with altitude of seed origin. Most others, such as Knudsen (1958) and Parrot (1960), found no consistent trends of geographic seed source with specific gravity.

It is difficult to generalize, but the following statement generally appears to hold. Genetic trends in wood properties related to geographic seed source are generally weak, but where a trend is present, specific gravity frequently decreases with increasing latitude and elevation of the parental seed source. In the southern pines, specific gravity in wood from natural stands increases from the northwest to the southeast in natural stands, but the genetic trend based upon seed origin appears to be in the opposite direction.

Clinal patterns of wood properties would be expected in the south, the northeast, and the boreal region, as described by Wright (1976). In the west, however, with its rugged topography, discontinuous variation is much more common, and broad trends, except for elevational differences, would be less frequent. These patterns are expected to be present in natural stands as well as in tests of the effects of seed origin.

8.2.2.2 Stand to Stand Variation in Specific Gravity

Very little information is available on genetic differences in wood from stand to stand within a given geographic area, since most of the reported genetic variation pertains either to geographic or tree to tree differences. Unpublished data from the Texas Forest Service and the experience of the authors indicate that the stand to stand genetic variation in specific gravity is rather small and in many instances none is detectable. When stand differences include greatly differing environments, the differences in wood can be considerable, but the individual tree variation is still large (Fig. 8.3).

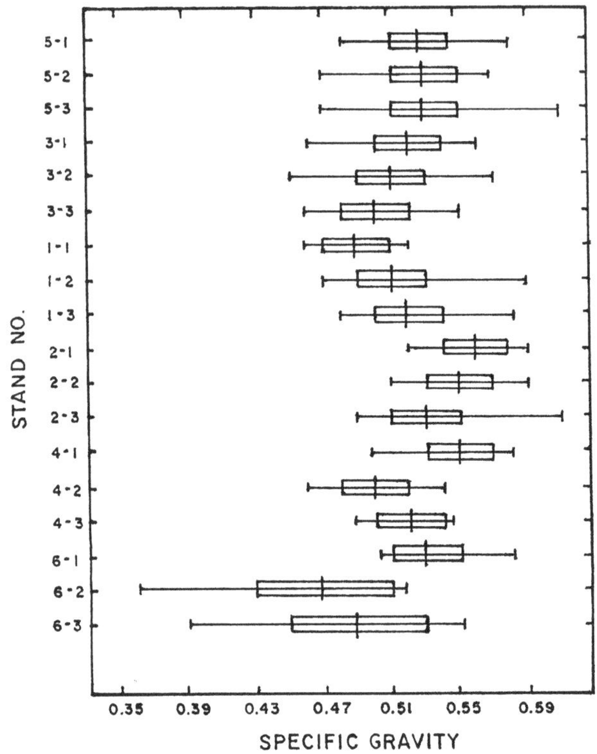

Fig. 8.3. Variation among and within stands of loblolly pine. The *vertical line* is the stand mean, the *rectangular area* is one standard deviation, and the *horizontal line* represents the range of values for individual trees within a stand. Note the size of the stand differences, much of which is environmental, while most of the tree to tree differences have a genetic origin (Courtesy Texas Forest Service)

8.2.2.3 Tree to Tree Variation in Specific Gravity

The literature on genetic differences of wood specific gravity among individual trees and their heritabilities is so voluminous that it is best to separate it into species or species groups. A summary of a limited number of papers is given in Table 8.2 to illustrate the magnitude of genetic control of specific gravity within different species or groups of conifers. Note that heritability is often referred to; its meaning was mentioned earlier in this chapter or is more extensively described by Wright (1976) and Zobel and Talbert (1984). The higher the heritability, the stronger is the genetic control.

The narrow sense heritability of wood specific gravity for individual conifer trees is generally high, ranging from 0.4 to 0.7. This is true for both juvenile and mature wood. For species with low wood specific gravity, such as Virginia pine, silver fir (*Abies alba*), and Norway spruce, the heritabilities tend to be somewhat lower than for the species that have denser wood, such as most hard pines, larch, and Douglas-fir.

Juvenile and mature wood specific gravities are generally very well correlated with each other. As expected, the gross heritability of wood specific gravity, determined from clonal materials, is higher, sometimes considerably so, than the narrow sense heritability, which is calculated based only upon the additive variance.

Table 8.2. A partial summary of genetic variation in wood specific gravity among individual trees and families in the conifers

Species	Reference	Inheritance pattern
The genus *Abies*		
Abies alba	Polge et al. 1971	Heritability based on parent/progeny regressions was 0.33. Based on between- and within-family comparison of 4-year-old seedlings, the heritability was 0.22
The genus *Larix*		
Larix leptolepsis	Isebrands and Hunt 1975	There is good potential to select for improvement of wood in Japanese larch. The possibilities of developing more uniform wood density within an annual ring were covered
Larix spp.	Lewark 1980	Heritability was highest for specific gravity, but the differences were too small for effective breeding
The genus *Picea*		
Picea abies	Birot and Nepveu 1979	A gross heritability of 0.29 to 0.49 was obtained
Picea abies	Ericson 1960a	There was a good correlation between ortet and ramet specific gravities
Picea abies	Kennedy 1966	A gross heritability of 0.84 was obtained for single growth rings
Picea abies	Kleinschmidt and Knigge 1967	Heritability for specific gravity was 0.84 for the whole stem
Picea abies	Lewark 1982	A high heritability was obtained
Picea abies	Mothe 1983	Heritability of wood density varied from 0.4 to 0.6
Picea abies	Rone 1970	A low heritability was obtained
Picea abies	Worrall 1975	The gross heritability varied from 0.51 to 0.70
Picea sitchensis	Brazier 1967	The high degree of variation between trees at comparable ages indicates the potential for improvement by selection
The genus *Pinus*		
Pinus caribaea	Burley et al. 1983	Heritability for mean specific gravity is 0.71, for minimum specific gravity 0.5, and for maximum specific gravity 0.5
Pinus contorta	Fries 1986	Seventy-four half-sib families showed individual heritabilities for wood density of 0.33, 0.34, and 0.76. There were no provenance differences in heritability. In breeding for biomass yield improvement, an indirect method using height growth is recommended
Pinus densiflora	Iwakawa et al. 1967	Specific gravity is only one of the traits with significant parent to progeny correlations
Pinus densiflora	Sudo 1983	Basic density and latewood percentage were chosen for selection for pulpwood development
Pinus echinata	Wiselogel and Tauer 1982	Heritability estimates for both juvenile and mature wood specific gravity are large enough to assure good genetic gains from selection
Pinus elliottii	Allen 1985	Wood density had very high heritabilities, which indicates their potential in a breeding program
Pinus elliottii	Einspahr et al. 1964	A gross heritability of 0.49 was obtained for specific gravity
Pinus elliottii	Goddard and Cole 1966	The parent to progeny correlation was 0.49. The heritability was 0.43
Pinus elliottii	Institute of Paper Chemistry 1962	Huge tree to tree variation makes large gains from selection and breeding possible. Specific gravity values varied from 0.47 to 0.65 for trees of similar age
Pinus elliottii	Sohn and Goddard 1974	Nine-year-old, open-pollinated families had a heritability of 0.68 for wood specific gravity

Table 8.2 (continued)

Species	Reference	Inheritance pattern
Pinus elliottii	Zobel et al. 1962	A gross heritability of 0.46 was obtained in an analysis pooled across two locations
Pinus kesiya	Geary 1967	The family with the heaviest wood was 13% denser than the family with the lightest wood
Pinus kesiya	Ladrach 1986a	Significant gains in wood density and wood uniformity should be obtained with a tree improvement program
Pinus nigra	Arbez and Millier 1972	A high heritability was found for wood specific gravity
Pinus nigra	Nanson et al. 1975a	Good genetic control was found. Heritability was 0.6 to 0.7. Greatest gain was by individual tree selection
Pinus oocarpa	Lima 1987	Specific gravity of *P. oocarpa* is under strong additive genetic control
Pinus patula	Kromhout and Toon 1977	There are great opportunities for breeding due to large selection differentials in breeding programs
Pinus pinaster	Keller 1973	Heritability is high for density components, particularly for differences between maximum and minimum specific gravity. Parent to progeny correlations for specific gravity are strong
Pinus pinaster	Polge and Illy 1968	Narrow sense heritabilities ranged from 0.39 to 0.76
Pinus radiata	Bannister and Vine 1981	Heritability ranged from 0.51 to 0.72 in 15-year-old open-pollinated progenies
Pinus radiata	Burdon and Harris 1973	The gross heritability was 0.75 for wood density
Pinus radiata	Dadswell et al. 1961	Gross heritability varied from 0.45 to 0.75 in clones. For seedlings, the estimates were from 0.43 to 0.57
Pinus radiata	Nicholls et al. 1964	Narrow sense heritability was only 0.16
Pinus radiata	Nicholls et al. 1976	Ramets had specific gravities similar to those of the ortets
Pinus radiata	Nicholls et al. 1980	Heritability of specific gravity was about 0.45
Pinus radiata	Shelbourne et al. 1972	Narrow sense heritability was quite high (0.55) contrary to some previous reports
Pinus taeda	Blair et al. 1975	Selection of trees with high density juvenile wood will increase the wood specific gravity of the trees in the plantation. High specific gravity trees (0.43) produced progeny of 9-year-old trees with higher pulp yields than the progeny from the lower density (0.36) parents
Pinus taeda	Goggans 1964	Heritability of wood specific gravity was about 0.9 for whole increment cores, 0.65 for latewood specific gravity, and 0.20 for earlywood specific gravity
Pinus taeda	Jett and Talbert 1982	One generation of nonintensive selection for specific gravity (one of each two individuals saved) are about 0.75 lbs/ft^3 of dry wood in 12-year-old trees
Pinus taeda	Kellison 1981	Progeny of parent trees selected for high juvenile wood density yielded 32 kg/m^3 higher basic density than progeny from unselected trees
Pinus taeda	Loo et al. 1985	Narrow sense heritabilities for age of transition from juvenile to mature wood were high enough to suggest that moderate gains are possible. Fast growth is probably related to early age of transition
Pinus taeda	McKinley et al. 1982	Analyses of plantations showed a strong degree of genetic control for wood specific gravity
Pinus taeda	Megraw 1985	Specific gravity is under a high degree of genetic control, with a narrow sense heritability of at least 0.5
Pinus taeda	Pearson et al. 1980	It is feasible to change the specific gravity of juvenile wood by selecting and breeding trees with a high specific gravity of the wood

Table 8.2 (continued)

Species	Reference	Inheritance pattern
Pinis taeda	Schmidtling 1973	Genetic variation was found to be related to intensive culture, including fertilization
Pinus taeda	Stonecypher et al. 1964	In 2- and 3-year-old open-pollinated progenies, narrow sense heritability was 0.56 to 0.72
Pinus taeda	Stonecypher and Zobel 1966	Narrow sense heritability was 0.73 for unextracted specific gravity in 5-year-old open-pollinated progenies
Pinus taeda	Talbert et al. 1982	A heritability of 0.82 was found for specific gravity in a 10-year-old control-pollinated genetic test. Juvenile wood properties can be altered through tree breeding
Pinus taeda	van Buijtenen 1964	Gross heritability of 0.64 to 0.84 occurred for 5-year-old grafts and over 0.64 on 6-year-old, open-pollinated progenies
Pinus taeda	Zobel 1970	Parents that lack "normal juvenile wood" produce progeny with the specific gravity of young wood very similar to that of the parents
Pinus taeda	Zobel 1984	The improvement in wood quality from genetics will partially offset some quality declines caused by young trees and thinnings
Pinus virginiana	Meier and Goggans 1977	Heritability varied from 0.38 to 0.41
Pinus virginiana	Rink and Thor 1973	Most wood variation was among seed sources. The narrow sense heritability of specific gravity varied from 0.13 to 0.19
The genus *Pseudotsuga*		
Pseudotsuga menziesii	McKimmy and Campbell 1982	Differences in wood density indicate that there are genetic differences in stem strength properties
Pseudotsuga menziesii	Silen 1964	Many wood properties are strongly inherited with about the same strength in different environments
Pseudotsuga menziesii	Wellwood and Smith 1962	There is significant genetic control of specific gravity

The range of wood specific gravity values in a species is frequently fairly limited but can sometimes be quite large (Fig. 8.4). These have been dealt with in other chapters. Most likely the physiological and structural requirements put limits on the range of values that are adaptive. This may be one reason why wood is a very conservative trait, so that it is often impossible to distinguish related species by their wood anatomy, while their foliage and flower morphology may be quite distinct.

8.2.3 Inheritance of Wood Specific Gravity in Hardwoods

Although the literature on the inheritance of specific gravity of hardwoods is not as abundant as for conifers, it is still substantial. A summary of some papers is given in Table 8.3. The inheritance of fiber characteristics in hardwoods was reviewed by Zobel (1965a) but there has been a great deal of information developed since then, especially for the genera *Eucalyptus* and *Populus*.

Fig. 8.4. One of the more strongly inherited wood properties is wood specific gravity. Differences in latewood content are shown for four high-density and four low-density loblolly pines from Texas. Progeny from these trees will strongly reflect the wood quality of the parents (Courtesy Texas Forest Service)

Most of the work on specific gravity of hardwoods has concentrated on the eucalypts and poplars. Although some of the studies were not as rigorous as could be desired, it appears that wood specific gravity in these two genera is inherited as strongly or stronger than in the conifers. This seems to be true for sycamore (*Platanus*) as well. The inheritance of specific gravity in *Swietenia* and teak (*Tectona*) is somewhat weaker, but results are based on fewer studies. There is a current upsurge of interest in inheritance of wood properties in hardwoods, and soon many results will be available for trees beyond the juvenile stage.

8.3 Genetic Control of Wood Properties Other than Specific Gravity

Assessment of the inheritance of fiber properties has been much less intensive than that related to the inheritance of wood specific gravity, with the exception of cell length. Evidence of inheritance patterns can be obtained from several lines of inquiry such as studies of actual genetic differences, studies of environmental effects on wood, tree to tree variation, and geographic variation.

The available evidence indicates that the inheritance of wood properties other than specific gravity is also very strong. This even includes such factors as pulp yield, tear factor and stretch of paper, which were found to differ greatly among families of the *Pinus elliottii* × *P. caribaea* var. *hondurensis* hybrid, *P. caribaea* var.

Table 8.3. A summary of some references on the inheritance of wood specific gravity in hardwoods

Species	Reference	Inheritance pattern
The genus *Betula*		
Betula pendula	Nepveu and Velling 1983b	The inheritance of basic density and shrinkage was fairly strong, while factors related to pulp yield had a low heritability. Specific combining ability was poor
The genus *Eucalyptus*		
Eucalyptus camaldulensis	Siddiqui et al. 1979	Large differences in specific gravity occur among 10-year-old provenance plantings
E. globulus	Dillner et al. 1970	Great variabilities were found in wood density and cellulose yield among 260 selected trees
E. grandis	Vital and Lucia 1980	Wood specific gravity is strongly dependent on the provenance in 5-year-old trees
E. grandis	Wang et al. 1984	Differences in density and moisture content among progenies was significant; variations in wood properties were largely due to genetic differences
E. saligna	King 1980	The total variation in specific gravity among coppice clones was 0.34 to 0.52. Selecting for higher specific gravity markedly increases wood uniformity. Gross heritability of wood density among 3-year-old clones was 0.65
E. tereticornis *E. grandis*	Hans 1976	Significant differences in specific gravity were found among open-pollinated families
E. viminalis	Otegbeye and Kellison 1980	A heritability of 0.61 for specific gravity was obtained for young trees
The genus *Gmelina*		
Gmelina arborea	Akachuku 1984	There was sufficient variation to encourage genetic improvement of wood qualities by breeding those with the best wood density
The genus *Platanus*		
Platanus occidentalis	Land et al. 1983	Heritabilities for specific gravity of stem wood were 0.68, and 0.14 for stem bark. Genetic variation among trees within stands was highly significant
Platanus occidentalis	McCutchan 1982	Selection for tree dry weight would result in gains of 29 to 37%, following roguing of the first generation seed orchard. Selection for tree dry weight will produce sycamore that has enhanced utility for energy, veneer, pulp, and sawtimber
Platanus occidentalis	Nebgen and Lowe 1982	The narrow sense heritability for specific gravity was 0.73 for 7-year-old open-pollinated families
Platanus occidentalis	Webb et al. 1973	Four-year-old, open-pollinated families had a narrow sense heritability of 0.78
The genus *Populus*		
Populus deltoides *P. nigra* *P. euramericana*	Avanzo 1964	Strong genetic differences in specific gravity occur within and between species
Populus deltoides	Farmer et al. 1966b	Narrow sense heritability was 0.62 in juvenile cottonwood
Populus deltoides	Farmer and Wilcox 1968	A heritability of 0.70 was found for specific gravity in cottonwood
Populus deltoides *P. trichocarpa*	Gabriel 1956	Clonal differences in specific gravity within species were statistically significant within the two species
Populus deltoides	Herpka 1979	A high gross heritability was found for specific gravity in 3-year-old trees

Table 8.3 (continued)

Species	Reference	Inheritance pattern
Populus deltoides	Posey 1969	Clones from eastern Oklahoma had lower specific gravity than clones from western Oklahoma
Populus nigra *P. euroamericana*	Nepveu et al. 1978	Juvenile to mature correlations of wood specific gravity of 0.72 and 0.60 occur in these two species
Populus spp.	Herpka 1973	Specific gravity decreased with decreasing latitude of provenance
Populus spp.	von Pechmann 1958	Comparison of poplar clones indicate the effect of heredity on wood structure
Populus spp. hybrids	Hamilton and Wendel 1967	There was a 25% variation in wood density among the 13 clones of hybrids studied
Populus spp. hybrids	Mohrdick 1979	Individual parents were more important in determining offspring specific gravity than species or hybrid groups in *P. alba* × *P. canescens* and *P. tremula P. tremuloides* × *P. grandidentata*
Populus hybrids	Reck and Dietrich 1976	There is a consistent difference in specific gravity between *P. robusta* and *P. alba* × *P. grandidentata* hybrids
Populus spp. hybrids	Stevens et al. 1983	Robusta had a higher specific gravity than *P. trichocarpa* and *P. deltoides* hybrids. Basic density is genetically improvable in a poplar selection program
Populus trichocarpa	Cech et al. 1960	Specific gravity is strongly inherited in this species
Populus trichocarpa	Reck 1974	A gross heritability of 0.56 was found
The genus *Quercus*		
Quercus petraea *Q. robur* *Q. rubra*	Nepveu 1984a	The gross heritability of specific gravity was 0.56 in *Q. petraea*, 0.59 in *Q. robur*, and 0.37 in *Q. rubra*; this might lead to high gains in breeding
Quercus robur	Nepveu 1984b	Genotypic control is strong for width of the earlywood zone and intermediate for vessel percentage, while environment more strongly influences fiber percentage
The genus *Swietenia*		
Swietenia macrophylla	Chudnoff and Geary 1973	Large differences in specific gravity were found among progenies, but there was no correlation with mother trees
The genus *Tectona*		
Tectona grandis	Purkayestha et al. 1973	The environment had a greater effect than seed source on the specific gravity of the mature wood

bahamensis, and *P. cubensis* families (Wright 1987). Calculating the genetics of pulping and papermaking traits is recommended by Wright, who reports that it can be done using micropulping. Such information integrates both the morphological and chemical properties of wood.

8.3.1 Latewood Percent

Although the average wood properties in conifers are important, the difference between earlywood and latewood can be so striking that one often needs to assess earlywood and latewood separately. A good starting point is the evaluation of the proportion of latewood. Latewood has been studied rather intensively in relation to wood specific gravity with which it is strongly correlated, but assessments of the actual inheritance of the proportion of latewood are not numerous.

Differences in the earlywood-latewood ratio among seed sources, when trees are planted in the same location, appear to be slight (Rees and Brown 1954; Saucier and Taras 1967; Kennedy 1971; Worrall 1975). In loblolly pine, the percent latewood increased with a more northerly origin of the parents, although the differences were not statistically significant. In the jack pine study reported by Kennedy (1971) there was no systematic trend in the ratio related to provenance, and provenances showed an interaction with test location. For example, the Smoky Falls Ontario Provenance developed significantly more latewood at Petawawa than at Acadia, largely because its earlywood formation was retarded, causing most of the wood to be formed during the summer.

Heritabilities of percentage latewood vary greatly; for example, Goggans (1962) reported heritability estimates for loblolly pine ranging from 0.25 to 0.92, values that were greater than the heritability of wood specific gravity. However, in slash pine, Squillace et al. (1962) found that the heritability of latewood percent was only moderate, having a narrow sense heritability of only 0.26 and a broad sense heritability of 0.48. These values are similar to the gross heritability of 0.53 reported for slash pine by Einspahr et al. (1964). Inheritance of latewood percent appears somewhat variable in radiata pine, with heritabilities ranging from 0.15 to 0.72 (Dadswell et al. 1961; Nicholls et al. 1964; Nicholls et al. 1980); these values are similar to those reported for loblolly pine by Goggans (1962). Kennedy (1966) found a gross heritability of 0.85 for latewood percent in Norway spruce.

As reported in detail in Chapter 7, the formation of latewood is influenced by environmental factors such as nitrogen fertilization (Paul 1959; Williams and Hamilton 1961; Posey 1965). Despite the effect of environment, percent latewood appears to be under fairly strong genetic control. This is important since it is one of the primary factors that control the average tracheid cross-sectional dimensions which are so important to product quality. Not only are cross-sectional cell dimensions affected by the proportion of latewood, but tracheid length is also influenced somewhat since the two properties are slightly correlated according to Goggans (1962). The correlation between cell length and latewood is sometimes reported as positive and sometimes negative, depending on the investigator (Jackson and Green 1957; Jackson 1959; Ericson 1960a; Zobel et al. 1961; Goggans 1962). The general concensus, however, is that latewood cells are slightly longer than those of earlywood.

8.3.2 Cell Length

Cell length differences among stands are sometimes considerable and are nearly always large among trees within stands (Fig. 8.5). Cell length is strongly inherited. A number of reviews related to the inheritance of cell length appeared in the 1960s (Dinwoodie 1961; van Buijtenen 1965; Smith 1967), but much work has been done since then. Evidence for hardwoods is somewhat scant, but those studies reported indicate that there is a moderate genetic control of cell length. A striking example of a large increase in fiber length was that obtained through

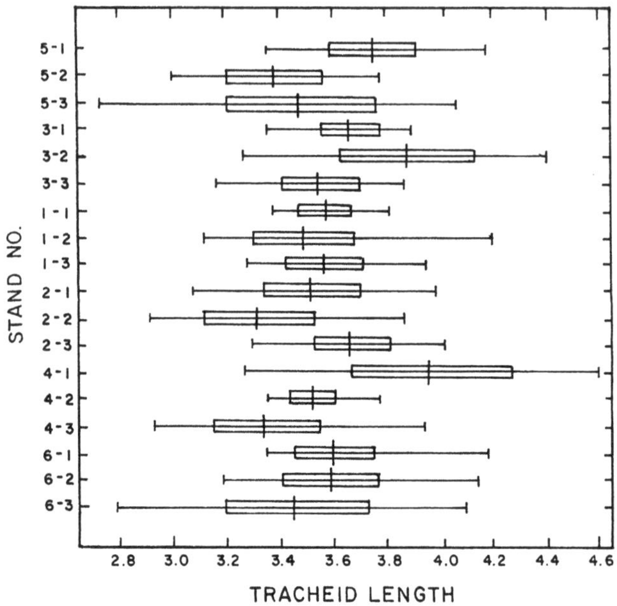

Fig. 8.5. Cell length can vary from stand to stand but it also varies a great deal from tree to tree within a stand, as indicated here for loblolly pine. Both of these types of variation are largely inherited, especially the tree to tree variability (Courtesy Texas Forest Service)

Table 8.4. A partial summary of the inheritance of cell length in hardwoods and conifers

Species	Reference	Inheritance patterns
Hardwoods		
Eucalyptus camaldulensis	Rudman et al. 1969	There are marked fiber length differences among clones
Eucalyptus viminalis	Otegbeye and Kellison 1980	The family heritability for fiber length was 0.42
Fraxinus americana	Armstrong and Funk 1979	The southern sources and the polyploids had the longest fibers. There are long vessel elements and fibers in polyploids
Populus deltoides	Farmer and Wilcox 1968	The gross heritability of fiber length is 0.36
Populus deltoides	Gabriel 1956	Statistically significant within-species differences were found among clones
Populus deltoides	Posey 1969	Strong clonal differences in fiber length occur. Clones from eastern Oklahoma had the longest fibers
Populus hybrids	Hamilton and Wendel 1967	There were no significant differences in fiber length among 13 clones studied
Populus hybrids	Mohrdick 1979	There was no significant difference in fiber length among a number of hybrid progenies
Populus polyploids	Bakulin 1980	Tetraploid poplars had larger cells
Populus polyploids	Einspahr et al. 1963	There is a gross heritability of 0.86 for ring 30 and 0.50 for average fiber length in *P. tremuloides* triploids

Table 8.4 (continued)

Species	Reference	Inheritance patterns
Populus tremuloides	Einspahr 1967	Fiber length is under moderate genetic control in 5-year-old clones
Populus trichocarpa	Gabriel 1956	Statistically significant within-species differences were found
Populus trichocarpa	Reck 1974	A gross heritability of 0.71 was found for cell length
Conifers		
Picea abies	Kennedy 1966	A gross heritability of 0.38 was reported for tracheid length
Picea abies	Rone 1970	A high heritability was found for cell length
Pinus contorta	Henderson and Petty 1972	Tracheid length is 20% shorter in the southern coastal provenances
Pinus echinata	Posey et al. 1970	Seedlings from different sources had different fiber lengths
Pinus echinata	Wiselogel and Tauer 1982	Substantial genetic gains in tracheid length can be expected from selection based upon the heritabilities obtained
Pinus elliottii	Allen 1985	There was no significant variation in tracheid length among families, showing that there is no appreciable genetic variation
Pinus elliottii	Einspahr et al. 1964	Gross heritability of cell length was 0.73
Pinus elliottii	Zobel et al. 1962	Gross heritability of cell length was 0.56
Pinus hybrid (*P. rigitaeda*)	Koo and Hong 1967	Tracheids are longer in the *P. rigida* × *P. taeda* hybrid than in the parents
Pinus pinaster	Keller 1973	Heritability of cell length is very low
Pinus ponderosa	Echols 1973	There was no effect of elevation of seed source on tracheid length
Pinus radiata	Dadswell 1960	Tracheid length of the parent was carried over to the new plant when rooted cuttings were grown. Gross heritability was 0.66
Pinus radiata	Harris 1961	Tracheid length differences among trees were 3.4 to 4.6 mm for mature wood; none of the differences could be attributed to environmental variation
Pinus sylvestris	Dorn 1969	The more northerly the origin, the shorter the tracheids
Pinus sylvestris	Miler et al. 1979	Trees from the Netherlands had the longest tracheids and those from Finland the shortest
Pinus taeda	Stonecypher et al. 1973	A narrow sense heritability of 0.44 was found for 3-year-old, control-pollinated progenies

polyploidy in *Populus* spp., where van Buijtenen et al. (1958) found fiberlength to be increased by 21–26%.

In conifers, the evidence for inheritance of tracheid length is more complete, with narrow sense heritabilities ranging from 0.28 to 0.97 and broad sense heritabilities from 0.56 to 0.86. A partial summary of some of the more recent information on inheritance of tracheid length in given in Table 8.4.

Differences in cell length in conifers related to source of seed occur regularly as demonstrated in Scots pine, lodgepole pine, and shortleaf pine. The patterns differ among species, and it is not possible to make any generalizations about the correlation between cell length and seed source. Tracheid length is sometimes reported to be positively related to specific gravity but Harris (1961) did not find a relationship between tracheid length and specific gravity in radiata pine.

The reported heritabilities of tracheid length vary from weak to strong, and seem to be somewhat weaker than those reported for wood specific gravity. In the hardwoods, the influence of polyploidy on cell size is confirmed for artificially produced tetraploids of a number of poplar hybrids.

8.3.3 Tracheid Diameter and Wall Thickness

Studies of open-pollinated progeny tests of loblolly pine by Goggans (1962, 1964, 1965) were made on the double cell wall thickness, radial tracheid diameter, radial cell width, and tangential cell width, in both the earlywood and latewood of the same annual rings. Heritability values ranged from nonsignificant to 0.84. Although there is definite evidence for genetic control for these cell dimensions, it does no appear to be as strong as the control of tracheid length. However, van Buijtenen et al. (1968) found significant differences in tracheid lumen diameter and wall thickness among 8-year-old loblolly pine clones, selected for high and low wood specific gravity. For the wood of hybrid larch, Chowdhury (1931) reported that the lumen diameter and wall thickness of the tracheids was intermediate between those of the parents, while Mergen and Furnival (1960) demonstrated genetic control of tracheid wall thickness in the hybrid between *Pinus thunbergii* and *Pinus densiflora*. The average wall thickness of the hybrid exceeds that of *P. thunbergii*, but falls within the range of *P. densiflora*.

In a recent study, Otegbeye and Kellison (1980) found family heritabilities of 0.82 and 0.94 for fiber diameter and fiber wall thickness in young *Eucalyptus viminalis*. These represent some of the highest values reported related to inheritance of fiber dimensions.

In summary, it appears that both wall thickness and cell diameter are rather strongly inherited, based upon the rather small number of species studied. Particularly in conifers, indirect evidence from the inheritance of wood specific gravity and percent latewood strongly supports the strong inheritance of cell dimensions.

8.3.4 Chemical Composition

The genetic control of the chemical composition of wood has not been studied as intensively as that of the specific gravity or the anatomical properties of wood. The main chemical properties that have been studied are lignin, cellulose, and extractives contents, although a small study on inheritance of the sugar content of the wood of loblolly pine was reported by Zobel (1971). The sugar content among families was essentially constant; polysaccharide differences were not closely related to specific gravity. It was suggested that breeding for chemical differences should be done only when they are related to yield improvement.

Knowledge of the inheritance of cellulose content is rather limited. Zobel et al. (1966) and Jett et al. (1977) found that considerable tree to tree differences in

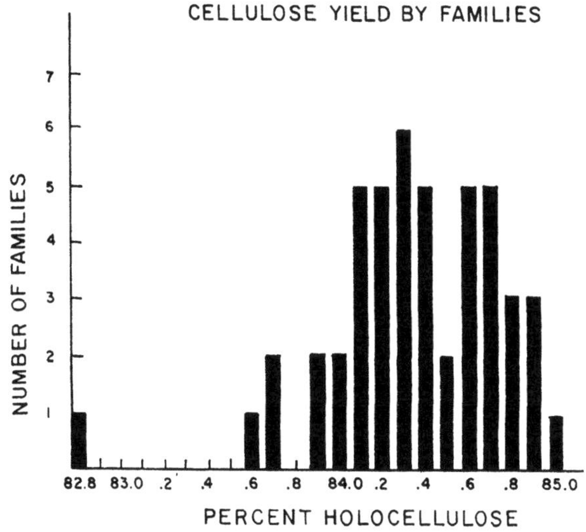

Fig. 8.6. Cellulose yield varies greatly by individual trees and by families. This includes both alpha- and holocelluloses. Percentage yield differences are shown for loblolly pine. Good gains can be made by using vegetative propagation, but because of its low narrow sense heritability only very small genetic gains are achievable using a selection and seed orchard program

holocellulose as well as alphacellulose existed in loblolly pine, but that family differences were small, and this characteristic was so poorly inherited that it is impractical to breed for increased cellulose content through a selection program (Fig. 8.6). Good gains have been made in genetic improvement of cellulose content using vegetative propagation in the eucalypts (Campinhos 1980). The Aracruz company was able to improve pulp yields from 47 to 51% just by using high holocellulose producers in their rooted cutting program. For Douglas-fir and western hemlock (*Tsuga heterophylla*) Wellwood and Smith (1962) found that almost all differences in percentage of cellulose appear to be caused by tree to tree differences, which probably are inherent (Fig. 8.7).

The accuracy of methods for microdetermination of cellulose is inadequate. However, one can make the assessment by determining the lignin content, for which a more time-consuming, but highly accurate technique is available. Using this method van Buijtenen et al. (1968) found significant differences in lignin among clones in loblolly pine which were highly correlated with pulp yield. The differences were small, however, i.e., 1 to 2%, and probably not sufficient to justify an intensive breeding program. Additional evidence for inheritance in lignin has been reported for slash pine and quaking aspen by Einspahr et al. (1964), who found that lignin is strongly inherited, with a broad sense heritability of 0.72.

Significant differences in cellulose and lignin contents among provenances of *Pinus contorta* grown in Germany were noted by Schütt (1958), while Dadswell et al. (1961) failed to find significant heritabilities of these chemical properties in radiata pine. Mixed results were also obtained with poplars by Schönbach (1960), who observed significant clonal differences, while Cech et al. (1960) did not detect a strong genetic component.

A large amount of rather intensive work has been devoted to the genetic control of extractives and oleoresins in wood. Studying variation among five different populations of radiata pine, Bannister et al. (1962) found no differences between

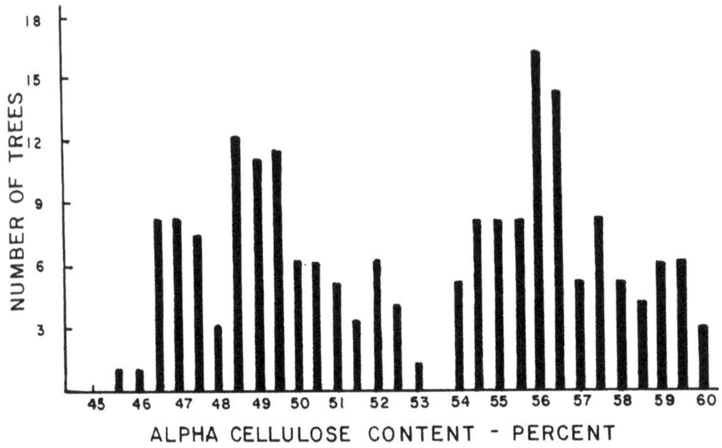

Fig. 8.7. Tree to tree differences in alpha-cellulose are very large, as indicated for this group of pine trees, all of which are the same age. As for holocellulose, these differences can mainly be exploited through use of a vegetative propagation program. There is no explanation for the bimodal pattern shown

two New Zealand populations but noticed significant differences in mean α-pinene content among the three California populations. Similarly, Zavarin et al. (1968) reported considerable difference in β-pinene, longifoline, and β-phellandrene among four provenances of *Pinus kesiya* when grown in Zambia. On the other hand, Barrett et al. (1964) found no differences in oleoresin yield among seven slash pine provenances grown in Louisiana.

Differences in oleoresin yield, viscosity, and the relationship with growth traits among open-pollinated families of slash pine were studied by Mergen et al. (1955) and Franklin (1974). They found strong differences among families. Surprisingly, it was reported that resin yields tended to be associated with large limbs.

In 1982, Strauss and Critchfield demonstrated that β-pinene was under the control of a small number of genes in the xylem of knobcone x Monterey pine hybrids (*P. attenuradiata*). High β-pinene content was partially dominant over low content. Since the mean levels of both 3-carene and α-pinene were increased when β-pinene was low, the β-pinene appeared to be under direct genetic control, rather than the result of differences in α-pinene production.

In summary, it appears that percentage lignin is under strong genetic control, but the range of variation is small. Cellulose content also has a strong genetic component but its inheritance pattern is such that selection is ineffective, although vegetative propagation can utilize the differences among clones (Jett et al. 1977; Brandão 1984) and meaningful genetic gains have been obtained in the eucalypts. Chemicals such as oleoresins are under strong genetic control of only a few genes and can be manipulated fairly readily.

8.3.5 Spiral Grain

Only limited studies have been made on the inheritance of spiral grain but there are large differences as shown in Fig. 8.8. It is difficult, however, to assess and measure accurately. Dadswell (1960) reported a broad sense heritability value of 0.66 for samples comprising the second to eighth ring inclusive from the center of the stem of vegetatively propagated *Pinus radiata*. When only the seventh and eighth rings were considered, the heritability decreased to 0.28. Narrow sense heritabilities of spiral grain were reported to be quite high by Nicholls et al. (1964) from data obtained with 10-year-old open-pollinated progeny of *Pinus radiata*. Using the fourth ring from the pith, a heritability of 0.44 was calculated. However, when spiral grain was measured at the seventh ring from the pith, the heritability was reduced to 0.35. The subject of spiral grain was thoroughly reviewed by Zobel (1965 b) who reported on the early studies such as those by Champion (1929). In his review and book, Harris (1984, 1988) reported that spiral grain is under moderate genetic control based upon research, e.g., by Nicholls et al. (1964) with radiata pine, Mikami (1973) with Japanese larch, and Birot et al. (1980) with *Eucalyptus dalrympleana*.

In loblolly pine, Zobel et al. (1968 b) found that spiral grain was not serious enough to cause economically important degrade in solid wood products. Most of the genetic variation appeared to be nonadditive, and it is most serious near the center of the tree. Arbez et al. (1978), working with maritime pine (*P. pinaster*) and Calabrian pine (*P. nigra* var. *calabrica*), found spiral grain oriented to the left

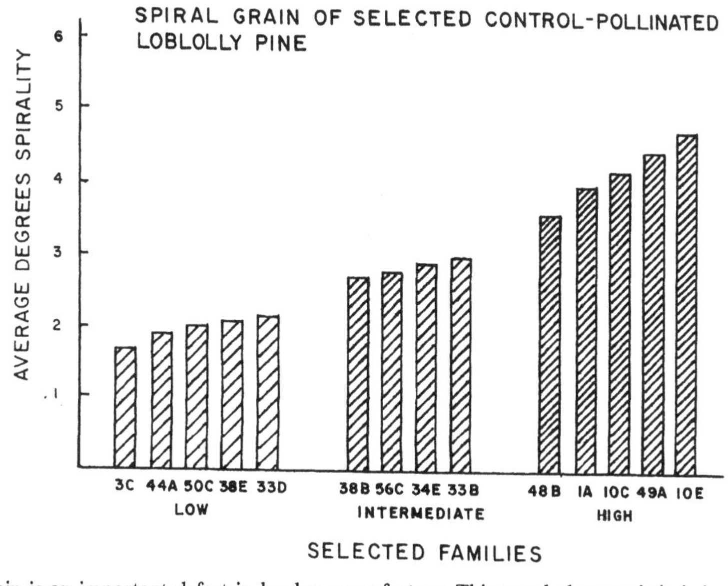

Fig. 8.8. Spiral grain is an important defect in lumber manufacture. This wood characteristic is inherited relatively strongly, based upon the limited number of studies made. Large differences are present among families, as indicated for 14 loblolly pine families, supporting the concept of the presence of a significant genetic component in this characteristic

in both seedlings and grafts. Broad sense heritabilities were 0.43 and 0.61, while narrow sense heritabilities were 0.16 and 0.58. Authors such as Allen (1985) suggest that genetic variation in slash pine can be exploited to improve spiral grain. Individual tree selection would be the most efficient to improve this characteristic.

The narrow sense heritability of spiral grain was 0.66 in beech (*Fagus sylvatica*) (Teissier du Cros et al. 1980). Grain angle was predominantly to the right, but individuals with left-handed grain or without spiral grain could also be found. Their work, like that of many others, emphasizes that individual tree variation of spiral grain is very large.

Other patterns of grain variation are occasionally important, such as the curly-grained individuals in some hardwoods (Harris 1988). Heikinheimo (1951) found that curly-grained parents of birch produced about 50% curly-grained progeny. Such trees can be of great value to the furniture industry, although their value drops rapidly if they become numerous from mass propagation. Some additional examples of inherited grain abnormalities were described in maple (*Acer pseudoplatanus*), and color variants were found in maple and *Ulmus glabra* (Conrad 1977).

For *Eucalyptus saligna*, Barr (1983) reported that parent trees with a decorative wavy grain produced progeny with the same feature. A major defect in species such as sweetgum (*Liquidambar styraciflua*) is interlocked grain (Harris 1988; see Fig. 1.12). In a study of 180 trees, Webb (1969) found important differences in interlocked grain among trees of the same age in a stand growing on a uniform site. Although it was not proven by tests, Webb reported that the pattern is such that a selection program to reduce interlocked grain might be successful.

8.3.6 Other Wood Properties

Several other wood properties of importance have been studied for their inheritance patterns, but not enough is known about them to warrant a separate section in this book. One of the most exhaustive studies was done by Shelbourne (1969) on compression wood of loblolly pine. It was inherited very strongly, yielding a heritability of 0.95 for the arc sine of the compression wood percent; this value is much higher than the heritabilty for stem straightness. Straightness was not measured on the same individuals as was compression wood, but the correlation with the straightness of family members was weak ($r = -0.17$), indicating that much of the strong inheritance of compression wood was not due to its correlation with the straightness of the tree. Burdon (1975) obtained similar results as Shelbourne (1969) with *Pinus radiata*, confirming the strong heritability of compression wood. The subject has been reviewed in detail by Timell (1986). In contrast, Einspahr et al. (1964) found that compression wood was inherited fairly weakly in slash pine. In eastern cottonwood, Posey (1969) found fewer gelatinous fibers in trees originating in eastern Oklahoma indicating less tension wood in this source. The increase in amount of gelatinous fibers further west is due, in part, to an increase in the amount of tree lean from east to west due to winds.

Heartwood formation in radiata pine was found to be inherited, with a narrow sense heritability of 0.20 and broad sense heritabilities ranging from 0.30 to 0.45

Fig. 8.9. Although the extractive content of wood usually has a strong genetic component, little is known about the propensity of trees to form excess amounts of heartwood, as in the *Pinus caribaea* shown. Resin infiltration usually is most severe in tropical climates and causes a severe defect in the wood produced (Courtesy of G. van Wyk, South Africa)

(Nicholls and Brown 1974). In tanoak (*Lithocarpus densiflora*) there was no evidence of a genetic effect on the occurrence of heart stain (Prestemon 1967). However, although the deposition of heartwood, as shown in Fig. 8.9 is common, very few documented inheritance patterns are known for this often severe defect.

Although few studies as such are available on the inheritance of fibril angle, it is shown to be highly correlated with tracheid length (Echols 1958) and wall thickness (Hiller 1964), making it probable that it is under moderate genetic control. Some evidence of genetic control of fibril angle was obtained by Mergen and Furnival (1960) in a study of the hybrid between (*Pinus thunbergii* and *Pinus densiflora*. In the only study we could locate, which reported directly on the inheritance of fibril angle, Keller (1973) found that in *Pinus pinaster* the heritability of fibril angle was not significant.

Other wood properties, such as shrinkage with collapse (Sesbou 1981) have been studied. Usually the variability among trees is very important; for example, over 40% of the total variability for shrinkage with collapse is of this type. A major problem encountered in a number of tropical hardwoods, and especially in the eucalypts, is end-splitting at the time of, or after harvesting. It has been possible to develop strains of trees, or clones, in which splitting was minimal. Genetic control was obvious, as discussed by Kromhout and Toon (1978) for *Eucalyptus*. They devised specialized methods to rate potential trees to overcome the splitting tendency. Another example has been reported by Barr (1983) who found that sawing quality of *Eucalyptus saligna* could be used as a guide to select trees for a breeding program. He reported this characteristic to be more associated with stands than with individual trees within stands.

Response to wood decay also appears to have a genetic component. A body of evidence is accumulating that the response of a tree to wounding and the subsequent infection by compartmentalizing the discolored wood, sets up barriers to further progress of the pathogen. Genetic differences in this capability have been demonstrated in poplar clones and maples (Garrett et al. 1976; Shigo et al. 1977; Santamour 1979).

In a recent study on yellow poplar (*Liriodendron tulipifera*), Lowerts and Kellison (1981) found a single-tree heritability of 0.49, indicating that discoloration and decay resistance were under moderately strong genetic control. In loblolly pine, Schmidtling and Amburgey (1982) determined that decay susceptibility of the wood is inversely related to its specific gravity. There is a strong genetic basis for the correlation between specific gravity and decay susceptibility; breeding for high specific gravity would reduce decay.

Heat of combustion of various woody materials has become of interest because of increased fuel costs. Howard (1973) examined the range of values for various southern pine materials which ranged from 8600 Btu/oven dry pound for stemwood to 5965 Btu for dried black liquor. In a study with American sycamore, Ezell et al. (1983) found considerable differences among families in BTU production over and above those due to differences in wood specific gravity.

Although not a wood property as such, the amount of bark has a major effect on utilization of a species. As reported elsewhere, the average percentage of bark

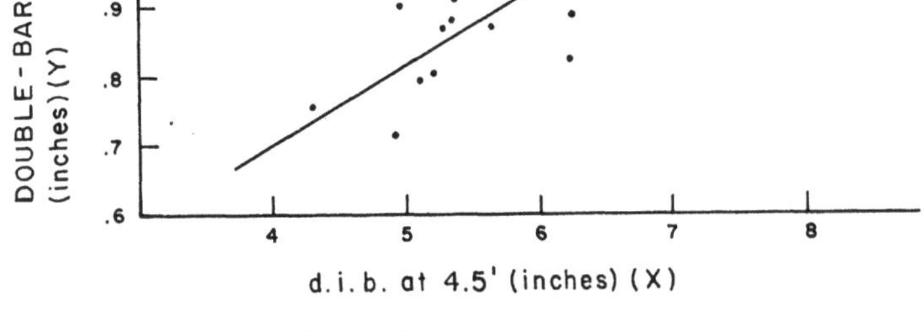

Fig. 8.10. Bark thickness is one of the more strongly inherited wood characteristics. To properly assess it, however, tree size must be taken into account. Note how closely bark thickness of the 8-year-old pine from the 20 trees within one control-pollinated family is related to *d. i. b.* (diameter inside bark). Faster growing families usually have thicker bark than slower growing ones, but there are many instances of the faster growing (large size) families having very thin bark. This trait is strongly inherited

varies with species and especially with tree size (Fig. 8.10). Normally, 10% bark is expected for mature trees but young trees have much more, especially the pines. According to Pederick (1970), 10-year-old loblolly pine has 31% bark in the coastal plain and 50% bark in the piedmont area. Bark thickness is strongly inherited, which is of considerable importance. Pederick (1970) states "bark thickness is one of the most strongly inherited tree characteristics... heritability was estimated to be about 0.65."

8.4 Interrelationships Among Traits

The interrelationships among different growth, form, and wood traits are not only of scientific interest, but are of crucial importance to the tree breeder. If, unknown to the breeder, two traits are related to each other, a change in one trait will cause an inadvertent change in the second. Some wood property traits are related but many are not; this was reported for tanoak by Prestemon (1967) and for *Pinus taeda* by Zobel et al. (1961a; see Fig. 1.15). Another example has been given by Harris (1961) who found that there was no correlation between tracheid length and specific gravity in radiata pine (Fig. 8.11). Several kinds of interrelationships have been mentioned throughout this chapter and in some detail in Chapter 1.

If one selects for high wood specific gravity, and if it were negatively correlated with growth rate, one would unintentionally select for slow growth as well. Such a correlated effect could be either favorable or unfavorable: if in the foregoing hypothetical example a newsprint mill would be interested in breeding for low

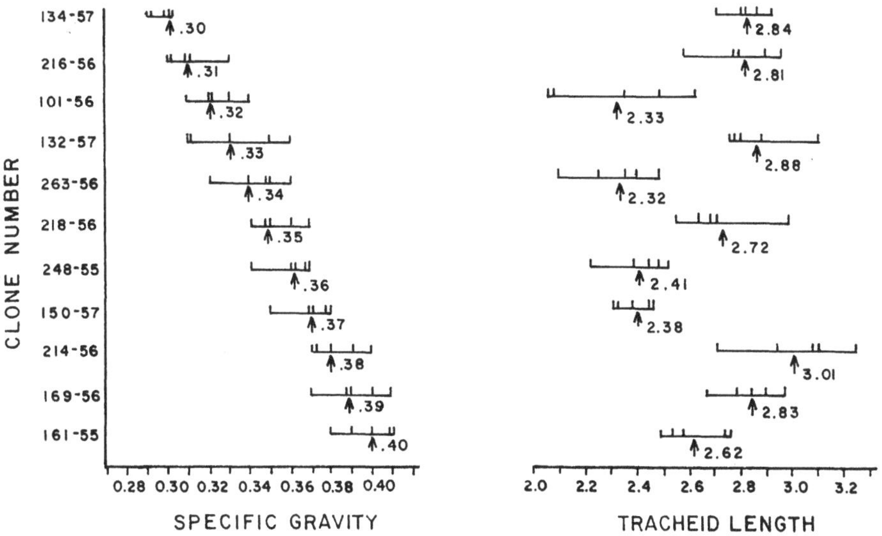

Fig. 8.11. The independence of specific gravity and tracheid length is shown for 11 clones of slash pine. The clones are arranged in order of increasing specific gravity; note the random pattern of tracheid length

wood specific gravity, they would obtain an increased volume production as well. In most tree improvement programs the situation is usually not that simple, since it is often necessary to improve several traits at once. In that case the progress made in each trait is highly dependent on the relationships among the traits.

Theoretically, index selection is the best way to achieve improvement in several traits simultaneously. It is accomplished by calculating a score or index for each tree and selecting the trees with the highest index. The index is obtained by multiplying each trait by a weighting factor and computing the index as the sum of the resulting values. This method was perfected by animal breeders (Hazel and Lush 1942), but it is equally applicable to trees. If two traits are strongly and negatively correlated with each other, it may not be possible to improve them both simultaneously, and one has to emphasize the trait of greatest value or settle for a compromise solution.

Sometimes gains in wood properties can be obtained by selecting for traits other than wood properties. One example of this was reported by Fries (1986) who states "indirect selection for dry stem biomass was more effective using height growth than wood density."

One of the most useful statistics for dealing with the so-called correlated responses is the coefficient of genetic prediction (CGP) described by Baradat (1976). The computational method is shown in Table 8.1. The CGP relates the selection intensity in one trait to the correlated response in a second trait in exactly the same way as the heritability relates the selection intensity to the gain in a single trait. In other words, if one selects a population that is one standard deviation above the average in specific gravity, the resulting change in growth rate will be the appropriate CGP times one standard deviation in growth rate. By looking at a table of CGPs one can readily see how selection in any one trait will affect the others. Examples are given in Tables 8.5 and 8.6, based on a study of Norway spruce (*Picea abies*) by Birot and Nepveu (1979).

8.4.1 Genetic Relationships Between Specific Gravity and Growth Traits

In Chapter 5 much of the literature relating growth rate to specific gravity has been summarized. No attempt was made, however, to separate genetic from environmental effects. It is important to make this distinction. Since it is easy to be overwhelmed by numbers, it might be best to take one study as an example. Let us therefore use Birot and Nepveu's (1979) tables (Table 8.5 and 8.6) on Norway spruce as the example, supplementing them with other data when appropriate. A cursory examination of Table 8.5 shows that for this species both the specific gravity of the inner wood and the weighted specific gravity are negatively correlated with measurements of growth rate, such as diameter. The genetic correlation is usually the strongest, the environmental correlation intermediate, and the phenotypic correlation the weakest. The correlations of wood properties with 1974 height were negligible. The CGPs show a similar pattern: the core specific gravities are weakly and negatively correlated to height and diameter.

How does this compare to the general literature? It seems to be fairly typical of some species, although it certainly is not an absolute rule. Negative genetic cor-

Table 8.5. Genotypic, environmental, and phenotypic correlations among traits (adapted from Birot and Nepveu 1979)

	Specific gravity outer wood		Nematode damage 1972		Bud break 1972		Circumference 1967		Diameter 1975		Total height 1968		Total height 1974	
	A	B	A	B	A	B	A	B	A	B	A	B	A	B
Specific gravity inner wood	0.90^{XX} 0.14^{XX} 0.42^{XX}	0.81^{XX}					-0.30 -0.12^{XX} -0.11^{X}	-0.10			-0.12 -0.19^{XX} -0.21^{XX}	-0.23		
Weighted specific gravity			-0.04 -0.06 -0.06	-0.21	-0.16 -0.10^{X} 0.07	-0.06			-0.56^{XX} -0.21^{XX} -0.16^{XX}	-0.17			0.19 -0.08 0.02	0.19
Nematode damage 1972					-0.39^{XX} -0.05 -0.16^{XX}	-0.32^{X}			0.00 0.14^{X} -0.08	-0.15			-0.40^{XX} 0.12^{XX} -0.02	-0.43^{XX}
Bud break 1972									0.50^{XX} -0.06 0.02	0.22			-0.23 -0.11^{X} -0.12^{X}	-0.19
Circumference 1967									-0.05 0.69^{XX} 0.71^{XX}	0.42^{XX}	0.86^{XX} 0.87^{XX} 0.86^{XX}	0.78^{XX}	0.62^{XX} 0.62^{XX} 0.60^{XX}	0.71^{XX}
Diameter 1975											-0.11 0.72^{XX} 0.70^{XX}	0.19	-0.31 0.67^{XX} 0.57^{XX}	0.01
Total height 1963											-0.63^{XX} 0.72^{XX} 0.70^{XX}	0.56^{XX}	0.59^{XX} 0.47^{XX} 0.44^{X}	0.39^{XX}
Total height 1968													0.78^{XX} 0.69^{XX} 0.71^{XX}	0.69^{XX}

[a] The three numbers in each cell are respectively the genetic correlation r_g, the environmental correlation r_e, and the phenotypic correlation r_p. A, analysis performed taking family effects into account; B, analysis performed ignoring family effects. Using method B only r_g is given, since r_e is identical for both methods and r_p is very close.

X Statistically significant at the 5% level.

XX Statistically significant at the 1% level.

Table 8.6. Coefficients of genetic predicition with family effects taken into account (first line), and all families confounded (second line)[a]

Ortets / Ramets	Circum-ference 1956	Circum-ference 1968	Nematode damage 1952	Ice damage	Bud break 1951	Bud break 1952	Bud break 1954	Core specific gravity 1956	Core specific gravity 1976	Height 1949	Height 1956
Diameter 1975	0.12 0.12	0.02 0.01	0.06 0.03	0.13 0.10	−0.02 −0.05	−0.08 0.04	−0.05 0.00	−0.12 −0.11	-0.19^{X} −0.15	−0.10 −0.12	−0.02 0.03
Nematode damage 1972	0.18 0.19	0.18 0.21^{X}	−0.07 −0.05	−0.03 −0.00	−0.03 −0.01	0.00 −0.01	−0.09 −0.06	−0.15 −0.08	-0.27^{X} -0.25^{X}	0.17 0.17	0.05 0.04
Bud break 1972	−0.18 −0.16	−0.10 −0.10	-0.34^{X} -0.37^{X}	0.51^{XX} 0.54^{XX}	0.57^{XX} 0.59^{XX}	0.69^{XX} 0.68^{XX}	0.63^{XX} 0.62^{XX}	−0.06 −0.04	−0.06 −0.04	-0.37^{X} −0.34	0.03 0.14
Specific gravity inner wood	−0.07 −0.12	−0.08 0.00	0.00 0.00	0.03 0.06	−0.02 0.04	−0.02 −0.04	0.03 −0.06	0.21^{X} 0.16	0.24^{X} 0.24^{XX}	0.03 0.00	0.04 0.07
Weighted specific gravity	−0.20 −0.25	−0.10 −0.19	0.00 0.02	0.04 −0.01	0.03 −0.02	−0.06 −0.06	0.03 −0.03	0.38^{XX} 0.24	0.42^{X} 0.44^{XX}	−0.06 −0.10	−0.09 −0.02
Height 1968	−0.11 −0.11	-0.18^{X} -0.18^{X}	−0.03 0.03	0.09 0.05	−0.03 −0.05	−0.06 −0.06	−0.04 −0.06	0.02 −0.04	−0.02 −0.02	−0.10 −0.01	−0.13 −0.08
Height 1974	−0.05 0.03	−0.10 −0.11	−0.16 −0.16	−0.07 −0.08	-0.20^{X} -0.19^{X}	-0.19^{X} −0.16	-0.24^{X} -0.21^{X}	−0.01 −0.08	−0.02 −0.02	−0.10 −0.05	0.07 −0.07

[a] After Birot and Nepveu (1979). Further explanation see Table 8.5

relations of growth to wood density have been reported for *Eucalyptus camaldulensis, Populus trichocarpa,* sycamore, loblolly pine, slash pine, radiata pine, Scots pine, and Douglas-fir (Squillace et al. 1962; van Buijtenen 1963; Stonecypher and Zobel 1966; Dorn 1969; Shelbourne et al. 1972; Webb et al. 1973; Reck 1974; Sohn and Goddard 1974; Siddiqui et al. 1979; Nicholls et al. 1980; McKimmy and Campbell 1982). Weak or moderately positive genetic correlations have been found in loblolly pine, slash pine, maritime pine, and poplar hybrids (Thor and Brown 1962; Polge and Illy 1968; Shelbourne et al. 1972; Stonecypher et al. 1973; Avanzo 1974; Sohn and Goddard 1974). It is not uncommon to find a negative correlation between specific gravity and diameter and a positive correlation between specific gravity and height (Squillace et al. 1962; Shelbourne et al. 1972; Sohn and Goddard 1974).

A physiological trait that has been shown to affect wood specific gravity is date of bud break. Most studies have been done with Norway spruce (Mergen et al. 1964; Lacaze and Polge 1970; Thiercelin 1970; Worrall 1970; Birot and Nepveu 1979; Nepveu and Birot 1979). Early flushing families tend to have a lower proportion of latewood, a lower wood specific gravity, and greater ring width. In a similar study of Douglas-fir, McKimmy (1959) found a positive correlation between early flushing and specific gravity.

The effect of time of bud burst was also studied in Douglas-fir by Kennedy (1970). Variation in time of flushing accounted for more than a quarter of the total variation in specific gravity, with early flushing trees having the greatest specific gravity. He summarizes that early flushing is associated with high specific gravity and this characteristic can be used to help locate potential high density trees. Exactly the same observation was made for spruce by Thiercelin (1970).

The genetic relation between growth rate and specific gravity can best be summarized as follows: commonly, there is a negative genetic correlation between growth rate and specific gravity, although variability is large among species and even within different geographic areas and races within the same species. The correlations are generally not strong, making it often possible to find desirable combinations of growth rate and specific gravity.

8.4.2 Relationships Between Wood Specific Gravity and Its Components

The components of wood specific gravity are quite different in conifers and hardwoods and will therefore be discussed separately.

Our interest here centers on the genetic relationships among wood properties which are best illustrated by the work of Goggans (1964), who made one of the most detailed analyses published to date. His pertinent relationships are summarized in Table 8.7.

The phenotypic correlations are somewhat lower than expected, but are otherwise reasonable. The importance of the specific gravity of the earlywood itself in affecting other traits is surprisingly high. This is probably a reflection of the young age of the material used. The genetic correlations tend to be much higher than the phenotypic ones, in some cases considerably more than 1.00. This probably reflects a bias in the estimation of the variance and covariance components.

Table 8.7. Phenotypic and genetic correlations between wood specific gravity and its components, based on a Georgia open-pollinated progeny test of loblolly pine (The genotypic correlations are in parentheses)[a]

	Whole core specific gravity	Latewood specific gravity	Earlywood specific gravity
Latewood traits			
Percent latewood	0.65	0.30	0.14
	(0.89)	(0.29)	(0.64)
Latewood specific gravity	0.47	–	0.04
	(0.67)	–	(0.69)
Double tracheid	0.21	0.40	−0.05
wall thickness	(NC)	(NC)	(NC)
Radial lumen	−0.04	−0.29	0.04
diameter	(−0.01)	(−0.57)	(1.72)
Radial tracheid	0.06	−0.04	0.01
width	(NC)	(NC)	(NC)
Tangential tracheid	0.03	0.00	−0.09
width	(−0.76)	(−0.85)	(2.50)
Earlywood traits			
Earlywood specific gravity	0.37	0.04	–
	(0.77)	(0.69)	–
Double tracheid	0.22	0.13	0.30
wall thickness	(NC)	(NC)	(NC)
Radial lumen	−0.12	0.18	−0.34
diameter	(−0.52)	(0.82)	(−1.13)
Radial tracheid	−0.09	0.19	−0.30
width	(−0.52)	(0.83)	(−1.10)
Tangential tracheid	−0.05	0.01	−0.06
width	(−0.37)	(0.20)	(0.59)

NC, Not calculated because of negative variance component.
[a] Adapted from Goggans (1964).

It appears that in this case the specific gravity of the earlywood itself is the most important component, which is in turn primarily determined by the diameter of the earlywood tracheids. Goggans' work confirms the negative phenotypic correlation of specific gravity with fiber width and the positive correlation with fiber wall thickness described by Zobel et al. (1961 a).

In a phenotypic study of slash pine, van Buijtenen (1964) obtained similar results. Percent earlywood was the most important factor affecting wood specific gravity, both in juvenile and mature wood, but the amount of cell wall material in the latewood, as measured by light transmission under the microscope, was also important in mature wood. According to Squillace et al. (1962), specific gravity in slash pine was only weakly correlated with latewood percent (r = 0.30 to 0.37). In radiata pine Nicholls et al. (1980) found genetic correlations between latewood percent and average wood specific gravity, ranging from 0.44 to 0.67.

Although it exerts its influence through its effect on tracheid dimensions, compression wood is sufficiently important to merit a separate consideration. Compression wood was reported to be strongly inherited (Shelbourne et al. 1969) and also to be somewhat negatively correlated with specific gravity. This is contrary to the author's experience and the reports by Timell (1986) which indicate a positive association between specific gravity and compression wood (Chap. 4).

The relationships are totally different in hardwoods. Here, the proportions of different cell types such as vessel elements, tracheids, libriform fibers, and medullary ray parenchyma are the important factors, as reviewed by Mottet (1965). At the time very little was known about the inheritance of the proportion of these cells, and only a few studies have been reported since then. Crist and Dawson (1975) found a distinct difference in percent vessels between *Populus "tristis* #1" and *P. "Northwest"*. The lower percent vessels was associated with a higher wood specific gravity. A full discussion of environmental effects related to wood properties is given in Chapter 6.

8.4.3 Relationships Among Other Wood Properties

Cell length is probably the second most important wood property after specific gravity. However, information on its relation to other wood properties is somewhat scarce, although some data are summarized in Chapter 1. In slash pine, tracheid length appears to be negatively correlated with growth rate and wood specific gravity (Strickland and Goddard 1966). In contrast, Jackson and Strickland (1962) report that tracheid length of loblolly pine was not correlated with specific gravity. In loblolly pine Zobel (1961a) found negative correlations of specific gravity with tracheid length and width; Goddard and Strickland (1964b) reported a negative, but low ($r = -0.12$) correlation between these two wood properties for slash pine. Some of the most detailed data for loblolly pine were reported by Goggans (1964) with the pertinent information summarized in Table 8.8.

Growth rate and wood specific gravity are negatively correlated with tracheid length. Tracheid length is positively correlated with various measurements of fiber width, such as lumen diameter, tangential cell width, and radial cell width. The relation with wall thickness is positive in latewood, but virtually absent in earlywood. A negative genetic correlation of tracheid length with specific gravity was reported by Matziris and Zobel (1973), but positive correlations were found with cell diameter and tree height. It is somewhat puzzling that the correlation of tracheid length with volume was almost zero.

Fibril angle is important because of its relation to the strength of the cells. The average fibril angle of loblolly pine earlywood tracheids is generally higher than that of latewood tracheids. The angle decreases from the pith to the cambium in latewood, but remains approximately constant in the earlywood (McMillin 1973). It increases with increasing wall thickness in the latewood of both slash pine and loblolly pine (Hiller 1964), and decreases in slash pine with increasing tracheid length (Echols 1955).

Not too much is known about juvenile to mature correlations of wood properties, since many progeny tests are either too young or have been sampled for wood properties only once. For specific gravity, the relationship is sometimes fairly strong, since the correlation between parent specific gravity and that of juvenile progeny is usually quite strong (van Buijtenen 1963; Matziris and Zobel 1973).

Table 8.8. Relation of fiber length to other wood properties based on a Georgia open pollinated progeny test of loblolly pine. (After Goggans 1964)

	1959 Latewood tracheids Length	1959 Earlywood tracheids Length
Entire core		
Specific gravity	0.02[a]	−0.15
	(0.01)[b]	(−0.17)
Percentage latewood	0.00	−0.05
	(−0.16)	(−0.42)
1959 Latewood		
Percentage in ring	0.00	0.06
	(−0.28)	(−0.19)
Width	−0.08	−0.01
	(−0.44)	(−0.49)
Specific gravity	0.33	0.17
	(0.98)	(0.98)
1959 Latewood tracheids		
Double wall thickness (DWT)	0.32	0.26
	(NC)[c]	(NC)
Radial lumen diameter	0.00	0.17
	(−0.90)	(−1.47)
Radial width (RW)	0.14	0.18
	(NC)	(NC)
Tangential width	0.23	0.23
	(0.50)	(0.10)
Length	–	0.68
	–	(0.89)
DWT/RW	0.20	0.08
	(0.88)	(0.94)
1959 Earlywood		
Width	−0.11	−0.11
	(−0.15)	(−0.37)
Specific gravity	−0.28	−0.37
	(−0.01)	(−0.33)
1959 Earlywood tracheids		
Double wall thickness	0.04	−0.02
	(NC)	(NC)
Radial lumen diameter	0.33	0.38
	(1.10)	(1.21)
Radial width	0.34	0.37
	(1.18)	(1.28)
Tangential width	0.14	0.16
	(1.22)	(0.90)
Length	0.68	–
	(0.89)	–
DWT/RW	−0.22	−0.29
	(−0.68)	(−0.86)

[a] Phenotypic correlations were based on 254 d.f.; $r_{0.05} = 0.12$; $r_{0.01} = 0.16$.
[b] Genotypic correlations are in parentheses.
[c] Not calculated because one $\hat{\sigma} m^2$ was negative.

Nepveu and Birot (1979) obtained a strong juvenile to mature specific gravity correlation in 23-year-old Norway spruce. They found that adult specific gravity could be predicted from the specific gravity of a single ring at the juvenile stage.

Wood moisture content is strongly inherited in *Eucalyptus viminalis* (Otegbeye and Kellison 1980) and loblolly pine (Matziris and Zobel 1973). It is also highly negatively correlated with wood specific gravity. This leads to the interesting situation that with increasing wood specific gravity, the green weight of the wood remains relatively constant, since water is being replaced by wood substance.

Wood brightness is also under genetic control (Wilcox 1975) and is correlated with low compression wood, low specific gravity, and low latewood percent. This relationship is of considerable practical importance for the production of mechanical pulp.

Changes of the inheritance of wood properties with age were studied by Nicholls (1967b) in radiata pine using individual growth rings. The trends were different for different traits. The gross heritability of tracheid length increased from about 0.2 near the pith to a maximum of 0.5 around age 7 gradually declined again to 0.2 by age 14. Wood specific gravity started with a heritability of 0.6 near the pith, declined to 0.13 around age 9 or 10, and then increased again to 0.6. The heritability of the deviation of the grain angle from the tree axis declined from 0.6 at the pith to 0.1 around the tenth ring. The meaning of such changes in heritability is not clear.

According to one theory (Franklin 1979), a stand of trees goes through three phases: a juvenile genotypic phase, a mature genotypic phase, and a codominance-suppression phase. The heritabilities are initially high, possibly as the result of differences in seed size and germination rate, but decrease as these effects wear off. When competition sets in (the mature genotypic phase), heritabilities rise again as inherent differences in competitive ability are expressed. Subsequently, heritabilities drop again as the stand begins to approach the carrying capacity of the site (the codominance-suppression phase). The theory is intriguing and has some interesting implications for early selection: it might be more effective under conditions of intense competition. Although this theory is still controversial and not strongly supported by available data, it is clear that if early selection is to be effective, there needs to be meaningful heritabilities at the age at which selection takes place and at the age at which the trees are harvested. In addition, the coefficient of genetic prediction needs to be large enough to allow economically attractive genetic gains. These considerations are important for predictions of the correlation of wood properties with the age of the trees.

8.5 Controlling Wood Properties by Breeding

As we have shown, most wood properties are moderately to strongly inherited and can be modified readily by breeding. Our enthusiasm must be tempered somewhat, however, by the realization that the range of variation among trees for some traits is not large, and that there is a large amount of within-tree variation and environmental variation to contend with. Defining the kind of wood desired can be a difficult task, as Paul (1953) expressed the problem:

"Objectives in tree breeding, therefore, should include the important factors that control the use requirements of the final product. Accentuation of a single phase of development may be insufficient. The apparent advantages in volume growth per tree may lead to a disadvantage in the usefulness of the lumber from it. A well-rounded background knowledge of environmental influences upon wood quality is basic to the formulation of objectives in tree breeding. A test of the quality of the wood resulting from experiments in forest genetics needs to be made at the earliest possible opportunity in order to find out whether new types of wood structure exist and, if so, which ones are most suited for further development."

Even though our knowledge of the environmental and genetic control of wood properties has advanced tremendously over the past 30 years, Paul's perception of the importance of the end use requirements and the interrelationships among the many factors involved in determining wood properties is as valid now as it was then.

Ordinarily a tree improvement program should progress through a number of fairly well-defined stages: definition of the breeding objectives, choice of species, delineation of breeding regions, identification of the best seed sources to use for each breeding region, initial tree selection, and the development of a long-range breeding program. Wood properties are an important consideration at each of these stages and a tree improvement program that does not include wood (many do not) is deficient if wood is to be the final product. During initial tree selection, wood quality deserves strong emphasis. Since wood properties are so strongly inherited, a substantial immediate gain can be made. Unfortunately, many breeding programs neglect wood quality at the initial stage or pay only lip service to it and later find that they have well-adapted, good growing trees that have undesirable wood.

The breeding objectives for wood quality have to be defined in terms of the proposed end use (Mitchell 1960; Fig. 8.12). The wood properties required for structural lumber, energy use, newsprint, and fine writing papers are quite different. The changing demands of the market-place put an additional burden on the decision maker who tries to define wood quality objectives. In the temperate zones there may be a 50–75 year lag between initiation of a breeding program and the harvest of the resulting mature tree. If the past is a guide to the future, the product requirements will have changed considerably over such a long period. The best strategy appears to be the development of a general purpose tree, that may not be best for any particular product, but is acceptable for a wide range of products. As was stressed in earlier chapters and by Blair and Olson (1984), wood uniformity is of key importance, and a breeding program that can achieve that objective is a success. They go on to state that "genetic improvement can be employed directly by selecting individuals that produce maximum dry weight..."

In the tropics and subtropics the problems of defining the desired end product are less serious, since a breeding cycle may be less than 10 years, as exemplified by the eucalypt breeding program of Aracruz in Brazil where the rotation age is 6 and 7 years (Campinhos 1980). For species such as *Gmelina arborea* the rotation age can be as low as 4 years. With such a short time, one can be more certain of the market demands.

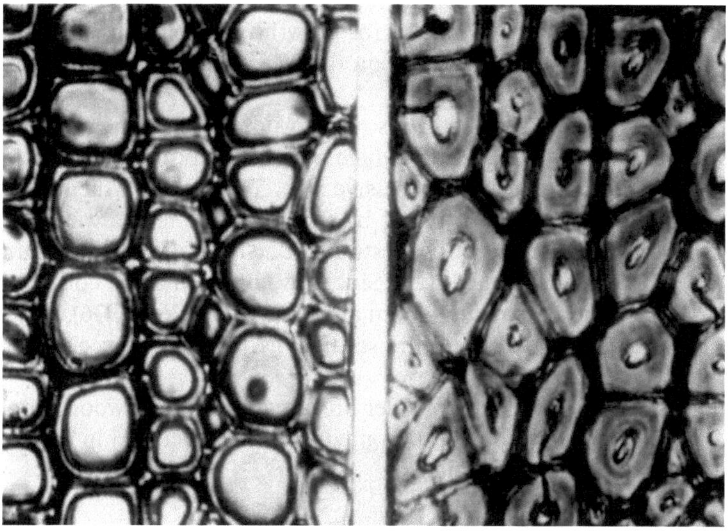

Fig. 8.12. Shown are woods of two *Eucalyptus grandis* trees of the same age growing side by side. Note the very thick cell walls of one and the thin walls of the other. Any breeding program must determine which type of cell is best. Wood properties are usually strongly genetically controlled (Zobel and Talbert 1984)

Delineation of breeding zones depends mostly on climatic and physiographic considerations, but the choice of seed sources should also be influenced by considerations of wood quality, since sometimes seed sources within a species may differ in wood quality (Nicholls et al. 1963). Particularly when exotic species are used, the properties of the wood may be unpredictable and can be determined only by actual testing in the new environment (see Chap. 2).

Developing a long-range breeding program for wood quality, or, more likely, incorporating wood quality into an existing breeding program requires two major decisions: how to cope with correlations among traits and how to deal with the selection phase and subsequent progeny testing.

Index selection is the best way to deal with selection for interrelated traits, as discussed in Section 8.4. Wilcox and Smith (1973) developed a number of such indices for different end uses of loblolly pine. However, it is necessary to approach index selection with a certain amount of caution and common sense, since the genetic and economic parameters used to calculate the index are not yet well known and are subject to considerable uncertainty. Particularly poorly known is the economics of changes in certain wood characteristics.

The progeny test and an advanced generation selection phase are considerably simpler than index development. Since wood quality traits are so strongly inherited, progeny testing is not necessary and simple mass selection is the most efficient breeding method. This method consists of selecting the best individuals, mating them, raising the next generation, and repeating the process. Because selection for wood quality is usually part of a broader program, this means that it may not be necessary to determine wood properties on all of the progenies, but only on a selected subsample, at a considerable reduction in cost. For example, Jett and Talbert (1982) propose to select for growth rate first and subsequently, after age 8, to determine specific gravity on the selected trees and to eliminate those not desired.

Another attractive strategy is to incorporate selection for wood properties in a stepwise screening program, where part of the population is removed from the breeding program at each successive step. The step at which a given property is determined is largely controlled by the cost. The least expensive tests are done first, while the more expensive tests are done later, after the size of the population has been considerably reduced. Since juvenile to mature correlations for several wood properties are high (Nepveu and Teissier du Cros 1976), this could also play a useful role in an early testing scheme, as part of an accelerated breeding program (van Buijtenen 1984).

It is of importance to consider breeding for specific wood properties. Because of its importance wood specific gravity will be discussed in detail and separately from other wood properties.

8.5.1 Breeding for Wood Specific Gravity

Of all the wood properties, specific gravity is without question the single most important one, because of its strong relationship to both yield and quality (Zobel 1963, Kano and Saito 1970; Harris et al. 1976, van Buijtenen 1982). The question now immediately arises whether its improvement should be the main objective or whether it should be a secondary trait. Zobel (1963) strongly recommended that wood be a secondary trait, and it seems that most tree improvement programs are taking this route. Similarly, Curró (1972) advocated selecting first for form, growth habit, and pest resistance, next for volume, and finally for wood specific gravity. However, the high economic value associated with wood quality improvement, and the high heritability of wood specific gravity certainly are strong arguments to put specific gravity high on the priority list. The ideal approach would be to include it in a selection index based on solid genetic and economic data.

Should one breed for high or low wood specific gravity (see Fig. 8.12)? High specific gravity is usually favored, because it contributes to strength in lumber and plywood, to high yield in pulp, and to strength properties of paper such as tear. However, the effect of pulp and paper quality is mixed (Namkoong et al. 1969). Some properties such as tear factor are increased by high specific gravity, while others such as burst, tensile strength, and apparent density are decreased. At the moment the newsprint, printing paper, and tissue industries are the main ones interested in low wood specific gravity. The trend toward selection from high specific gravity is reinforced by the increased use of thinnings and shortening of the rotation age, which are common worldwide (Harris et al. 1976). To cope with shortening rotations it is particularly attractive to select for high juvenile specific gravity (Zobel et al. 1978).

A study done by Tappi (van Buijtenen et al. 1974, 1975) showed that wood of high specific gravity produces a stronger sheet, if both high tear and high tensile strengths are needed. This can be shown most clearly by plotting tear versus tensile strengths for different wood specific gravities, using the equations gleaned from the literature by the TAPPI Forest Biology Committee. The line moves up as specific gravity increases, indicating that a more favorable combination of tear and tensile strengths can be obtained from wood of high specific gravity. Since wood of high specific gravity requires more refining, this result comes at the expense of increased energy cost. The maximum tensile strength that can be achieved is also less for wood of high specific gravity.

Breeding for increased wood uniformity is valuable regardless of the trait selected for. Uniformity can be translated into greater efficiency and better quality control at all stages of the manufacturing process. Zobel et al. (1983) have been spectacularly successful in achieving this, using vegetatively propagated eucalypts in Brazil. The initial range of specific gravity was from 0.35 to 0.85. After three stages of selection, 25 clones were chosen ranging in specific gravity from 0.52 to 0.63 and with high cellulose yields. Volume yield on the best ten clones has been increased from 36 m^3/ha/year at 7 years of age to 70 m^3/ha/year at 5.5 years of age, while pulp yields are improved per m^3 of wood by nearly 25% (Brandão 1984).

The relation between per acre yield and wood specific gravity is complex. When growth rate and specific gravity are positively correlated, there is no problem. If they are negatively correlated, the best strategy is determined by economic considerations. A smaller landowner, who sells his wood to the highest bidder has no incentive to grow trees with increased specific gravity and decreased growth rate, since he sells his wood either by volume or by green weight, which is not directly related to wood specific gravity. A large industry, however, which strives to provide as large a portion of its timber supply as possible to its own mill has a powerful incentive to achieve the best possible balance between growth rate and specific gravity. If total dry matter production is the overriding consideration, maximum volume production is usually the best strategy (Goddard and Cole 1966; Webb et al. 1973; Avanzo 1974). If wood quality is also a major consideration, a compromise solution should be used. The best answer will be unique for each case, and needs to be based on a thorough economical and biological analysis.

8.5.2 Breeding for Other Wood Properties

Wood properties other than specific gravity are rarely included in breeding programs, since the cost of assessment is higher and their importance less. Tracheid length is used as one of the criteria for selection in radiata pine in Australia (Nicholls and Dadswell 1961). Strickland and Goddard (1966) discussed the possibility of selecting for tracheid length, pointing out the problems associated with the negative correlation with growth rate and specific gravity, but a program of selecting for tracheid length was never implemented. The other major cooperative tree improvement programs in the USA do not include tracheid length in their selection criteria. In poplar hybrids, tracheid length is occasionally included as a trait to be selected. For example, Son and Chung (1970) found a hybrid between *Populus alba* and *P. glandulosa* that possesses longer fibers and higher wood specific gravity than the *P. euramericana* hybrid I-476 commonly used.

Chemical characteristics (except oleoresins) are even more difficult to manipulate. Chemical characteristics can be manipulated indirectly by selecting for morphological characteristics, particularly bole straightness and branch size (Zobel 1971). These traits have a strong influence on the amount of reaction wood, which in turn changes the amount of lignin, cellulose, and hemicelluloses. A detailed discussion of this is given in Chapter 4. Direct selection for reduced compression wood can achieve a similar result (Shelbourne et al. 1969) but no known programs use this as a major selection criterion.

Active breeding programs for increased oleoresin production exist (Mergen 1954; Franklin 1974), but their scope has been greatly reduced, due to the competition from by-products of the pulp and petrochemical industries. Work has been done on the inheritance of terpenes with several species such as *Pinus pinaster* (Bernard-Dagan et al. 1971).

8.5.3 How Are Genetic Gains Obtained Operationally?

The common, and usual, method of transferring desired wood properties is through selection and breeding of trees with the preferred characteristics. As described in this chapter, most wood properties are inherited in such a way and are

Fig. 8.13. Use of vegetative propagation greatly increases the quality of wood produced, as shown for this 5-year-old *Eucalyptus* at Aracruz, Brazil. Wood uniformity is greatly improved and wood properties, including cellulose yields, can be easily manipulated in the desired direction. The use of vegetative propagation makes maximum use of genetic differences easy to obtain

strong enough (high heritabilities) that significant gains can be obtained through standard breeding practices, such as the use of seed orchards. It is, in fact, quite remarkable how high the heritabilities are for most wood properties. They are among the highest found in all characteristics of forest trees. A prime example is the often cited specific gravity (or wood density) where meaningful changes can be obtained from a breeding program. As always, there are some exceptions such as cellulose yield (Jett et al. 1977) where very little gain can be obtained by standard selection and breeding methodology.

The situation is very hopeful with the newly developing methods of regeneration using vegetative propagation rather than seed regeneration. This trend was emphasized by Kellison (1981) and Ikemori et al. (1986). Since it enables the transfer of all the genetic potential from a selected tree to a new tree, it even allows large gains in those wood properties, such as cellulose yield, which do not respond well to conventional selection and breeding (Zobel et al. 1983; Ikemori et al. 1986). It has been possible to improve pulp yields from 47 to nearly 52% in the best rooted cuttings now being planted. That, plus specific gravity improvement from 0.48 to 0.52, enabled the wood consumption at the mill to be reduced from 4.87 m^3/ton of pulp to 3.71. The emergence of vegetative propagation will enable the rapid production of large quantities of uniform and highly desirable wood on a large scale (Fig. 8.13).

One area rarely discussed is the relative cost of changing wood by genetic or other methods, and, as a result, which should be used. The most comprehensive study has been reported by van Buijtenen et al. (1974), in which is was found that genetic control of wood specific gravity was a major component of cost effectiveness. In 1972, Zobel reviewed the cost of changing wood properties and concluded "the costs and returns from changing wood qualities are most difficult to assess." Zobel calculated (in 1972 terms) the costs of including wood in a breeding program and indicated that it is feasible in many tree improvement programs.

Sometimes gains in wood quality can be obtained by improving wood uniformity, as discussed earlier. This area was emphasized by Zobel et al. (1983) and the gains in *Eucalyptus* were highlighted (see Fig. 8.13). From 3677 clones tested at Aracruz, Brazil, 25 were selected as having desired and very uniform wood density and high cellulose yields. Uniformity is much easier to achieve when vegetative propagation is used than when seed generation is required.

8.6 Summary

Genetic control of wood properties is generally strong and can be found at the provenance as well as the individual tree level. Interaction with environmental factors is usually present, but generally not strong except in the case where exotic species are growing on sites that differ greatly from their indigenous range.

Control of wood properties (primarily wood specific gravity) through breeding is generally achieved as part of broader tree improvement programs, where improvement in wood properties is an important but secondary objective. One is thus faced with the necessity for multiple trait breeding, which can be accomplished most effectively through index selection, if the necessary background information is available. The use of a stepwise screening procedure is a viable and cost-effective alternative.

High wood specific gravity is the most universal breeding objective, followed distantly by high tracheid length. Selection for chemical composition of the wood can best be done indirectly, by selecting for form, branch habit, and the acceptable amount of reaction wood.

Chapter 9

Future Importance of Knowing and Manipulating Wood Variability

9.1 General Concepts

The importance of wood variability is closely related to the product and quality needs of the future. Numerous symposia, such as "Impacts of the Changing Quality of Timber Resources" (Porterfield and Crist 1978) or "Utilization of the Changing Wood Resource of the Southern United States" held in Raleigh North Carolina at N.C. State University in June, 1984, deal with the subject, which is of great concern to the forest industry. In 1980, Preston stated it as: "increasingly inadequate supplies of preferred species is accompanied world wide by a decreasing supply of trees of large size and high quality ... industry is continuously facing the necessity of substitution, often simultaneously, of non-traditional species, size and quality for preferred raw material. This is a powerful driving force for technological advancement which moves us ever closer to ... produce needed wood products from an indiscriminate mix of forest biomass."

In addition to the importance of wood for fiber, solid wood products and energy uses, which have already been discussed, there are many good publications, e.g. Elliot (1984), which emphasize that an understanding of variation in macrofeatures, such as color, grain deviation, and knots, is of major importance in the future utilization of wood. The encouraging thing about it is, as emphasized by Ladrach (1986b) for plantation forestry, that "there exists a great opportunity to modify wood properties to suit forest product needs ..."

The changing quality of the world wood supply is recognized and is of concern to many in ways similar to those discussed by Erickson (1978), Fahey and Starostovic (1978), Hart (1978), Youngs and Erikson (1984), and Zobel (1984).

Changes in wood must always be paired with both known and anticipated needs. This was expressed by Bond (1985) as "... of major concern is the fact that we can expect the South's raw material to be quite different from today's. The pine resource will be younger, smaller, lower density, and of inherently lower grade." The latter may be very difficult to assess accurately. As a rough estimate, Kulp (1982) suggested that in the future, 27% of the woody biomass used for industrial purposes would be for high value solid wood products, 38% for wood chips, and 35% for processing energy. These anticipated industrial needs must be assessed in relation to the changing and variable future wood supply.

The needs, possibilities, and trends of the future of wood use was clearly spelled out by Hillis (1984) who made predictions for future wood requirements. The data he presents take into account the population increase and the resultant demands for forest products. Hillis states that by the year 2000 the demand for wood will exceed the supply of wood, especially that of coniferous woods and high quality timber. The only way that a future shortfall of needed wood can be avoided is to use wood with properties different from those now available; the in-

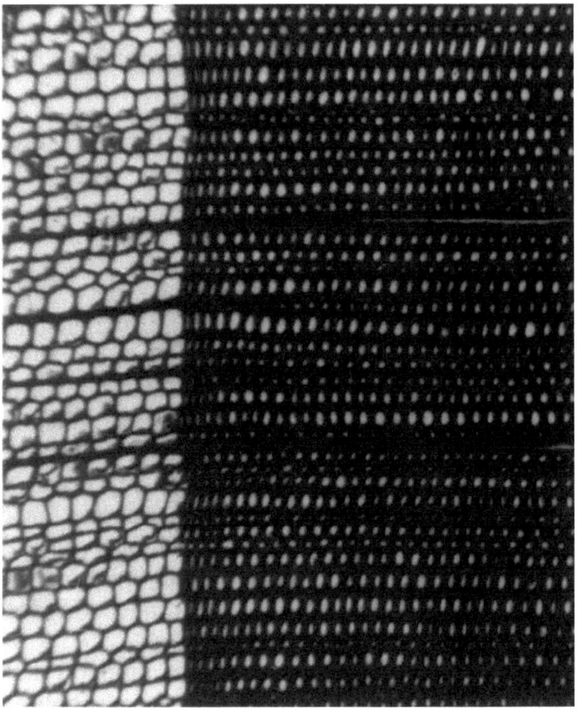

Fig. 9.1. One of the greatest needs for wood in the future is to make it more uniform. Both genetically and silviculturally, there has been some success in doing this from tree to tree. However, to reduce the extreme variation between earlywood and latewood shown within a tree is a real challenge for which there currently is no answer

dustry will need to adjust to the use of differing kinds of wood in the future, which will not only be different but in many instances will also be more variable. Hillis (1984) summarizes as follows: "consequently continuing technological developments within the forest products industries will be required to maintain the competitiveness of timber ..." In addition, we must emphasize that efforts to produce wood with more uniform properties will also require a major emphasis in the future (Fig. 9.1).

Changes in wood that will have a major effect on utilization and product properties need not be large. For example, Wilson (1964) showed that in the southern pines, a variation of 10% in earlywood content within any one species will produce greater differences in pulp quality than are usually found among the four major southern pine species. The relationship of wood properties to a changing wood supply has had wide coverage. For example, the summary by Zobel (1976) emphasized the effects of the current movement toward shorter rotations which will result in lower density wood and shorter cells as the proportion of juvenile wood becomes greater. Added to this is the trend for more complete utilization, such as use of top wood and young thinnings and the strong trend for harvesting of species once considered to be nonutilizable or of marginal value.

As the use of more variable wood becomes widespread, and technologies become more sophisticated, two apparently opposite trends appear to be developing relative to the need for wood uniformity. More uniform wood properties lead to better products and yields which result in industry becoming more efficient in producing a competitive product (Anon. 1962). On the other hand, new technologies such as production of oriented strandboard from inferior quality trees or reconstituted lumber, such as Scrimber (Coleman 1981 a, b, 1984) or Parallam (Barnes et al. 1976; Forgacs 1984) enable industry to use relatively low quality wood for high value solid wood products of wide usage. In the pulp and paper field, methodologies such as thermomechanical pulp (TMP) or chemothermomechanical pulp (CTMP) enable widespread employment of "low quality" wood, such as juvenile wood of pines or wood of certain less desired hardwood species, to make a number of highly desired or specialized paper products. Higgins and Puri (1976) report that CTMP can be used successfully with hardwoods such as young *Eucalyptus* as well as with the more commonly used conifers. In his review, Bond (1985) notes the need to develop an alternate manufacturing methodology to use the large amounts of juvenile wood that is becoming available. This must be accompanied by a review of product standards.

Even though improved technologies are being developed, the wood that will be available to the forest industry in the future often will be of marginal value for a given product, and its uniformity and utility will need to be improved. For example, juvenile wood of pines, which is available in increasing quantities, gives neither good tear strength nor good yields when pulped by the standard sulfate process. It also produces poor sawn products for *Pinus ponderosa* and for *P. taeda* (Pearson et al. 1980).

No matter what the future trends will be, it is certain that wood specific gravity will continue to be the major wood property of concern. This has been repeated many times in earlier chapters and was well stated by Veal et al. (1986): "specific gravity is a reliable index of wood quality for loblolly pine because it is highly correlated with the earlywood/latewood ratio and average fiber coarseness." Although not used too broadly, the coarseness of fibers is recognized by many pulp mill personnel as a key wood property.

One frequently quoted reason for wood variability in future utilization is a predicted greater usage of tropical hardwoods. Although we emphasized in earlier chapters that this highly variable raw material will not be as widely used as often thought, and that the exotic species which are grown in the grasslands and scrub-brushlands will be the main source of supply, numerous studies have been made on the properties of the greatly variable wood from the tropics; the results indicate that serious problems will be encountered (Mayer 1971; Fig. 9.2). For example, Palmer and Gibbs (1978) report that certain tropical hardwood species, in addition to having a very variable wood density, also have a high lignin content, and other species have a high silica content. Still other tropical species have been reported to have a variable wood, difficult to use in a standard pulping program (Palmer et al. 1983). Despite this variability, some tropical hardwoods are being pulped operationally and produce a usable paper (Gomez and Mondragon 1974). Several organizations are using mixed tropical hardwoods with considerable success. Some, such as Bratt (1971), make the point that paper and linerboard made from mixed tropical hardwoods are usually inferior in tear strength. Use of such highly variable wood is aided by new methodology such as computer-based wood identification (Wheeler et al. 1986). This reference manual enables grouping of similar woods for more effective utilization and can conceivably become very im-

Fig. 9.2. The stack of nice logs from a tropical hardwood forest looks very impressive. But the wood variation among species is enormous. Some of the logs have very good and usable wood, some have marginal wood, and a few have wood with a high silica content that make them almost impossible to use

portant in the future. Working with mixed tropical hardwoods is the ultimate in using wood with variable properties.

9.2 The Need for Wood Improvement

Wood improvement is beneficial in all forestry but it is especially critical for exotic forestry. Often the trees grown in exotic environments will produce wood quite different from that formed by the same species in their indigenous environment. This was stressed for tropical pines by Palmer and Tabb (1968) when they stated, "the outstanding finding is the variability of the pulping characteristics which, pending a deeper understanding, appears to preclude the possibility of these being forecast without making pulp from wood grown ... on the actual site ultimately to be planted." The same authors (Palmer and Tabb 1971) emphasize that wood variation caused by site factors and genetic differences is great enough to affect the final product significantly.

The increasing use of exotic species is a certain trend for the future, especially in the tropical and subtropical areas. This subject was treated in some depth in the books related to growing exotics by Zobel and Talbert (1984) and Zobel et al. (1987). Many recent publications are available on the wood of a number of species when grown as exotics, such as those by Worrall et al. (1977) on *Pinus caribaea*, by Ladrach (1986a) for *P. kesiya*, and for *P. radiata* by Bamber and Burley (1983).

The results of the studies on exotic conifers generally showed lower specific gravity, lower yield, and lower tearing strength of paper compared to the wood

Fig. 9.3. Wood from the indigenous range of most species, such as the pine shown here, is usually superior to that when the species is grown as an exotic. Notwithstanding, better forest management and better utilization technologies usually make the wood from exotic plantations competitive

now used from these species in their indigenous ranges (Fig. 9.3). Despite this, the wood of the exotic tropical pines is generally useful, as reported in many publications. Numerous similar studies on *Eucalyptus, Gmelina*, and other hardwoods indicate that the wood of young exotic hardwoods is frequently better than that from older trees growing in the indigenous range of the species. The supply of young wood of exotic hardwoods will become much larger in the near future. Such wood is easier to work with than wood of conifers because of the greater uniformity within the tree and among sites where the trees may be grown. However, individual tree differences within sites are still large, creating uniformity problems in utilization.

With the exotics, the wood is different not only because it is grown in a different environment, but also because of the high proportion of juvenile wood, knots, and reaction wood which are prevalent in fast-grown, short-rotation forest crops. Since the exotic trees often grow to a merchantable size in a few years, much of the wood harvested is juvenile wood. This is especially frustrating in the utilization of most conifers, but the consequences are less serious with hardwoods.

In the authors' opinions, even though manipulation of trees toward producing more desired properties has had relatively little emphasis in the past, it has a great opportunity in future forestry operations. For example, most tree improvement programs throughout the world emphasize the development of fast-growing, straight, adaptable, pest-tolerant trees. Frequently, however, these programs put little or no emphasis on wood properties of the trees to be produced. It is clear that foresters and those who use wood are not aware of the magnitude and importance of the variation patterns in wood, their effect on the final product, and the ways in which they can be manipulated. One objective of this book has been to try to help in such an understanding and to indicate how wood properties might be managed and controlled in the future.

9.2.1 A Specific Need – Uniformity

It is especially important to reduce the variability of wood used in the manufacturing operation. Both the yields and quality of the final product will respond positively to greater uniformity in the desired wood properties (Zobel 1984). The financial benefits from producing a uniform and predictable final product is evident from the current rapidly expanding demand for *Eucalyptus* fiber. The most frequently cited reasons for the desirability of eucalypt pulp are its consistently high quality and its predictable characteristics. The manufacturer of paper can rely upon the good quality and uniformity of this kind of pulp which results in a strong preference for eucalypt fiber.

The basic and primary need for improving wood for the future is the reduction of its variability by silvicultural manipulation and genetic control of wood properties. Much of this can be accomplished by the proper choice of species and provenances. This aspect has been thoroughly discussed in earlier chapters. The objective here is to remind the reader of their importance and especially to emphasize that improvements, some dramatic, are possible using the three methods listed above, and some improvements are already being made. For example, Palmer (1977) in response to criticisms of the quality of wood from pines grown in the tropics states that if the forest stands are handled correctly, the tropical pines will produce pulps similar to those from the southern pines in the USA. Their uniformity will be improved by selection of better trees as a source of seed; work of this type is already well underway. However, the problem of nonuniformity is still largely unsolved, both in plantation-produced wood and in individual trees from natural regeneration which often have extreme variablity among them.

Fig. 9.4. One of the greatest hopes for improving wood properties is the use of vegetative propagation. Great gains in wood properties and uniformity is currently available with some hardwoods. Use of vegetative propagation, as in the pine stand shown, is just starting but we expect it to become a standard practice. It is easy to obtain better and more uniform wood using vegetative propagation (Zobel and Talbert 1984)

Just one among hundreds of examples illustrating a high nonuniformity was given by Palmer et al. (1983), who reported on seven tropical hardwoods that had been grown in plantations in Fiji. Yields and wood properties varied so greatly among the species as to make it impossible to produce acceptably uniform products. There will always be significant variations among the tropical hardwood species and the only way to reduce it is to work with species groups with similar kinds of wood and to concentrate operations on a few of the most desirable ones. Currently, very few tropical hardwoods are being grown in plantations, but the future will show a definite increase in this activity (Cardenas 1971).

Another method of increasing wood uniformity is vegetative propagation, the importance of which is only now becoming widely realized in forestry (Fig. 9.4). Vegetative propagation makes it possible to reduce dramatically the variability in wood obtained from a plantation and to produce whole stands of trees with wood that is relatively uniform from tree to tree. Vegetative propagation not only makes possible large and rapid improvements in uniformity of the more important wood properties, such as specific gravity, but also properties such as cellulose yield (Jett et al. 1977). It is the best way of producing trees with a grain suitable for specialty, high-quality decorative products. A good example of this is the reproduction of straight-grained eucalypts which make beautiful lumber and veneer from this otherwise very crooked-grained genus. Through the use of vegetative propagation, it is possible to develop straight-grained clones which produce commercial quantities of high-quality decorative woods from species that have interlocked grain.

Although much effort is now being exerted on developing efficient methods of vegetatively propagating forest trees for operational planting, much more is needed in the future. Use of rooted cuttings is now operational for certain species of *Eucalyptus*, *Picea*, *Pinus radiata*, *Cryptomeria japonica*, *Populus*, and *Gmelina* and is in the advanced developmental stage for a few other coniferous and hardwood species.

Generally, however, the methodology of vegetative propagation is not yet operational in forestry. After it does become widespread, the wood-using industry will have a "uniformity bonanza" enabling it to produce much better and uniform quality products efficiently.

9.2.2 Possibilities and Trends for Changing Wood

At first it would appear that any attempt to make wood less variable by changing it in the desired direction might be quite impossible because there are such a large number of interacting forces which determine the kind of wood that a tree will produce. Yet, when done with care and with knowledge of the factors that influence wood variability, much improvement is possible, both by changing the environment in which the tree is grown and by genetic changes resulting from a breeding program (Fig. 9.5). Even when the raw wood is not ideal, the final product can be improved by new technologies in manufacture now available and by those under development.

Wood properties among individual trees within a species or provenance are very variable; only a couple yet studied have shown otherwise, although the amount of variability differs among them. There is always variation in wood properties within an individual tree which is difficult to change; this was emphasized in several earlier chapters. Variation in wood can also occur among geographic sources of seed or locations where the trees are grown. Sometimes the wood produced differs among sites within a location. These causes of variation must be understood and manipulated if wood with greater uniformity is to be produced in the future.

One major cause of differences in wood produced in forestry is related to the age of the tree when it is harvested; this results from the different proportions of juvenile wood. The importance of age was illustrated by Plumptre (1983) who clearly describes how wood from mature *Pinus caribaea* is much denser than that from young plantations. Thus, one most important improvement in future wood properties will result from rotation age control of the trees harvested. This one silvicultural action can greatly improve the utility and uniformity, and thus, the value of the wood produced.

Although well covered in earlier chapters, it is important to reemphasize the future and continued importance of wood density or wood specific gravity as a

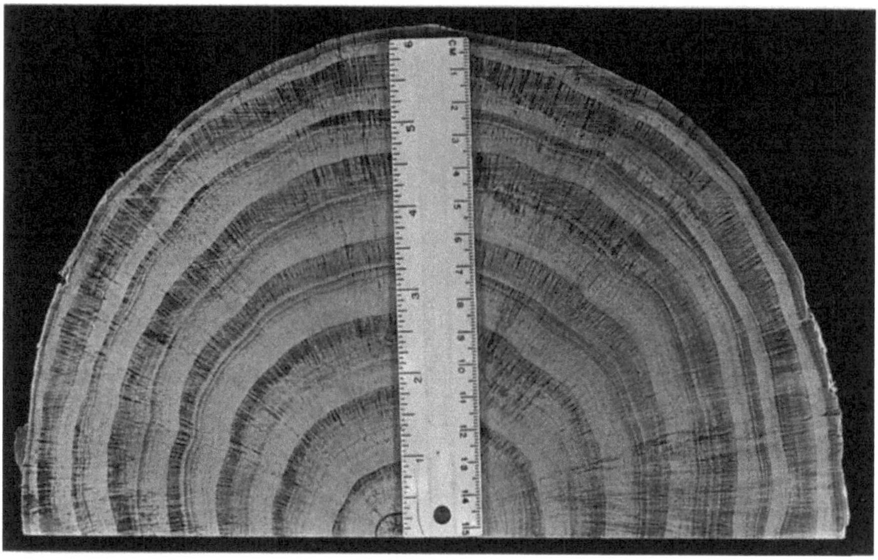

Fig. 9.5. It is essential to develop juvenile wood with higher specific gravity, as shown for the 7.5-year-old pine. Such an increase improves yields and certain strength properties for some papers and is good for some solid wood products. Despite its adverse characteristics, much wood of this type will be available to the forest industry in the future (Zobel et al. 1987)

key wood property. This idea was positively expressed by Mottet (1965) when he stated: "... among all the physical properties of wood, it is logical to select density as the reference criterion. Density is associated with the hardness of the wood, its shrinkage, its heat conductibility and its capacity for becoming impregnated ... It can be measured without undue difficulty." The future trend is to handle and buy wood by dry weight based upon species, site, and general age of stand. In progress, and to be greatly expanded in the future, is the need for dry wood weight-yield tables, such as the one developed by Zobel et al. (1969b). Specific gravity is an integral part of the determination of dry weight tables. A number of companies are switching from a volume or a green weight measuring system to inventories based upon dry weight. Thus, variations in wood density become of increasing importance.

Despite a continuing emphasis on wood density, we can expect certain other wood properties, such as number of fibers per unit volume or weight, ratios such as length to width of cells or cell diameter to lumen size, to be more greatly emphasized in the future. Greater knowledge and more sophisticated methodology may result in somewhat less emphasis on wood density per se; as a consequence, other important wood properties, some of which are components of wood density, will be studied in greater depth.

Without doubt, the easiest way to improve general wood uniformity and value is to control the amount of juvenile wood through silvicultural or other methods. This was described extensively in earlier chapters but again deserves mention as a hope for the future. It is quite clear that for economic reasons and because so

Fig. 9.6. The industries prefer wood from mature stands as represented by the photograph. The change to fast growing, short-rotation age wood is being made with reluctance. However, new technologies are making the transition easier

much fast-growing wood will be harvested from plantations of young ages in the future, the amount of juvenile wood will steadily become greater. Numerous problems related to this increase have been discussed earlier but in addition, to quote Pearson et al. (1980) "mills with limited digester volumes will encounter difficulties in maintaining production as the proportion of juvenile wood is increased." This trend was recognized by McKee (1984) who stated" there has been a recognition for several years of a general trend to increasing amounts of juvenile wood . . ." As recently emphasized by Zobel (1985): "one categorical statement can be made – juvenile wood is here to stay and will continue to make up a large proportion of the wood available from plantations."

Especially in the tropics, conifer silviculture essentially dictates thinnings, which produce practically all juvenile wood. The proportion of young trees from fast-grown plantations of both conifers and hardwoods from the tropics and subtropics is dramatically increasing and will always be great, supplying a large inventory of juvenile wood.

In the past, some industries have been strongly antagonistic to the use of large amounts of juvenile wood. Numerous articles have been published to that effect; one example is the article "Weak Wood" by Senft et al. (1985). They point out the many problems related to the use of juvenile wood (Fig. 9.6). Yet, with advanced technology and a good knowledge of the characteristics of juvenile wood, it is possible to use it for numerous products and it is even preferred for some technologies, such as TMP, CTMP, or for use in oriented strandboard. In the future, the poor connotation raised by the term "juvenile wood" will be considerably lessened.

Manufacturing techniques are being modified to take advantage of the useful qualities of juvenile wood. In hardwoods the industry has already adjusted to the use of juvenile wood, and in the conifers the technology for using juvenile wood is rapidly improving (Kellison and Zobel 1978). For example, Palmer and Ganguli (1982) state that although *Pinus caribaea* from plantations, now considered to be inferior because of its high proportion of juvenile wood, will likely yield a pulp with properties similar to those from Douglas-fir and the southern pines because of a reduction in juvenile wood as the plantations age. Also, various mixtures of pine and eucalypt wood can produce pulps especially useful for some products (Palmer et al. 1982b).

Two things are needed to properly utilize juvenile wood in the future. The first is to recognize what it is, its characteristics, and the amount present. The second is to capitalize on its good qualities by appropriate techniques and by making products for which it is best suited. Great strides are being made in this area, especially for solid wood products, but also for paper, using improved pulping methodology. In New Zealand, Bunn (1981) points out that the new wood becoming available is intrinsically more suitable for the rapidly developing mechanical pulps than the wood from older trees which is best for chemical pulps. For solid wood, methods such as high temperature drying, cutting to shorter lengths, and end gluing are making a greater and more efficient use of young wood possible. Other methods produce reconstituted products such as oriented strandboard. Scrimber and Parallam are useful products made from wood previously considered to be of little value (Barnes et al. 1976; Hart 1978; Coleman 1981a, 1984; Forgacs 1984).

Efforts are also needed to change product standards to accommodate juvenile wood and to educate the consumer about its utility. Such changes are now beginning, and quality standards for paper as well as boards are under review (Fig. 9.7). In pulp and paper manufacture, the newly emphasized pulping methods, such as TMP and CTMP, and products derived from it, such as newsprint, are opening up vast new potential uses for juvenile wood. This trend was discussed by Kellison and Hitchings (1984) who state "juvenile wood of medium density species such as the southern pines are preferred for mechanical pulps. A high proportion of earlywood, which is common in the juvenile zone of loblolly and slash pine, is preferred in the process because latewood tracheids are more often broken into stiff, poorly fibrillated fragments which compact poorly and yield weak, low density, rough sheets that are prone to lint." Some organizations now find juvenile wood satisfactory for writing grades of paper and for newsprint. Marton (1986) emphasizes that different wood species respond differently in refining energy and pulp properties, and that hardwoods give poor quality mechanical pulp.

Control of wood quality by segregation of different classes of wood in the mill was discussed by Veal et al. (1986). They describe a "3-pile" system of storing and using chips for loblolly pine. The low gravity pile (< 0.40) is from the tops of trees and low gravity plantation wood, for example thinnings, that are predominantly

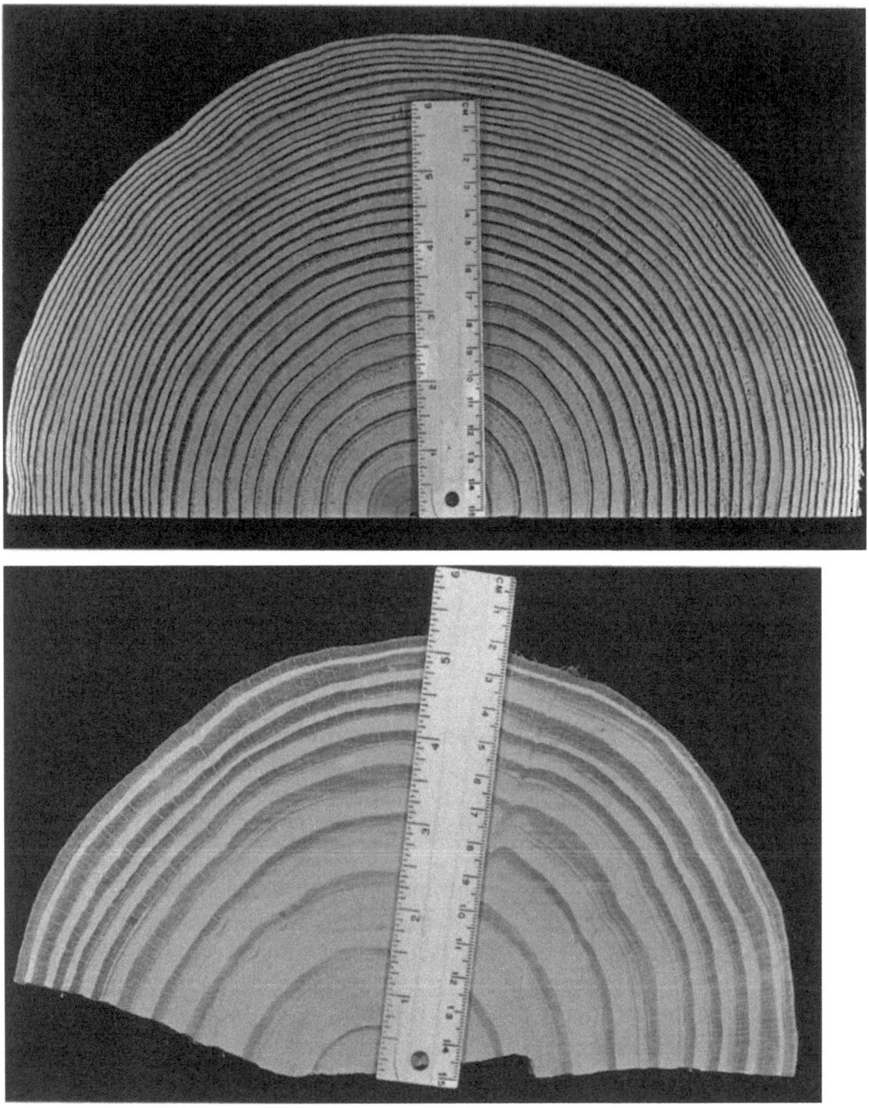

Fig. 9.7. Quality standards developed for mature timber as shown (*above*) must be drastically changed to accommodate the emerging type of fast-grown timber (*below*). Somewhat of a revolution in utilization criteria is taking place

juvenile wood. The intermediate pile, with a specific gravity of 0.40 to 0.45 is from intermediate stems and sawmill culls. The third, high density pile (>0.45) is from sawmill residuals and the butt ends of older trees. Although not yet extensively used in the industry, such a system of segregating wood will be of great value in the future to reduce variability of the furnish going to the mill. It will have a greater utility as plantation forestry increases, as emphasized by Andrews (1986) who closed his discussion on the effect of fiber morphology and chemical composition on the kraft process with the following, "... more rigid controls are strongly recommended for wood zone segregation during woodyard slashing/chipping operations during raw material procurement and preparation for paper-making." More wood segregation of this type will occur in the future.

In discussing the use of juvenile wood McKee (1984) states that the amount must either be minimized or its delivered cost must be held in line with its true worth. He notes that a closer cooperation is needed between wood procurement and the mills if this type wood is to be efficiently utilized.

9.3 Wood Properties Related to Product Quality – The Future

The large and most important subject relating wood properties to product quality has frequently been referred to throughout this book and needs to be handled in a separate book. In this short section we will merely try to point out the increasing importance of the effect of wood properties on product quality and a few actions that will be of importance in making forest products competitive in the future.

It can be stated with no contradiction that if the forest industry is to maintain a competitive position, it must be able to produce superior quality products at a modest cost. The key to achieving this is to use wood with properties that are uniform and predictable so that the quality of the final product will be well known and desired. The methods to do this have already been discussed in earlier chapters. Now the question is "what about the future?"

The prediction of product quality from a change in wood properties has attracted much attention (see Fig. 9.7). There are so many variables, some quite uncontrollable, that progress has been slow in achieving this goal. This idea was stressed by Keith (1985) in his discussion as how to define what is important in wood quality. One of the most ambitious attempts to relate wood properties with paper quality was the large study by a TAPPI subcommittee to assess quality control in the forest and in the mill (van Buijtenen et al. 1974). A systems model was developed to examine the relative efficiencies of tree improvement, silviculture, and technology in modifying the quality of pulp and paper. For southern pine kraft pulp, wood specific gravity was most important. Breeding for high wood density was desired, especially at short rotations, when high amounts of juvenile wood are harvested. Modifying the paper properties by refining and use of additives was more expensive than achieving the same quality of paper through the use of genetics and silviculture.

Such in-depth studies are needed as a guide for the effective modification of wood but they are so expensive and difficult that few have been done. Thus, the industry is not always sure of the kind of wood desired and certainly does not know the best way of obtaining it. A detailed report of the same TAPPI study by van Buijtenen (1975) describes the best way to produce linerboard and multiwall sack paper. Linerboard was not found to be as sensitive to specific gravity as multiwall sack paper, in which there is a critical lower specific gravity of the wood. In this study this specific gravity appears to be about 0.45.

One of several studies assessing product quality of future wood that will be available was carried out on "new crop" *Pinus* radiata (Corson and Foster 1981).

There has been widespread concern about the value and use of the wood available from young plantations of radiata pine. The results of comprehensive trials were reported as having "provided detailed evidence of the good pulp qualities that can be obtained from radiata and suggests that it could be rated with spruce in its potential properties ... it has confirmed that, within the limits of the wood densities of the samples examined, the influence of variations in wood density resulting from geographic location should have no significant effect on pulp quality."

Another example is Kibblewhite and Lloyd (1983) who did a study on wood, fiber, and pulping properties of 12-year-old thinnings of radiata pine. They reported the expected low tear and high burst, tensile strength, and apparent densities related to the lower wood basic densities and fiber dimensions found in young wood. As early as 1972, Kirk et al. stressed the increasing use of juvenile wood and that the "... unique properties of juvenile wood pulp could prove useful for a variety of specialty products ..." This idea was based upon the assumption that there is an economic problem: "the manufacturing cost of 1 ton of juvenile wood pulp is estimated to be 65% more than the cost of 1 ton of pulp from mature wood." Most of the cost differential results from wood procurement related to tree size and use of thinnings but part of it is directly related to the wood properties involved. This kind of information is essential as a guide to the modification of wood for future use.

Another example related to paper is the report by Hitchings (1984) on the use of juvenile wood. He states "the relationships between fiber length, cell wall thickness, fiber lumen diameter, fibril angle, and combinations of these basic measurements have demonstrated high statistical correlations with paper sheet strength properties such as burst strength, tear strength, and breaking length. The correlations between wood properties such as specific gravity and the pulp yield by a specific pulping process have not been as readily found and under some pulping conditions, no significant correlations were detected." Hitchings goes on to emphasize that from an economic point of view, an accurate prediction of pulp yield and quality based upon oven-dry weight or volume is of key importance to the industry.

In a discussion about the future wood supply and the coming high volumes of juvenile wood, McKee (1984) shows that small 4-in. diameter (young) trees contain 13.5% less fiber than 10-in. diameter pine. Pulp yield per unit dry weight of wood is less for juvenile than mature wood, making the economic penalty of using juvenile wood about 25%. Some paper properties will be improved by more juvenile wood which has increased burst and tensile strength, fold edurance, and apparent density, as reported by Semke (1984). The uniformity of formation and printing are improved by increasing amounts of juvenile wood.

The future effect of juvenile wood is especially relevant as shown by Semke (1984), who reported on yields and qualities of pulp and paper when woods of different types of loblolly pine were pulped. He states that "whether higher contents of juvenile pine fibers should be viewed as commercially good, bad or inconsequential depends on many factors... . Each paper producer should individually evaluate the relative importance of these..." In a short course sponsored in 1986 for the Forest Products Research Society, 15 papers were presented on the use and problems of juvenile wood as was a special symposium at N.C. State University in 1984. There is a large upsurge of interest in this subject related to the growing awareness by the industry that the supply of juvenile wood will become ever increasing in the future.

In their paper on the use of pine thinnings, Pearson et al. (1980) assessed the quality of wood available for solid wood products in conventional lumber production. They show the weakness and instability of juvenile wood when used for that purpose. The more recent technologies used to make oriented strandboard, Parallam and Scrimber, broaden the potential usage of the young wood but it still is a major problem for the producers of solid wood products. The manufacturer and seller of solid wood products made from the wide variety of wood properties available in the future will really need to use ingenuity and develop methodologies especially suited to the different types of wood available.

As stated several times in this book, the use of vegetative propagation in forestry is increasing and will eventually become quite standard. The opportunities provided by this system for greater change and uniformity in wood are considerable. A preview of what is possible was described in the results for *Eucalyptus* by Aracruz Florestal in Brazil. The recent paper by Ikemori et al. (1986) describes the changes in wood that have been made for better wood and more uniform wood through rooted cuttings. As a result of the changes those organizations who use rooted cuttings to supply the desired wood will be in a good competitive situation.

9.4 Nonconventional Wood Products

Another future and increasing use for wood is for what might be termed nonconventional products. This includes the use of wood for firewood in households. Firewood is really a most conventional use of wood since it has historically accounted for the bulk of wood usage. In some areas, it continues to be most important today with half the world wood consumption being for fuel, and it appears that firewood use will increase in the future as the nonrenewable energy sources become more expensive or scarce.

The common belief is that most any kind of wood can be used for firewood and that it is generally obtained from natural stands using wood unwanted for other purposes (Fig. 9.8). Although this is still largely true, it is evident that in the future more firewood will be obtained from higher quality wood and especially from plantations established for that purpose. When this occurs, the desired wood will have properties different from those for fibers or solid wood products. The best fuelwood has a relatively high density, a low moisture content, and burns slowly with little smoke and sparking.

Extensive usage of "nonconventional" wood is for industrial energy. Such use is rapidly escalating and is already employed on a large scale. In 1974, Smith made calculations of the potential for use of wood for energy in the pulp and paper industry in the southern USA. He reported that during the period of high petroleum prices, wood was cheaper for energy than oil, especially if waste wood, "green junk", was available. The value of wood energy to the pulp mills is now reduced

Fig. 9.8. Firewood usage is increasing. In the past, degenerate stands of waste wood, as shown, were used. This kind of wood is becoming short in supply in some areas, especially in the developing countries, and good wood is sometimes now being used for firewood (Zobel and Talbert 1984)

with lower oil prices but will rise again as fossil fuel prices increase. In some of the developing countries, wood and its residues are still the major source of energy for operating the conversion plants.

One prime example of the use of wood for industrial energy is in Brazil where a huge amount of charcoal is used to energize the production of steel. Desired is wood of high density, low moisture content, and a high BTU from trees that grow very rapidly. Huge plantations are being made yearly for energy production. In Brazil alone, about 100000 ha are planted yearly primarily to be used for industrial energy.

Considerable effort has been expended to develop wood with better properties for energy production (Zobel 1980a). The methods used to convert wood for industrial energy were described by Goldstein (1980a). He also describes the potential use of wood for organic chemicals. Conversion methodologies have been worked out but what the future holds for use of industrial wood energy and chemicals is highly dependent on the availability and cost of petroleum. Many researchers feel that the use of wood for charcoal will expand with time, especially as improved methods of production come into use (Paddon 1984a, b) and as waste wood can be more fully utilized. For example, Brito (1984) reports on the potential use of young pine thinnings for charcoal briquettes. Compared to the standard eucalypts, the pine produces charcoal with an inferior yield and bulk density. However, briquettes were obtained with properties suitable for both industrial and domestic use.

There are other nonconventional uses for wood in addition to a direct use for energy. Many useful organic chemicals can be made from wood; methods are known but the economics are not now favorable (Goldstein 1980a, b).

As of now, hardwoods are desired for both production of energy for industry and for chemicals. Frequently, waste and unused wood is used for energy purposes regardless of kind or quality. There is a preference for oak, higher density *Eucalyptus*, and other dense hardwoods. There is currently little movement toward expansion of specialization in developing wood especially useful for energy production, but we expect this activity to develop soon.

9.5 Summary

It is certain that wood properties available to the industry will change in the future; some changes will be dramatic (Zobel 1975b). This stems from more complete utilization of the tree, from the use of what were formerly considered "junk" species, from expansion of forestry into new environments, from extensive utilization of new species, from intensive silviculture, from extensive use of exotics grown in extreme environments where trees become of a merchantable size at young ages, and from breeding for more uniform and desirable wood. Economic pressures resulting in shorter forest rotations will have a major effect on the kind of wood available in the future as will land use changes such as replacement of forest lands with agricultural crops forcing trees to be grown on poorer quality lands, on which different species and provenances may be required.

The effect of a change of the wood resource on industry has been widely discussed by many people and each has his own ideas and opinions. One opinion was expressed by Zobel (1975b) as : "other than its effect on the total amount of cellulose, how will the changes in form of wood affect the industry? In the short term it will mean a change in methodology and revised product and quality standards. Some persons have told me that in the long term, the characteristics of the fibers and their form will not be of great importance. They argue that industry is looking to the time when the main usage of cellulose will not be in a fiber form as produced in the tree but as reconstituted cellulose and lignin products in which fiber identity will have been lost. Products will be made to the desired size and form with suitable strength, fire, pest, and water-resistant properties, much as the present day chipboard, hardboard, and flakeboard can be tailor-made for different end uses. I feel that this is an accurate assessment, and that the needs in the future will not be as dependent on fiber length, wall thickness, or cell size as in the past but more on the useful chemical constituents of wood. If timber is reduced to the form of reconstituted cellulose, large amounts of energy will be required; therefore the speed with which such conversion takes place will be dependent on the cost and abundance of energy available as well as on the timber supply. One of the brightest spots of all is that the timber supply is renewable and the industry can be largely self-supporting in energy requirements, using a high proportion of waste, and still be within limits of environmentally acceptable methods."

Another opinion was expressed by Rennel (1984) in a discussion of the future of paper in a telematic world. He states: "in summary, it is the considered opinion of the Jaakko Poyry organization that paper has a future and, what is more, quite a good one. Paper has unique advantages as a medium for communication and will be used as such even if the same functions could be performed more efficiently by electronic means. Paper use in offices and homes will be stimulated rather than depressed by the growth of computer-based office and home use systems – paper is made from a renewable resource, wood, and there are no good universal substitutes for paper in sight..."

No matter what the reasons are for changes in wood properties, the forest industry needs to determine the magnitude and causes of the changes and how they will affect the final product. From this arises the greatest need of all, how to use the changed wood to efficiently produce a product such that competing materials will not replace wood. The common platitude that "forests are renewable" is very real and very important and must never be forgotten. Handled correctly forest land can be made permanently productive. Construction materials made from wood are the most energy efficient of all so use of wood is one of the very best ways of reducing energy consumption. However, it is mandatory that the forester put an allout effort into producing uniform and high quality wood and the manufacturer must learn to efficiently use the wood that is available.

In pulping, Kellison and Hitchings (1984) state that in the future mechanical pulping will increase more rapidly than chemical pulping. This trend is already well established. Its rapidity will depend on the wood variations that are found within the various species. As shown by Palmer et al. (1982b) certain mixtures of conifer and hardwood pulps produce very desirable products.

Another trend is a greater use of wood for energy, both firewood and industrial. There may be an increased use of wood for chemicals; much depends on the cost of petroleum.

One major change in wood utilization currently evident is the inroads being made by *Eucalyptus* pulp in Europe, Canada, and the southern USA (Fig. 9.9). This trend will probably somewhat moderate but is currently very strong. For example, the 1985 Paper Tree Economics Letter, which gives analyses of forest products and economics, published an article "Move Over Softwood Kraft." They stated "the world pulp market is changing... this time, there's an important new dimension being added... Its name is *Eucalyptus*... Many softwood kraft producers feel...that rising *Eucalyptus* production is a very negative development... In terms of market shares, there are steady inroads being made by hardwoods in competition with the long dominant softwoods; the specific hardwood species taking all the growth is *Eucalyptus*". *Changes such as this caused by an emphasis on differing species will continue to occur.*

As emphasized by Laarman (1980), and stated a number of times relative to competition from South America in papers such as Zobel (1986), the competition from woods grown in the tropics and subtropics will continue to increase with

Fig. 9.9. Exotic plantations, such as the beautiful *Eucalyptus grandis* growing in South Africa, supply an ever larger amount of the wood needs of the world. Wood quality and uniformity can be controlled in such plantings (Courtesy of G. van Wyk, South Africa)

time and the industry must adjust to use of the type of wood grown there. Manufacturing methods must be adjusted, as reported by Palmer and Greenhalgh (1983) where they outlined the production of pulp and paper on a small scale, i.e., 30 tons per day or less. They list the arguments and advantages of small-scale production and its suitability in developing countries. If such proves to be workable, it will open up a whole horizon for the use of wood and other nonwoody fibrous materials now underutilized or wasted, and will make for more complete usage of the world wood resources in the future.

There has been concern expressed for the future need for use of wood and its possible replacement by competing materials. It seems proper to summarize this by a quotation by Elliot (1984) "the present century has seen the birth of a nonwood materials revolution... Light weight metals and plastics can be manufactured to defined engineering and design specifications, and this has inevitably changed the overall uses of wood. Only paper and board products have so far remained relatively immune from such competition. Despite these changes, the world demand for industrial wood has continued to rise, so that present demand. . . is equivalent on a weight basis to the annual world production of iron and steel." It is up to the forest industry to adapt to, and efficiently use, the changing types of wood that will be available in the future. As stated by Kellison et al. (1985): "the use of genetically improved planting stock, a change in the type of manufactured products and changes in log separation and processing will likely be the alternatives used to meet the challenge of the changing wood resource."

References

Aho P E, Cahill J M 1984 Deterioration rates of blowdown timber and potential problems associated with product recovery. U S For Serv Rept PNW-167, 11 pp

Akachuku A E 1984 The possibility of tree selection and breeding for genetic improvement of wood properties of Gmelina arborea. For Sci 30: 275–283

Aldridge F, Hudson R H 1955 Growing quality softwoods. II: A critical examination of the Turnbull hypothesis. Q J For 49: 260–270

Aldridge F, Hudson R H 1958 Variation in strength and density of Picea abies – specimens taken from a commercial consignment. Q J For 52: 107–114

Aldridge F, Hudson R H 1959 Growing quality softwoods. Q J For 53: 210–219

Allen P J 1977 Genotypic and phenotypic correlations of wood and tree characteristics. In: Variation of growth, stem quality, and wood properties in relation to genetic improvement of tropical forest trees. IUFRO Workshop Brisbane, Australia, 13 pp

Allen P J 1985 Estimation of genetic parameters for wood properties in slash pine in southeast Queensland. Res Note 41 Dept For Queensland, Australia, 15 pp

Amidon T E 1984 Hardwood as a substitute for juvenile pine. Util Chang Wood Res South U S N C State Univ Raleigh 1 p (Summary)

Amman G D 1970 Distribution of redwood caused by the balsam woolly aphid in Fraser fir of North Carolina. U S For Serv Res Note SE-135, 4 pp

Amos G L, Bisset I J, Dadswell H E 1950 Wood structure in relation to growth in Eucalyptus gigantea. Aust J Sci Res 3: 393–413

Andersson S 1967 Pruning investigations. II: Pruning in Scots pine. Dept For Yield and Res Royal Coll For Res Note 15, 57 pp

Ando T, Hirozami K 1961 Bulk density of branches of Cryptomeria. J Jpn For Soc 43: 280–283

Andrews E K 1986 Impact of fiber morphology and chemical composition on the kraft process and subsequent handsheet properties. R & D Conf TAPPI Proc, Raleigh, North Carolina, pp 111–119

Andrews I A, Hughes J F 1973 Variation in wood properties in 12 year-old trees of P. caribaea var. hondurensis in Trinidad – a summary report. Trop Prov Prog Res Int Coop Nairobi, Kenya, pp 532–535

Anonymous 1962 The influence of environment and genetics on pulpwood quality – an annotated bibliography. TAPPI Monograph Series 24, 316 pp (883 Abstracts)

Anonymous 1968 Population studies of Norway spruce in Poland. Pol For Res Inst, 180 pp

Anonymous 1978 Leucaena: the miracle tree. Africa 86: 75 pp

Anonymous 1979 Eucalypts for planting. FAO Series #11 Rome, 680 pp

Anonymous 1983 Progeny testing. Proc Genetics Workshop U S For Serv, Charleston, South Carolina, 597 pp

Arbez M, Baradat Ph, Birot Y, Azoeuf P, Hoslin R 1978 Variabilité et hérédité de l'angle du fil du bois mesuré à l'aide d'un traceur radioactif chez le Pin maritime et le Pin laricio de Calabre. Can J For Res 8: 280–289

Archer R R 1986 Growth stresses and strains in trees. Springer, Berlin Heidelberg New York Tokyo, 240 pp

Arganbright D G, Bensend D W 1968 Relationship of gelatinous fiber development to tree lean in soft maple. Wood Sci 1: 37–40

Arita M 1967 [Studies on the crown broadness in Cryptomeria japonica in view of genetical and ecological characteristics.] Res Bull Fac Agr Gifu Univ Japan, 25, 46 pp

Armstrong J E, Funk D T 1979 Genetic variation in the wood of Fraxinus americana. Wood Fiber 12: 112–120

Artuz-Siegel E A, Wangaard F F, Tamalong F N 1968 Relationships between fiber characteristics and pulp-sheet properties in Philippine hardwoods. Tappi 51(6): 261–267

Auchmoody L R 1972 Epicormic branching; seasonal change, influence of fertilization, and frequency of occurrance in uncut stands. U S For Ser Res Pap NE-228, 8 pp
Aytug B 1962 [Notes on wood anatomy research.] Istanbul Univ Drm Fak Derg 11A: 88–93
Avanzo E 1974 [Relative importance of basic density and basal area in the selection of clones of poplar of section Aigeiros.] Cellul Carta 25: 29–39
Baas P 1982a Leeuwenhoek's contributions to wood anatomy and his ideas on sap transport in plants. In Antoni van Leeuwenhoek 1632–1723. Editions Rodopi B V Amsterdam, pp 79–107
Baas P 1982b New perspectives in wood anatomy. W Junk Publishers, The Hague, Netherlands, 252 pp
Baas P 1984 Acid rain, wood structure and wood quality. IAWA Bull 5: 316
Bailey I W 1920 The cambium and its derivative tissues. II: Size variation of cambial intials in gymnosperms and angiosperms. Am J Bot 7: 355–367
Bailey J K, Feret P O, Bramlett D L 1974 Crown character differences between well-pruned and poorly-pruned Virginia pine trees and their progeny. Silvae Genet 23: 181–184
Baker G 1968 Effect of proximity to limb whorls on wood density. For Prod J 18(9): 112–113
Baker G, Shottafer J E 1967 Specific gravity relationships in plantation-grown red pine. 8th Lake States For Tree Impr Conf U S For Serv Res Pap NC-23, 15–19
Bakulin V T 1980 [Wood anatomy of induced polyploids of poplars.] Izvestiya Sibirgskogo Otdeleniya Akademii Nauk SSSR. Biologicheshikh Nauk 1: 25–30
Balch R E 1955 Effects of balsam woolly aphid on quality of pulp. Can Dept Agr Div For Biol Bimonthly Prog Rept 11(4): 1
Bamber R K, Burley J 1983 The wood properties of radiata pine. Commonw Agr Bur, England, 84 pp
Banch J 1980 Variation of wood structure due to secondary changes. IUFRO Meet Oxford, England, 16 pp
Banks C H 1957 A comparison of the strength of timber from the major and minor radius of eccentric trees. Commonw For Conf Aust and N Z, 10 pp
Banks C H, Schwegmann L M 1957 The physical properties of fast- and slow-grown Pinus patula and Pinus taeda from South African sources. For Res Inst Pretoria, S Africa, pp 45–49
Bannan M W 1967 Anticlinical divisions and cell length in conifer cambium. For Prod J 17: 63–69
Bannister M H, Burdon R 1969 Genetical survey of P. radiata. Rep For Res Inst New Zealand, pp 20–21
Bannister M H, Vine M H 1981 An early progeny trial in Pinus radiata 4. Wood density. N Z J For Sci 11: 221–243
Bannister M H, Williams A L, McDonald I R, Forde M B 1962 Variation of turpentine composition in five population samples of Pinus radiata. N Z J Sci 5: 489–495
Baradat Ph 1976 Use of juvenile mature relationships and information from relatives in combined multitrait selection. Proc IUFRO Joint Meet Genetic Work Parties Advan Breed, Bordeaux, France, pp 121–138
Barefoot A C 1963a Abnormal wood in yellow poplar. For Prod J 13: 16–22
Barefoot A C 1963b Selected wood characteristics in young yellow poplar (Liriodendron tulipifera). For Prod J 13: 233–239, 443–448
Barefoot A C, Hitchings R G, Ellwood E L 1965 Wood characteristics and kraft paper properties of four selected loblolly pines. III: Effect of fiber morphology on pulp. Tappi 49: 137–147
Barefoot A C, Hitchings R G, Ellwood E L, Wilson E 1970 The relationship between loblolly pine fiber morphology and kraft paper properties. Bull N C Agr Exp Sta N C State Univ Raleigh, North Carolina Tech Bull 202, 88 pp
Barger R L, Ffolliott P F 1971 Effect of extractives on specific gravity of southwestern ponderosa pine. U S For Serv Res Note RM-205, 4 pp
Barker J A 1973 Location effects on heritability estimates and gain predictions for 10-year-old loblolly pine. PhD Thesis School For Res N C State Univ Raleigh, North Carolina, 105 pp
Barner H 1966 Classification of seed sources. Seminar For Seed and Tree Improv Denmark, 19 pp
Barnes D, Admiraal L, Pike R L, Mather V N 1976 Continuous system for drying lumber with microwave energy. For Prod J 26(5): 31
Barnes D P, Sinclair S A 1983 Time-related changes in specific gravity and moisture content of spruce budworm killed balsam fir. Can J For Res 13: 257–263
Barnes R D, Woodend M A, Schweppenhauser M A, Mullin L J 1977 Variation in diameter growth and wood density in six-year-old provenance trials of Pinus caribaea on five sites in Rhodesia. Silvae Genet 26: 163–167

Barr N A 1983 Sawing qualities of Eucalyptus saligna as a guide to selection of trees for breeding and seed collection. Fast Growing Trees Silvicultura 32: 800–801

Barrett J P, Bengston G W 1964 Oleoresin yields for slash pines from seven seed sources. For Sci 10: 159–164

Barrichelo L E, Brito J O 1976 Wood of Eucalyptus with prime material for the pulp and paper industry. PRODEPEF 13: 145 pp

Barrichelo L E, Brito J O 1979 Variabiliadade radial da madeira de Pinus caribaea var hondurensis. IPEF 18: 81–102

Barrichelo L E, Brito J O, do Couto H T, Campinhos E C 1983 Densidade básicá, teor de holocelulose e rendimento em celulose de madeira de Eucalyptus grandis. Fast Growing Trees Silvicultura 32: 802–808

Barrichelo L E, Kageyama P Y, Speltz R M, Bonish H J, Brito J O, Ferreira M 1977 Studies on provenances of Pinus taeda established for industrial improvement. IPEF 15: 1–14

Barse R C, Laidly P R 1980 Wood specific gravity of plantation red pine little affected by spacing. U S For Serv Res Note NC 251, 2 pp

Bauch J 1980 Variation of wood structure due to secondary changes. IUFRO Conf Div 5 Oxford, England, p 16 (Abstracts)

Bauch J 1983 Alterations in stem and root tissue of conifers due to pollution influence. IUFRO Conf Div 5 Madison, Wisconsin, p 21 (Abstracts)

Bay R R, Boelter D H 1963 Soil moisture trends in thinned red pine stands in northern Minnesota U S For Serv Res Note LS-29, 3 pp

Beineke W F 1977 Corrective pruning of black walnut for timber form. Purdue Univ For Nat Res 76 West Lafayette, Indiana, 5 pp

Bendtsen B A 1978 Properties of wood from improved and intensively managed trees. For Prod J 28(10): 61–72

Bendtsen B A 1984 Mechanical and anatomical properties in individual growth rings of plantation-grown eastern cottonwood and loblolly pine. Util Chang Wood Res South U S N C State Univ, Raleigh, North Carolina, 1 p (Summary)

Bennett F A 1969 Spacing and slash pine quality timber production. U S For Serv Res Pap SE-53, 9 pp

Bernard-Dagon C, Fillon C, Pauly G, Baradat P, Illy G 1971 Les terpènes du Pin maritime: espects biologiques et génétiques. I. Variabilité de la composition monotérpenique dans un individu, entre indidivus et entre provenances. Ann Sci For 28: 223–258

Bentum A L, Côté W A, Day A C, Timell T E 1969 Distribution of lignin in normal and tension wood. Wood Sci Technol 3: 218–231

Bertoloti G, Simoes J W, Nicolielo N, Garnica J B 1983 Efaitos de differentes metodes e intensidades de desbaste na produtivadade de Pinsu caribaea var. hondurensis. IPEF 24: 47–54

Besley L 1964 The significance of fiber geometry and distribution in assessing pulpwood quality. Tappi 47(11): 183A–184B

Bethel J S 1977 Wood as a renewable material. S A F Ann Meet – For for Am Future: Beyond the Bicentennial New Orleans, Louisiana, 77–80

Bhat K M, Bhat K V 1983 Wood properties of 1-year-old Eucalyptus tereticornis. IUFRO Conf Div 5 Madison, Wisconsin, 1 p (Summary)

Biblis E J 1969 Tensile properties of loblolly pine growth zones. Wood Fiber 1: 18–28

Bingham C W 1983 Faster growth: greater utilization. IUFRO Conf Div 5 Madison, Wisconsin, 3–11

Birot Y, Dufour J, Ferrandes P, Teissier du Cros E, Azoeuf P, Hoslin R 1980 Variabilité de l'angle du fil du bois chez quelques feuillus: Hêtre, Chêne et Eucalyptus dalrympleana. Ann Sci For 37: 19–36

Birot Y, Nepveu G 1979 Variabilité clonale et liaisons ortets-ramets dans une population d'Epicéa. Silvae Genet 28: 37–47

Birt D V, Harris J M 1970 Preliminary assessment of wood density in provenance trials of Pinus taeda, P. elliottii, and P. caribaea. N Z For Serv FRI Rotorua New Zealand For Prod Rep 295, 7 pp

Bisset I J, Dadswell H E 1949 The variation of fibre length within one tree of Eucalyptus regnans. Aust For 8: 86–96

Bisset I J, Dadswell H E, Wardrop A B 1951 Factors influencing tracheid length in conifer stems. Aust For 15: 135–146

Blair R L, Zobel B J, Franklin E C, Djerf A C, Mendel J M 1974 The effect of tree form and rust infection on wood and pulp properties of loblolly pine. Tappi 57(7): 46–50

Blair R L, Zobel B J, Barker J A 1975 Predictions of gain in pulp yield and tear strength in young loblolly pine through genetic increases in wood density. Tappi 58(1): 89–91

Blair R, Zobel B, Hitchings R G, Jett J B 1976 Pulp yield and physical properties of young loblolly pine with high density juvenile wood. Applied Polymer Symp 28: 435–444

Boden D I 1982 The relationship between timber density of the three major pine species in the Natal midlands and various site and tree factors. 35th Ann Rep Wattle Res Inst S Africa, 120–126

Bond B C 1985 Summary and observations. Util Chang Wood Res South U S N C State Univ Raleigh, North Carolina, 3 pp

Books D J, Tubbs C H 1970 Relation of light to epicormic sprouting in sugar maple. U S For Serv Res Note NC-93, 2 pp

Boone R S, Chudnoff M 1970 Variations in wood density of the mahoganies in Mexico and Central America. Turrialba 20: 369–371

Boone R S, Chudnoff M 1972 Compression wood formation and other characteristics of plantation-grown Pinus caribaea. U S For Serv Res Pap ITF-13, 16 pp

Borsdorf W 1965 Über den Einfluß mineralischer Düngung auf Wuchsleistung und Darrdichte junger Pappelpflanzen. Arch Forstwes 14: 61–78

Bosshard H H 1951 Variabilität der Elemente des Eschenholzes in Funktion von der Kambiumtätigkeit. Schweiz Z Forstwes 102: 648–665

Boulding K E 1977 Wood in the energy economy. S A F Ann Meet. – For for Am Future: Beyond the Bicentennial New Orleans, Louisiana, pp 1–7

Boutelje J B 1965 The anatomical structure, moisture content, density, shrinkage and resin content in and around knots of Swedish redwood (Pinus silvestris) and Swedish whitewood (Picea abies). IUFRO Sect 41 Congr Melbourne, Australia, 13 pp

Boyce S G, Kaiser M 1961 Environmental and genetic variability in the length of fibers of eastern cottonwood. Tappi 44: 363–366

Boyd L 1967 Genetic variation in packing density within a selected population of loblolly pine (Pinus taeda). MS Thesis Texas A & M Univ College Station Texas, 36 pp

Boyd J D 1977 Basic causes of differentiation of tension wood and compression wood. Aust For Res 7: 121–143

Boyd C W, Koch P, McKean H B, Morschauser C R, Preston B, Wangaard F F 1976 Wood for structural and architectural purposes. Wood Fiber 8: 11–72

Brandaõ L G 1984 The new eucalpyt forest. Marcus Wallenberg Symp Proc 1, Falun, Sweden, 3–15

Brasil M A, Veiga R A de Arruda, Ferreira M 1977 Variation of specific gravity in transverse sections of the stem from base of trunk to crown in Eucalyptus. IPEF 15: 73–82

Brasil M A, Veiga R A de Arruda, Mello H de Amoral 1979 Wood basic density of 3-year-old Eucalyptus grandis. IPEF 19: 63–76

Bratt L C 1971 Trends in pulping of mixed tropical hardwoods. U N Ind Dev Org, Vienna, Austria, 19 pp

Brazier J D 1967 Timber improvement I. A study of the variation in wood characteristics in young Sitka spruce. Forestry 4: 117–128

Brazier J D 1983 Fast grown woods of the future. IUFRO Conf Div 5, Madison, Wisconsin, 13–31

Brewbaker J L, Hutton E M 1979 Leucaena – versatile tropical legume. New Agr crops – Am Assoc Adv Sci Press, Boulder, Colorado, 207–259

Briscoe C B, Harris J B, Wyckoff D 1963 Variation of specific gravity in plantation-grown trees of bigleaf mahogany. Carib For 24: 67–74

Brito J O 1984 Estudo tecnologico do madeira de Pinus spp para a producao de carvao vegetal e briquetagem. IPEF 26: 25–30

Brito J O, Barrichelo L E, do Couto H T, Mendez C J, de Rezende G C 1984 Study of the behavior of wood of Eucalyptus in the process of dry distillation. Bol Tec 8: 5–36

Brito J O, Barrichelo L E, Gutierrez L E 1980 Quality of the rosin and turpentine of tropical pines. IPEF 21: 55–63

Brito J O, Ferraz E S, Barrichello L E, do Couto H T 1986 Effects of mineral fertilization upon growth rings of Pinus caribaea var. bahamensis. IPEF 32: 5–20

Britt K W 1970 Handbook of pulp and paper technology: 2nd Ed. Van Nostrand Reinhold, New York, 723 pp

Britwun S P 1977 Species and provenance selection of tropical hardwood species with special reference to West Africa. 3d World Consul For Tree Breed, Canberra, Australia

Brix H, Mitchell A K 1980 Effects of thinning and nitrogen fertilization on xylem development in Douglas-fir. Can J For Res 10: 121–128

Brown C H 1965a The pruning of radiata pine in Chile. IUFRO Sect 41, Melbourne, Australia, 4 pp

Brown C H 1965b Improvement of the quality of timber from coniferous plantations in New Zealand by silvicultural means. IUFRO Sect 41, Melbourne, Australia, 10 pp

Brown G A 1971 A statistical analysis of density variation in Pinus caribaea grown in Jamaica. 15th IUFRO Congress, Gainsville, Florida, 17 pp

Brown G A 1973 A statistical analysis of density variation in Pinus caribaea grown in Jamaica. Selec Breed Improve Some Tropical Conifers CFI England Vol 2: 70–72

Brown G S 1965 The yield of clearwood from pruning: some results with radiata pine. Commonw For Rev 44: 197–221

Brown H P, Panshin A J 1940 Commerical timbers of the United States. McGraw-Hill, New York, 554 pp

Brown H P, Panshin A J, Forsaith C C 1949 Textbook of wood technology Vol I. McGraw-Hill, New York, 554 pp

Brown G A, Pawsey C K 1969 A study of stub length and occlusion depth following pruning in Pinus radiata. Aust For 23: 63–80

Buckingham F M 1967 The growth of loblolly pine (Pinus taeda) as influenced by soil moisture and other environmental factors. Diss Abstr 27(4): 1010

Bunn E H 1981 The nature of the resource. N Z J For 26: 162–169

Burdon R D 1975 Compression wood in Pinus radiata clones on four different sites. N Z J For Sci 5: 152–164

Burdon R D, Harris J M 1973 Wood density in radiata pine clones on four different sites. N Z J For Sci 3: 286–303

Burdon R D, Harris J M, Bannister M H, Birt D V, Low C B 1972 Variation between and within populations of Pinus radiata in wood density, ring width, and bark thickness. N Z For Serv FRI Rotorua New Zealand Gen Tree Impr Rep 60, 13 pp

Burdon R D, Thulin I J 1965 The improvement of pinus radiata. N Z For Serv Symp 6, 92 pp

Burley J 1973 Variation of wood properties of P. patula in Malawi. Trop Prov Prog Res Int Coop Nairobi Kenya, pp 574–583

Burley J, Nikles D G 1973 Selection and breeding to improve some tropical conifers. Dept For Queensland, Australia, 456 pp

Burley J, Nuevo C C, Franklin P G 1983 Preliminary estimates of heritability of wood density parameters in Pinus caribaea var. hondurensis from Fiji and Queensland, Australia. Meet IUFRO Division 5. Madison, Wisconsin

Burley J, Palmer E R 1979 Pulp and wood densitometric properties of Pinus caribaea. CFI Occ Pap 6, 66 pp

Burley J, Wood P J 1976 A manual on species and provenance research with particular reference to the tropics. CFI Trop For Pap 10, 226 pp

Burton J D, Smith D M 1972 Guying to prevent wind sway influences, loblolly pine growth, and wood properties. U S For Serv Res Pap SO 80, 8 pp

Burton J D, Shoulders E 1974 Fast grown, dense loblolly pine sawlogs: a reality. J For 72: 637–641

Byram T D, Lowe W J, McKinley C R, Robinson J F, Stander A F, van Buijtenen J P 1984 Thirty-second progress report of the cooperative tree improvement program. Texas A & M University, College Station, Texas, 27 pp

Byrd V L, Ellwood E L, Hitchings R G, Barefoot A C 1965 Wood characteristics and kraft paper properties of four selected loblolly pines. II: Wood chemical constituents and their relationship to fiber morphology. For Prod J 15: 313–320

Callaham R Z 1964 Provenance research: investigation of genetic diversity associated with geography. Unasylva 18: 40–50

Campbell R K 1965 Phenotypic variation and repeatability of stem sinuosity in Douglas-fir. Northwest Sci 39(2): 47–59

Campinhos E 1980 More wood of better quality through intensive silviculture with rapid-growth improved Brazilian Eucalyptus. Tappi 63(11): 145–147

Cardenas P R 1971 Practical experiences on pulping mixed tropical hardwoods. U N Ind Dev Org, Vienna, Austria, 48 pp

Carlson T C, Nimlos T J 1966 Using soil series to predict site index and wood specific gravity in Western Montana. Northwest Sci 40(2): 56–67
Carmichael A J 1969 Quality wood program. 11th Meet Commit For Tree Breed Can Quebec, 3–9
Cech M Y, Kennedy R W, Smith J H 1960 Variation in some wood quality attributes of one-year-old black cottonwood (Populus trichocarpa). Tappi 43: 857–859
Champion H G 1929 More about spiral grain in conifers. Indian For 55(2): 57–58
Chidester C H, McGovern J N, McNaughton G C 1938 Comparison of sulphite pulps from fast-growth loblolly, shortleaf, longleaf and slash pines. TAPPI Conf New York, 1–4
Choong E T, Box B H, Fogg P J 1970 Wood properties of young loblolly pine in response to cultural management. Louisiana State Univ For Notes 88, Baton Rouge, Louisiana, 3 pp
Chowdhury K A 1931 Anatomical studies on the wood of a hybrid larch. J For 29: 797–805
Chudnoff M 1961 The physical and mechanical properties of Eucalyptus camaldulensis. Isr J Agr Res 66, 39 pp
Chudnoff M 1976 Density of tropical timbers as influenced by climatic life zones. Commonw For Rev 55: 203–217
Chudnoff M 1980 Tropical timbers of the world. U S For Serv FPL, 826 pp
Chudnoff M 1984 Tropical timbers of the world. U S For Serv Agr Handbook 607, 464 pp
Chudnoff M, Geary T F 1973 On the heritability of wood density in Swietenia macrophylla. Turrialba 23: 359–361
Clark A, Taras M A 1969 Wood density surveys of the minor species of yellow pine in the eastern United States, II – Sand pine (P. clausa). U S For Serv Res Pap SE-51, 14 pp
Clark J 1962 Effects of fiber coarseness and length. I. Bulk, burst, tear, fold, and tensile tests. Tappi 45: 628–634
Clarke S H 1939 Stresses and strains in growing timber. For 13: 68–79
Cleaveland M F, Wooten T E, Saucier J R 1974 Annotated bibliography on the effects of fertilization on wood quality. Dept For Clemson Univ Clemson South Carolina For Res Series 29
Clifton N C 1969 Resin pockets in Canterbury radiata pine. N Z J For 14: 38–49
Cole D E, Zobel B J, Roberts J H 1965 Slash, loblolly, and longleaf pine in a mixed natural stand; a comparison of their wood properties, pulp yields, and paper properties. Tappi 49: 161–166
Coleman J D 1981a A "reconsolidated" wood for structural purposes. CSIRO Div Chem Tech Rev, pp 1–10
Coleman J D 1981b A "reconsolidated" structural wood substitute from radiata pine. 35th Appita Gen Conf Rotorua, New Zealand, 11 pp
Coleman J D 1984 Further developments with Scrimber. Proc 21st For Prod Res Conf Melbourne, Australia, 8 pp
Colley J 1973 Influence of vessel elements on the picking tendency of eucalypts pulps. Pap Technol Ind 14: 293–296
Conover D F 1958 Early thinnings have increased forking in a northern hardwood stand. U S For Serv Tech Note 532 Lake States For Expt Sta, 2 pp
Conrad J 1977 Ein Indiz für die Vererbbarkeit gewisser Holzstrukturen. Forst Holzwirt 32(12): 217–218
Cooper G A 1960 Specific gravity of red pine as related to stem- and crown-formed wood. Iowa State J Sci 34: 693–708
Corson R C, Foster R S 1981 Refiner pulping of "new crop" pinus radiata. Appita 35: 412–416
Cown D J 1971 Wood properties of Pinus nigra. In 1970 report. N Z For Serv For Res Inst, 7 pp
Cown D J 1973a Effects of severe thinning and pruning treatments on the intrinsic wood properties of young radiata pine. N Z J For Sci 3: 379–389
Cown D J 1973b Resin pockets: their occurrence and formation in New Zealand forests. N Z J For 18: 233–251
Cown D J 1974 Comparison of the effect of two thinning regimes on some wood properties of radiata pine. N Z J For Sci 4: 540–551
Cown D J 1977 Summary of wood quality studies in fertilizer trials. The Use of Fertilizers in New Zealand Forestry. FRI N Z Symp 19 Rotorua, New Zealand
Cown D J 1981 Wood density of Pinus caribaea var. hondurensis grown in Fiji. N Z J For Sci 11: 244–253
Cown D J, McConchie D L 1981 Effects of thinning and fertilizer application on wood properties of Pinus radiata. N Z J For Sci 11: 79–91

Cown D J, McConchie D L 1982a Wood density prediction for radiata pine logs. FRI Bull 9 Rotorua, New Zealand, 10 pp
Cown D J, McConchie D L 1982b Rotation age and silvicultural effects on wood properties of four stands of Pinus radiata. N Z J For Sci 12: 71–75
Cown D J, McConchie D L, Treloar C 1984 Timber recovery from pruned pinus radiata butt logs at Mangatu: effect of log sweep. N Z J For Sci 14: 109–123
Cown D J, McConchie D L, Young G D 1983 Wood properties of Pinus caribaea var. hondurensis grown in Fiji. FRI Bull 17 N Z For Serv Rotorua, New Zealand, 54 pp
Cown D J, Parker M L 1979 Densitometric analysis of wood from five Douglas-fir provenances. Silvae Genet 28: 48–53
Crist J B, Dawson D H 1975 Anatomy and dry weight yields of two Populus clones grown under intensive culture. North Cent For Exp Stat U S For Ser, St Paul, Minnesota, 6 pp
Cromer R N, Barr N J, Williams E R, NcNaught A M 1985 Response to fertilizer in a Pinus radiata plantation. Above ground biomass and wood density. N Z J For Sci 15: 59–70
Cromer R N, Hansen N W 1972 Growth, nutrient uptake and pulping characteristics of young Eucalyptus globulus. Appita 26: 187–190
Crow A B 1962 Site, silviculture, and wood quality. 11th Ann For Symp Louisiana State Univ Baton Rouge, Louisiana, 16–34
Curró P 1960 Technological investigations on the wood of some euro-american poplar hybrids – physical and mechanical properties. Agr For Expt Center 3, Italy, 28
Curró P 1972 [Variations of basic density in wood of Populus×euramericana (P.×canadensis).] Cellul Carta 23(3): 41–49
Cutter B E, Murphy W K 1978 Effects of potassium on growth and wood anatomy of a populus hybrid. Wood Fiber 9: 282–288
Dadswell H E 1957 Tree growth characteristics and their influence on wood structure and properties. 7th British Commonw For Conf CSIRO, Australia, 19 pp
Dadswell H E 1958 Wood structure variations occurring during tree growth and their influence on properties. J Inst Wood Sci 1: 1–24
Dadswell H E 1960 Tree growth-wood property interrelationships. Proc Spec Field Inst For Biol Sch For N C State Univ Raleigh, North Carolina, 86 pp
Dadswell H E 1972 The anatomy of eucalypt wood. CSIRO Melbourne, Australia, 36 pp
Dadswell H E, Fielding J M, Nicholls J W, Brown A G 1961 Tree to tree variations and the gross heritability of wood characteristics of Pinus radiata. Tappi 44: 174–179
Dadswell H E, Nicholls J W 1959 Assessment of wood qualities for tree breeding. I. Pinus elliottii from Queensland. Div For Prod Tech Rept 4 Commonw Sci Ind Res Org Melbourne, Australia
Dadswell H E, Wardrop A B 1959 Growing trees with wood properties desirable for paper manufacture. Appita 12: 129–136
Dadswell H E, Wardrop A B, Watson A J 1958 The morphology, chemistry and pulping characteristics of reaction wood. In: Bolam F (ed) Fundamentals of papermaking fibres. Tech Sect Brit Pap Makers' Assoc, London, 187–219
Dargavel J B 1968 Variations in the basic density of mountain ash. Aust For Res 3(3): 25–30
Darrow W K, Roeder K R 1983 Provencance trials of Eucalyptus urophylla and E. alba in South Africa: Seven year results. S Afr For J 125: 20–28
Davey C B, Wallum A G 1979 Nitrogen inputs to forest ecosystems through biological fixation. Proc Impact Inten Harvest for Nutrient Cycling SUNY Coll For Syracuse, New York, 62–74
Davis E M 1961 The effect of density upon wood properties. Wood Work Dig 63(9): 52–53
de Guth E B 1970 Variation of physical and chemical characters of several sub-tropical Pinus species in northeastern Argentina. IDIA 6: 33–40
de Guth E B 1973 Density and tracheid morphology in species of pine planted in deep red soils in the province of Misiones. IDIA 8: 1–7
de Guth E B 1980 Relationship between wood density and tree diameter in Pinus elliottii of Misiones Argentina. IUFRO Conf Div. 5 Oxford, England, 1 p (Summary)
De Jesus R M, Rossman N C, Brouard J S 1988 Eucalyptus – Leucaena mixture experiment. II Wood properties. (In press – School For Res North Carolina State Univ Raleigh, North Carolina, 4 pp)
De Souza V R, Carpim M A, Barrichelo L E 1986 Wood basic density of different provenances of Eucalyptus grandis and E. saligna from different diameter classes and different heights along the trees. IPEF 33: 65–72

de Villiers A M 1973 Observations on the timber properties of certain tropical pines grown in South Africa and their improvement by tree breeding. Sel Breed Impr Trop Pines CFI Oxford, 95–116

Deleporte P 1983 The variability in inter- and intraprovenance of Picea sitchensis. Improvement strategy by vegetative propagation. AFOCEL, 283–337

Delgado GA 1983 Wood quality of six species from young experimental plantations in Venezuela. IUFRO Conf Div 5 Madison, Wisconsin, 6

Delwaulle J C 1985 Paper production from clonal eucalypt stands in Congo. 26th FAO Comm Pulp Pap Conf Rome, Italy, 11 pp

Denne M P 1971 Tracheid length in relation to seedling height in conifers. Wood Sci Technol 5: 135–146

Derr J H, Enghardt H 1960 Is geographic seed source of slash pine important? South Lumberm 201(2513): 95–96

Dietrichson J 1963 Some results from an anatomic investigation on Norway spruce provenances in Sweden and Norway. 1st World Consul For Gen Tree Improv Stockholm, 9 pp

Dietrichson J 1964 The selection problem and growth-rhythm. Silvae Genet 13: 178–184

Dietrichson J, Rognerud P A, Haveraaen O, Skrøppa T 1985 Stem cracks in Norway spruce (Picea abies (L) Karst). Medd Norsk Inst Skogsforskn 38.21, 32 pp

Dillner B, Ljunger A, Herod O A, Thune-Larsen E 1970 The breeding of Eucalyptus globulus on the basis of wood density, chemical composition and growth rate. Prod Ind Util Eucalyptus Lisbon Vol 23, 120–155

Dinwoodie J M 1961 Tracheid and fibre length in timber: A review of literature. For 34: 124–144

Dinwoodie J M 1965 Growth stresses in timber – a review of European investigations. IUFRO Sect 41 Melbourne, Australia, 10 pp

Dodd RS 1985 Within-tree variation in wood production and wood quality in sycamore (Acer pseudoplatanus): its relation to crown characteristics. Can J For Res 15: 56–65

Doerksen A H, Mitchell R G 1965 Effects of balsam woolly aphid upon wood anatomy of some western true firs. For Sci 11: 181–188

Doimo L 1984 What is wood? Tech Pap 38 Dept For Queensland, Australia, 20 pp

Doran J C 1974 Genetic variation in wood density of Eucalyptus regnans. Res Work Group Aust For Commonw For Gen Melbourne, Australia, 99–101

Dorn D 1969 Relationship of specific gravity and tracheid length to growth rate and provenance in Scotch pine. 16th Northeastern For Tree Improv Conf Quebec, Canada, 8 pp

Drow J T 1957 Relationship of locality and rate of growth to density and strength of Douglas-fir. U S For Serv Rep FPL-2078, 56 pp

Durand P Y 1983 Variation in the technological quality of wood of framire in plantations in the Ivory Coast. Fast Growing Trees Silvicultura 32, pp 811–815

Echols R M 1958 Variation in tracheid length and wood density in geographic races of Scotch pine. Yale Univ Sch For Bull 64, 52 pp

Echols R M 1960 Effects of growing space on wood specific gravity in loblolly pine. Proc Soc Am For Ann Meet, 140–143

Echols R M 1971 Patterns of wood density distribution and growth rate in ponderosa pine. Symp Effect of Growth Accel on Wood Prop U S For Serv FPL Madison, Wisconsin, 12 pp

Echols R M 1973 Effects of elevation and seed source on tracheid length in young ponderosa pine. For Sci 19:47–49

Echols R M, Conkle M T 1971 The influennce of plantation and seed source elevation on wood specific gravity of 29-year-old ponderosa pines. For Sci 17:388–393

Echols R M, Dorman K W 1962 Heritability of specific gravity and sumerwood percent and relation to other factors in slash pine. Tappi 45:599–601

Eguiluz-Piedra T, Zobel B J 1984 Geographic variation in wood properties of Pinus tecunumanii. Wood Fiber Sci 18: 68–75

Ehrenberg C 1970 Breeding for stem quality. Unasylva 24(3): 23–31

Eichorn F 1895 Untersuchungen über das Holz der Roteiche. Forst-Naturwiss Z 4: 233–264

Einspahr D W 1967 Variation and heritability of wood and growth characteristics of five-year-old quaking aspen. Inst Pap Chem Genet Physiol Note. 6 pp

Einspahr D W, Benson M K 1967 Geographic variation of quaking aspen in Wisconsin and upper Michigan. Silvae Genet 16:106–112

Einspahr D W, Benson M K, Harder M L 1971 Influence of irrigation and fertilization on growth and wood properties of quaking aspen. Symp Effect Growth Accel on Wood Prop U S For Ser FPL Madison, Wisconsin, 11 pp
Einspahr D W, Benson M K, Harder M L 1972 Within-tree variation in specific gravity of young quaking aspen. Rept Inst Paper Chem 15, 8 pp
Einspahr D W, Goddard R E, Gardner H S 1964 Slash pine wood and fiber property heritability study. Silvae Genet 13:103–109
Einspahr D W, van Buijtenen J P, Peckham J R 1963 Natural variation and heritability in triploid aspen. Silvae Genet 12:51–58
Elliott G K 1970 Wood density in conifers. Commonw For Bureau Tech Comm, 44 pp
Elliott G K 1984 Wood properties and future requirements for wood products. Attributes of trees as crop plants. Midlothian, Scotland, 18 pp
Encev E, Kostov P 1959 [The influence of resin tapping on the physical and mechanical properties of Pinus sylvestris and P. nigra.] Nanc Trud Lesoteh Inst 7 Sofija, pp 221–235
Erdmann G G, Peterson R M 1972 Crown release increases diameter growth and bole sprouting of pole-size yellow birch. U S For Ser Res Note NC-130, 4 pp
Erickson J R 1978 Changing resource quality – its impact in harvesting and transportation. For Prod Res Soc Proc No P-28-21: 14–17 (from Impacts on the changing quality of timber resources)
Erickson H D, Arima T 1974 Douglas-fir wood quality studies. Part II: Effects of age and stimulated growth on fibril angle and chemical constituents. Wood Sci Technol 8:255–265
Erickson H D, Harrison A T 1974 Douglas-fir wood quality studies. Part I: Effects of age and stimulated growth on wood density and anatomy. Wood Sci Technol 8: 207–226
Erickson H D, Lambert G M 1958 Effects of fertilization and thinning on chemical composition, growth, and specific gravity of young Douglas-fir. For Sci 4: 307–315
Ericson B 1960a Latewood percentage, density and volumetric shrinkage in wood of Picea abies (L) Karst f virgata Jacq. A comparison with Picea abies. Stat Skogsforskninginst Stockholm Rapp 2, 15 pp
Ericson B 1960b Studies on the genetical wood density variation in Scots pine and Norway spruce. Stat Skogsforskningsinst Stockholm Rapp 4, 52 pp
Ericson B 1966 Gallringens inverkan på vedens torr-råvolymvikt, höstvedhalt och kärnvedhalt hos tall och gran. (Effect of thinning on the basic density and content of latewood and heartwood in Scots pine and Norway spruce). Skogshögsk Stockholm Inst Skogsprod Rapp Upp 10, 330 pp
Ezell A W, Lowe W J, Murphy W K, Wright J A 1983 Plantation culture for energy production – the importance of within species selection. Wood Fiber Sci 15: 69–73
Fahey T D, Starostovic E 1978 Changing resource quality – its impact on the forest product and construction industries. For Prod Res Soc 21: 19–25 (from Impacts on the changing quality of timber resources)
Falkenhagen E R 1979 Provenance variation in growth; timber and pulp properties on Pinus caribaea in South Africa. Tech Bull 59 Dept For S Afr, 65 pp
Farmer R E, Wilcox J R 1966a Specific gravity variation in a lower Mississippi Valley cottonwood plantation. Tappi 49: 210–211
Farmer R E, Wilcox J R 1966b Variation in juvenile growth and wood properties in half-sib cottonwood families. U S For Serv Res Pap NC-6, 4 pp
Farmer R E, Wilcox J R 1968 Preliminary testing of eastern cottonwood clones. Theor Appl Gen 38: 197–201
Farr W A 1973 Specific gravity of western hemlock and Sitka spruce in southeast Alaska. Wood Sci 6: 9–13
Farrar J L 1964 Tracheid development in certain conifers as affected by soil drought. 10th Int Bot Congr, 367
Farrington A, Hansen N W, Nelson P F 1977 Utilization of young plantation E. globulus. Appita 30: 313–319
Ferguson R B, Land S B, Cooper D T 1977 Inheritance of growth and crown characters in American sycamore. Silvae Genet 25: 180–182
Fernow B E 1896 Southern pine – mechanical and physical properties. U S For Serv Circ 12, 20 pp
Ferrari G, Scaramuzzi G 1980 Investigations on the effects of site on wood characteristics in Populus×euroamericana. IUFRO Conf Div 5 Oxford, England, 40 (Summary)
Ferraz E S 1983 Growth rings and climate in Eucalyptus. Fast Growing Trees Silvicultura 32, 821, 822

Ferreira C A, Freitas M, Ferreira M 1979 Densidade basica da madeira de plantações commericais de eucaliptos, na regiâo de Mogi-Guaçú (São Paulo). IPEF 18: 106–117
Ferreira M 1968 Studies of the variation of basic density of the wood of Eucalyptus alba and Eucalyptus saligna. Univ São Paulo Piracicaba Brasil, 71 pp
Ferreira M 1972 Variation of wood basic density in commerical plantations of Eucalyptus grandis at ages of 11, 12, 13, 14 and 16 years. IPEF 4: 65–75
Ferreira M 1978 Wood dry weight yield of pine plantations. IPEF 17: 78–79
Fielding J M 1960 Branching and flowering characteristics of Monterey pine. For Timber Bureau Bull 37 Canberra, Australia, 59 pp
Fielding J M 1965 Pruning Pinus radiata in Australia, with particular reference to the wood produced. IUFRO Sect 41 Melbourne, Australia, 8 pp
Fielding J M, Brown A G 1960 Variations in the density of the wood of Monterey pine from tree to tree. For Timber Bur Leaflet 77 Canberra, Australia, 1–28
Foelkel C E, Barrichelo L E, do Amaral A C, do Valle C F 1975 Variations in the wood characteristics and sulphate-pulp properties of Pinus oocarpa as a function of age of the forest stand. IPEF 10: 81–87
Foelkel C E, Barrichelo L E, Garcia W, Brito J O 1976 Kraft cellulose of juvenile and adult wood of Pinus elliottii. IPEF 12: 127–142
Foelkel C E, Ferreira M, Nehring J H, Rohm M B 1975 Variability in slash pine wood in the radial direction. IPEF 10: 1–11
Foelkel C E, Zvinakevicius C 1980 Hardwood pulping in Brazil. Tappi 63:39–42
Foil R R 1961 Late season soil moisture shows significant effect on width of slash pine summerwood. Louisiana State Univ For Note 44 Baton Rouge, Louisiana, 10 pp
Forest Products Research Society 1986 "Expletive Deleted" – Juvenile Wood. Shortcourse 15 papers Savannah, Georgia
Forgacs O L 1984 Parallam project – a case history. Celebration 1000th PhD in Chemistry McGill Univ, Montreal, Canada, 6 pp
Foulger A N 1968 Effect of aphid infestation on properties of grand fir. For Prod J 18: 43–47
Foulger A N, Freese F, Harris J 1972 Effect of nutrient concentration on the stem anatomy of white ash seedlings. Wood Fiber 4: 112–114
Foulger A N, Hacskaylo J 1968 Stem anatomy variation in cottonwood growth under nutrient deficient conditions. 8th Lake States For Tree Impr Conf U S For Ser Res Paper NC-23, 41–47
Franklin E C 1974 Phenotypic and genetic variation of sulfate naval stores yields in loblolly pine. TAPPI For Biol Conf, 99–102
Franklin E C 1979 Model relating levels of genetic variance to stand development of four North American conifers. Silvae Genet 28: 207–212
Franklin E C, Bridgwater F E 1981 Genetics of fast growth and wood properties. Proc TAPPI Meet Atlanta, Georgia, 445–448
Franklin E C, Callaham R Z 1970 Multinodality, branching and forking in lodgepole pine (Pinus contorta var murrayana). Silvae Genet 19: 180–184
Franklin E C, Meskimen G 1975 Wood properties of some eucalypts for the southern United States. National Conv Soc Am For, Washington DC, 454–458
Frederick D J, Madgwick H A, Oliver G R 1982 Wood basic density and moisture content of young Eucalyptus regnans grown in New Zealand. N Z J For Sci 12: 494–500
Fries A 1986 Volume growth and wood density of plus tree progenies of Pinus contorta in two Swedish field trials. Scand J For Res 1: 403–419
Fukazawa K 1983 Juvenile wood of hardwoods. Div 5 IUFRO Conf Madison, Wisconsin, 28–29 (abstracts)
Fukazawa K, Aoki A 1960 Studies on the relation between physical properties and growth conditions for planted Sugi (Cryptomeria japonica) in central district of Japan VI. On hardness test. Res Bull Fac Agr Gifu Univ Japan 12: 109–117
Gabriel W J 1956 Preliminary report on clonal differences in the wood and phloem of Populus deltoides and P. trichocarpa. 3 rd Northeast For Tree Improv Conf Ithaca, New York, pp 33–34
Gajda T P, Lamb F M, Ifju G 1982 How to use beetle-killed southern pine timber. Res Dev Div Conf TAPPI Asheville, North Carolina, 111 (Summary)
Gansel C R 1965 Inheritance of stem and branch characters in slash pine and relation to gum yield. 8th South Conf For Tree Improv Savannah, Georgia, 57–63

Garges L D, Labosky P, Blankenhorn P R, Rishel L E 1984 Lumber recovery from gypsy moth-killed red and white oak trees. For Prod J 34(7–8): 45–50

Garrett P W, Shigo A L, Carter J 1976 Variation in diameter of central columns of discoloration in six hybrid poplar clones. Can J For Res 6: 475–477

Geary T F 1967 Variation in density and diameter between open-pollinated families of Pinus khasia. Rhod Zamb Mal J Agr Res 5(3): 297–300

Gentle SW 1970 Incidence of boron and sulfur deficiency in Pinus radiata plantations in NSW. Aust Plant Nutrit Conf Mt. Gambier, South Australia, 2 pp

Gentle S W, Bamber R K, Humphreys F R 1968 Effects of two phosphate fertilizers on yield, financial yield and wood quality of radiata pine. For Sci 14: 282–286

Geyer W A, Gilmore A R 1965 Effect of spacing on wood specific gravity in loblolly pine in southern Illinois. Agr Expt Stat Note 113 Univ Illinois, Urbana, Illinois, 5 pp

Gibson E J 1980 Product development and quality. IUFRO Conf Div 5 Oxford, England, 15–29

Gilmore A R 1968 Geographic variation in specific gravity of white pine and red pine in Illinois. For Prod J 18: 49–51

Gilmore A R 1969 Soil moisture stress and tracheid length in loblolly pine. 10th South Conf For Tree Improv, Houston, Texas, 205–207

Gilmore A R 1971 Specific gravity of plantation-grown yellow poplar is not related to site. Wood Fiber 3: 182–183

Gilmore A R, Boyce S G, Ryker R A 1966 The relationship of specific gravity of loblolly pine to environmental factors in southern Illinois. For Sci 12: 399–405

Gilmore R C, Pearson R G 1969 Characterization of the strength of juvenile wood of loblolly pine (Pinus taeda). Tech Rept 42 Sch For Res N C State Univ Raleigh, North Carolina, 35 pp

Giordano G 1959 Influenza della resinazione sulle caratteristiche del legno di Pinus pinea. Anal Acad Ital Sci For 8: 147–163

Gladstone W T, Barefoot A C, Zobel B J 1970 Kraft pulping of earlywood and latewood from loblolly pine. For Prod J 20: 17–24

Gladstone W T, Gray R L 1972 Effects of forest fertilization on wood quality. South For Res Cen Weyershaeuser Co, Hot Springs, Arkansas, 18 pp

Goddard R E, Cole D E 1966 Variation in wood production of six-year-old progenies of select slash pines. Tappi 49: 359–362

Goddard R E, Strickland R K 1962 Geographic variation in wood specific gravity of slash pine. Tappi 45: 606–608

Goddard R E, Strickland R K 1964a Crooked stem form in loblolly pine. Silvae Genet 13: 155–157

Goddard R E, Strickland R K 1964b Interrelationships among tree and site characteristics. Tree Impr Tips No 1 Univ Florida, Gainesville, Florida, 2 pp

Goebel N B, Taras M A, Smith W R 1960 Tension wood and its relation to splitting in hickory. South Carolina Agr Exp Stat Bull 480: 3–21

Goggans J F 1961 The interplay of environment and heredity as factors controlling wood properties in conifers with special emphasis on their effects on specific gravity. Tech Rep 11 School For Res N C State Univ Raleigh, North Carolina, 56 pp

Goggans J F 1962 The correlation, variation and inheritance of wood properties in loblolly pine. Tech Rep 14 School For Res N C State Univ Raleigh, North Carolina, 155 pp

Goggans J F 1964 Correlation and heritability of certain wood properties in loblolly pine (Pinus taeda). Tappi 47: 318–322

Goggans J F 1965 Variation of tracheid width and wall thickness within and between trees of southern pine species. IUFRÓ Section 41, Melbourne, Australia, 10 pp

Göhre K 1960 Die Verteilung von Rohwichte im Pappelstamm. Wiss Abh Deutsch Akad Landwiss Berlin 44: 51–79

Göhre K, Götze H 1956 Untersuchungen über die Rohwichte des Rotbuchenholzes. Arch Forstwes 5: 716–748

Goldman S E, Cleveland G D, Parker J A 1979 Lightwood induction and associated beetle attacks on slash pine. For Sci 25: 80–83

Goldstein I S 1979 Energy sources derivable from wood on an industrial scale for individual, commercial and manufacturing operations. Spec Rept FAO N C State Univ Raleigh, North Carolina, 42 pp

Goldstein I S 1980a New technology for new uses of wood. Tappi 63(2): 105–108

Goldstein I S 1980b Chemicals from biomass: present status. Alternate feedstocks for petrochemicals. Proc Am Chem Soc San Fransisco, California, 19 pp
Gomez C H, Mondragon I 1974 The pulping of Colombian hardwoods for linerboard. Tappi 57(5): 140–142
Gooding J W, Smith W H 1971 Effects on fertilization on stem, wood properties and pulping characteristics of slash pine (P. elliottii). Symp Effect Growth Accel on Wood Prop U S For Serv FPL Madison, Wisconsin, 10 pp
Götze H 1959 Vergleichende Untersuchungen über die Beziehung zwischen Baumalter and Rohwichte. Wiss Z Humboldt Univ Berlin, Math-Naturwiss Reihe 8: 738–752
Govette R L, Sinclair S A, Bowyer J L 1983 Lumber grade yields from spruce budworm damaged balsam fir. For Prod J 33(5): 46–50
Grah R F 1961 Relationship between tree spacing, knot size, and log quality in young Douglas-fir stands. J For 59: 270–272
Grant D J, Anton A, Lind P 1984 Bending strength, stiffness, and stress-grade of structural Pinus radiata: effect of knots and timber density. N Z J For Sci 14: 331–348
Gray R L, Kyanka G H 1974 Potassium fertilization effects on the static bending properties of red pine wood. For Prod J 24(9): 92–96
Gray R L, de Zeeuw CH 1975 Effects of potassium fertilizer on the wood density and related anatomical characteristics of red pine. 22d Northeastern For Tree Impr Conf Syracuse, New York, 77–110
Greaves A 1982 Pinus oocarpa. Annot Bibliogr No F 22 CFI Oxford, England, 70 pp
Gregory R A, Wong B L 1983 The effect of defoliation on the production and differentiation of wood elements following defoliation and reflushing in Acer saccharum. IUFRO Conf Div 5 Madison, Wisconsin, 6–7 (Abstracts)
Grigal D F, Sucoff E I 1966 Specific gravity variation among thirty jack pine plots. Tappi 49: 497–498
Grisez T J 1978 Pruning black cherry in understocked stands. U S For Serv Res Pap NE-395, 10 pp
Guiher J K 1965 Effect of rings-per-inch on specific gravity of red oak. For Prod J 15: 409–411
Guiher J K 1968 Relation of specific gravity to rings-per-inch in cottonwood construction lumber. For Prod J 18: 22
Gupta V K, Kukreti M C 1983 A note on physical and mechanical properties of Acacia mearnsii (A. mollissima) from Tanil Nadu. Indian For, 395–400
Haigh R W 1961 The effect of provenance and growth rate on specific gravity and summerwood percentage of young Douglas-fir. Univ Brit Col For Club Res Commit 19, 57 pp
Hakkila P 1969 Geographical variation of some properties of pine and spruce pulpwood in Finland. Commun Inst For Fenn, 66.8, 60 pp
Hall P J 1984 The relationship between wood density and growth rate and the implications for the selection of black spruce (Picea mariana) plus trees. Info Rept N-X-244 Newfoundland For Res Ctr Can For Serv, 22 pp
Hamilton J R 1961 Variation of wood properties in southern red oak. For Prod J 11: 267–271
Hamilton J R, Harris J B 1965 Influence of site on specific gravity and dimensions of tracheids in clones of Pinus elliottii and P. taeda. Tappi 48:330–333
Hamilton J R, Cech F C, Shurtliffe C E 1976 Estimating bole specific gravity from limbs of mature black cherry and northern red oak trees. Wood Fiber 7: 281–292
Hamilton J R, Litwin P J, Tryon F H 1978 A note on the influence of soil parent material on northern red oak specific gravity. Wood Fiber 10: 2–5
Hamilton J R, Wendel G W 1967 Specific gravity and fiber length of some hybrid poplars growing in West Virginia. West Virginia Agr Expt Sta Bull 556T, 6 pp
Hans A S 1976 Variation in wood density of Eucalyptus grandis and E. tereticornis. J Sci Technol 1(4): 109–112
Hans A S, Burley J 1972 Wood quality of Eucalyptus grandis in a fertilizer trial at Siamambo, Zambia. E Afr Agr For J 38(2): 157–161
Hans A S, Burley J, Williamson P 1972 Wood quality in Eucalyptus grandis grown in Zambia. Holzforschung 26: 138–141
Harrington C A, DeBell D S 1980 Variation in specific gravity of red alder (Alnus rubra). Tappi 10: 293–299
Harris J M 1961 A survey of the wood properties of radiata pine grown in the Kaingaroa Forest. For Prod Rep 76 For Res Inst N Z For Serv, 1–16

Harris J M 1963a The effect of pruning on the resinification of knots in radiata pine. N Z For Res Notes 34: 1–7

Harris J M 1963b The influence of environment on the wood density of radiata pine grown in New Zealand. 1st World Consul For Gen Tree Breed Stockholm, Sweden, 8 pp

Harris J M 1965 A survey of wood density, tracheid length and latewood characteristics of radiata pine grown in New Zealand. N Z For Serv Tech Rep 47, 31 pp

Harris J M 1967 Latewood, earlywood and wood density. 14th IUFRO Congr Sect 41, 56–69

Harris J M 1971 FAO project on tropical pines. Rep FRI 1970 N Z For Serv Rotorua, New Zealand, 78–79

Harris J M 1973a The use of beta rays to examine wood density of tropical pines grown in Malaya. Sel Breed Improve Some Trop Conifers. CFI Oxford, England, 86–94

Harris J M 1973b Physical properties, resin content, and tracheid length of lodgepole pine grown in New Zealand. N Z J For Sci 3: 91–109

Harris J M 1973c Spiral grain and xylem polarity in radiata pine; microscopy of cambial reorientation. N Z J For Sci 3:363–378

Harris J M 1977a Variation of growth, stem quality, and wood properties in relation to genetic improvement of tropical forest trees. IUFRO Workshop Brisbane, Australia, 9 pp

Harris J M 1977b Note on wood density of Pinus caribaea grown under temperate, subtropical and tropical conditions. Queensland Dept For Tech Pap 4, 19 pp

Harris J M 1981 Wood quality of radiata pine. N Z For Serv Reprint 1474 Rotorua, New Zealand, 211–215

Harris J M 1983 Variation in wood quality: research requirements and research strategy. IUFRO Conf Div 5 Madison, Wisconsin, 117–134

Harris J M 1984 Non-destructive assessment of spiral grain in standing trees. N Z J For Sci 14: 395–399

Harris J M 1988 Spiral grain and wave phenomena in wood formation. Springer, Berlin Heidelberg New York Tokyo

Harris J M, James R N, Collins M J 1976 Case for improving wood density in radiata pine. N Z J For Sci 5: 347–354

Harris J M, Meylan B A 1965 The influence of microfibril angle on longitudinal and tangential shrinkage in Pinus radiata. Holzforschung 19: 144–153

Hart T L 1978 Coping with changing timber quality – the future. For Prod Res Soc 21: 44–49 (from Impacts on the changing quality of timber resource)

Hatton J V 1984 Chemical pulping of budworm-killed balsam fir and white spruce. 70th Ann Meet Tech Sect Can Pulp Pap Assoc, 43

Hawthorn J C 1961 An assessment of the use of growth rate as an indicator of specific gravity in hemlock and balsam. Thesis School For Univ Brit Columbia, Vancouver, Canada

Hazel L N, Lush J L 1942 The efficiency of three methods of selection. J Hered 33: 393–399

Heger L 1974 Longitudinal variation of specific gravity in stems of black spruce, balsam fir and lodgepole pine. Can J For Res 4: 321–326

Heikinheimo O 1951 Kokemuksia visakoivun kasvatuksesta. (Experiences in the growing of curly birch.) Comm Inst For Fenn 39.5, 53 pp

Henderson J, Petty J A 1972 A comparison of wood properties of coastal and interior provenances of lodgepole pine, Pinus contorta. Forestry 45(1): 50–57

Herpka I 1965 On the variability of the length of wood fibers and the specific gravity in a natural population of white willow (Salix alba) in the inundation area of the Danube. Inter Poplar Comm Novi Sad, Yugoslavia 12th Session, 1–9

Herpka I 1979 Genetic variability of basic density of juvenile wood in Populus deltoides and its significance for selection. IUFRO Meet Conc Poplars, France and Belgium. Wageningen, Netherlands, 12 pp

Higa A R, Kageyama P Y, Ferreira M 1973 Variation in the wood specific gravity of Pinus elliottii var elliottii and Pinus taeda. IPEF 7: 79–89

Higgins H G 1985 Letter from Australia. Tappi 68(2): 24

Higgins H G , Puri V 1976 Chemithermomechanical pulp from young eucalypts. 1st Latin Am Congr on Pulp Pap, Buenos Aires, Argentina, 15 pp

Higgs M L, Rudman P 1973 Effects of fertilizing and thinning on the wood properties of Eucalyptus regnans. Appita 27: 51–55

Hildebrandt G 1954 Study of the increment and pure wood content of spruce stands. Deutscher Verlag der Wissenschaften, Berlin, 133 pp
Hildebrandt G 1960 The effect of growth conditions on the structure and properties of wood. 5th World For Congr Seattle, Washington, 12 pp
Hiley W E 1955 Quality in softwoods. Q J For 49: 159–164
Hiller C H 1964 Correlation of fibril angle with wall thickness of tracheids in summerwood of slash and loblolly pine. Tappi 47: 125–128
Hiller C H, Brown R S 1967 Comparison of dimension and fibril angles of loblolly pine tracheids formed in wet or dry growing seasons. Am J Bot 54: 453–460
Hillis W E 1972 Properties of eucalypt woods of importance to the pulp and paper industry. Appita 26: 113–122
Hillis W E 1980 The efficient use of the wood resource. Proc Inter Union For Res Org Div 5 Conf Oxford, England, 14 pp
Hillis W E 1984 Forest products Then – Now – Future? 21st For Prod Res Conf CSIRO Victoria, Australia
Hillis W E 1987 Heartwood and tree exudates. Springer, Heidelberg, 268 pp
Hillis W E, Brown A C 1978 Eucalyptus for wood production. CSIRO Griffin Press Ltd, Adelaide, Australia, 434 pp
Hilmi H A 1960 Density variation in Pinus brutia. Extr Rep Imp For Inst Oxford, England, 21
Hirai S 1958 Studies on weight growth of forest trees VI: Chamaecyparis obtusa in the Tokyo University Forest in Chiba. Bull Tokyo Univ For 54: 199–217
Hitchings R G 1984 The effect of juvenile wood of loblolly pine on the kraft pulp yield (Pinus taeda). Util Changing Wood Res South U S N C State Univ Raleigh, North Carolina, 13 pp
Hodges J F 1985 Genotype by fertilizer interaction in slash pine and its effect on family selection for seed orchard roguing. MS Thesis School For Res N C State Univ Raleigh, North Carolina, 79 pp
Holley D L, Veal M A 1977 Economic impact of fusiform rust. Conf. Fusiform Rust Gainesville, Florida, 14 pp
Holst M J 1960 Forest tree breeding and genetics at the Petawawa Forest Experiment Station. 7th Meet Commit For Tree Breed, Vancouver, Canada, K1–K27
Holubcik M 1969 [Share of spruce summerwood (Picea abies) in High Tatras.] Lesn Cas 4(2): 115–124
Höster H R, Liese W 1966 Über das Vorkommen von Reaktionsgewebe in Wurzeln und Ästen der Dikotyledonen. Holzforschung 20: 80–90
Howard E T 1973 Heat of combustion of various southern pine materials. Wood Sci 5:194–197
Howe G E 1969 Early results of a sugar maple improvement study. 11th Meet For Tree Breed Can For Branch Dept Fish For Canada, Ottawa, Canada, 286
Howe J P 1968 Influence of irrigation on ponderosa pine. For Prod J 18: 84–93
Howe J P 1970 Effect of irrigation on wood formed at various heights in ponderosa pine. Wood Sci 3: 126–128
Howe J P 1974 Relationship of climate to the specific gravity of four Costa Rican hardwoods, an exploratory study. Wood Fiber 5: 347–352
Howell R A, Leach G N, Nutter D E, Steltenkamp M S 1984 Some age associated trends of planted slash pine as related to kraft pulping. Util Changing Wood Res South U S Raleigh, North Carolina, 16 pp
Hsu J K, Walters L S 1975 Effect of irrigation and fertilization on selected physical and mechanical properties of loblolly pine (Pinus taeda). Wood Fiber 7: 192–206
Huang T C, Liu Y J 1959 [The effect of growth and development on the timber quality of fast-grown Pinus koraiensis grown in Tsaohokow (Liaoning Prov NE China).] For Sci (Peking) 6: 489–496
Hughes F E 1965 Tension wood; a review of literature. For Abstr 26(1): 2–9
Hughes J F 1973 A comparison of the wood properties of P. caribaea var. hondurensis from two plantation sites in Jamaica. Trop Prov Prog Res Int Coop Nairobi, Kenya, 544–546
Hughes M, Gertjejansen R 1984 Spruce budworm killed balsam fir as a raw material for flakeboard and waferboard. For Prod J 34(2): 42–46
Hughes R E 1968 Wood quality for fast growing tropical species; a new approach to species trials. Commonw For Inst Oxford, England, 11 pp
Hughes R V, Mackney A W 1949 Density and moisture content of New Zealand Pinus radiata. Appita 3: 387–407
Hunt K 1977 Kraft pulp yields from fertilized and thinned jack pine. Tappi 60(8): 136–137

Hunter A G, Goggans J F 1968 Variation in specific gravity, diameter growth, and colored heartwood of sweetgum in Alabama. Tappi 51:76–79

Hunter A G, Goggans J R 1969 Variation of fiber length of sweetgum in Alabama. Tappi 52: 1952–1954

Hurley D W, Volkman D A, Andrews R S, Schmidt J J 1976 Pulping and by-product yield of paraquat treated slash pine. Proc Lightwood Res Coord Council, Savannah, Georgia, 143–144

Ibrahim S 1972 Variation of wood properties of Pinus caribaea var hondurensis. MS Thesis Sch For N C State Univ Raleigh, North Carolina, 73 pp

Ifju G 1964 Tensile strength behavior as a function of cellulose in wood. For Prod J 14: 366–372

Ifju G 1969 Within-growth-ring variation in some physical properties of southern pine wood. Wood Sci 2: 11–19

Ifju G, Ferguson P C, Oderwald R G 1977 Pulping and papermaking properties of southern pine harvested from beetle-infested forests. TAPPI For Biol Comm Madison, Wisconsin, 169–176

Ikemori Y K, Martins F C, Zobel B J 1986 The impact of accelerated breeding on wood properties. 18th IUFRO World Congr Ljubljana, Yugoslavia, 359–368

Inglis C S, Cleland M R 1982 Predicting final branch size in thinned radiata pine stands. FRI Bull 3 For Res Inst N Z For Ser, 17 pp

Inokuma T, Shimaji K, Hamaya T 1956 [Studies on poplars (I). Measurement of fiber-length and specific gravity of Japanese giant poplar (Populus japano-gigas).] Misc Infor Tokyo Univ For 11: 77–86

Institute Paper Chemistry 1962 Application of the pulping and paper-making sciences to the genetic improvement of southern pulpwood. Base-line values for judging wood quality of slash pine. Prog Rept 7 (to three companies), 46 pp

Isebrands J G, Einspahr D W, Phelps J E, Crist M B 1982 Kraft pulp and paper properties of juvenile hybrid larch grown under intensive culture. Tappi 65(9): 92–96

Isebrands J G, Hunt C M 1975 Growth and wood properties of rapid-grown Japanese larch. Wood Fiber 7: 119–128

Iwakawa M, Watanabe M, Sato T, Mikami S, Inuma M, Kida S 1967 (Inheritance of some characters in open-pollinated progenies of Pinus densiflora Sieb et Zucc.) Bull For Exp Sta Meguro, Japan 207: 31–67

Jackson D S 1963 On the homocline of slash pine. Dissertation Abst 24(1): 11

Jackson L W 1959 Loblolly pine tracheid length in relation to position in tree. J For 57:366–367

Jackson L W, Greene J T 1957 Method for fiber measurements. J For 55: 300

Jackson L W, Parker N 1958 Anatomy of fusiform rust galls on loblolly pine. Phytopathology 48: 637–640

Jackson L W, Strickland R E 1962 Geographic variation in tracheid length and wood density of loblolly pine. Georgia Res Counc 8, 1–4

Jacobs M R 1965 Stresses and strains in tree trunks as they grow in length and width. Dept Nat Dev Leafl 96 Australia, 15 pp

Jain K K, Seth M K 1979 Intra-increment variation in specific gravity in wood of blue pine. Wood Sci Technol 13: 239–248

Jayme G, Branscheid F, Harders-Steinhäuser M, Eser L 1943 Eignungsvergleich verschiedener Schwarzpappelhölzer zur Zellstoffgewinnung. Holz Roh-Werkst 6: 231–240

Jayme G, Krause T 1963 Über die Packungsdichte der Zellwände in Laubhölzern. Holz Roh-Werkst 21: 14–19

Jayne B A 1958 Effect of site and spacing on the specific gravity of wood of plantations-grown red pine. Tappi 41: 162–166

Jeffers J W 1959 Regression models of variation in specific gravity in four provenances of Sitka spruce. J Wood Sci 4: 44–59

Jefferson P A, Yanchuk A 1985 Heritability and genetic relationships for wood quality characteristics – the key. Forintek Canada Corp Workshop, Quebec, Canada, 1–19

Jett J B, Talbert J T 1982 Place of wood specific gravity in the development of advanced generation seed orchards and breeding programs. South J Appl For 6:177–180

Jett J B, Weir R J, Barker J A 1977 The inheritance of cellulose in loblolly pine. TAPPI For Biol Comm Meet Madison, Wisconsin, 4 pp

Jett J B, Zobel B J 1975 Wood and pulping properties of young hardwoods. Tappi 58(1): 92–96

Johnson J W, McElwee R L 1967 Geographic variation in specific gravity and three fiber characteristics of sweetgum. 9th South For Tree Improv Conf Knoxville, Tennessee, 8 pp
Johnson L P 1942 Studies on the relation of growth rate to wood quality in Populus hybrids. Can J Res 20: 28–40
Kaeiser M, Boyce J G 1964 Averages and correlation coefficents between specific gravity and a number of wood properties on Eastern cottonwood, including gelatinous fibers. Am J Bot 51:673
Kaneski L 1938 Über die Länge der Buchenholzfasern von Fagus crenata. Nippon Rungakakai Shi 20, 263 pp
Kanô T 1961 [The position in stems indicating the average values of bulk density.] Bull Govt For Expt Sta Meguro, Japan 134: 141–148
Kanô T, Saiki H 1970 [On the selection of the basic density in the Japanese red pine (Pinus densiflora Sieb et Zucc) wood.] Mokuzai Gakkaishi 16: 305–309
Karami P K 1973 Improvement of stem form in Eucalyptus grandis. Proc Meet Trop Prov Prog Res Inter Cooper Nairobi, Kenya, 413–416
Karki L 1980 Genetically narrow crowned and find-branched trees are valuable in forestry. Found For Tree Breed Finland Info Sheet 3, 3 pp
Karnik M G, Bhatia K, Lal J 1970 Terminalia tomentosa – its commercial properties. Indian For 96:612–620
Kazumi 1983 Juvenile wood of hardwoods. IUFRO Conf Div 5 Madison, Wisconsin, 28–29 (Abstract)
Keating W G 1983 Property needs for the wood user. IUFRO Conf Div 5 Madison, Wisconsin, 81–94
Keating W G, Bolza E 1982 Characteristics, properties and uses of timbers. Texas A & M Univ Press, College Station, Texas
Kedharnath S, Chacko V J, Gupta S K, Mathews J D 1963 Geographic and individual tree variation in some wood characteristics of teak (Tectona grandis). Silvae Genet 12: 181–187
Keiding H, Lauridsen E B, Wellendorf H 1984 Evaluation of a series of teak and gmelina provenance trials. Dan For Seed Cent Humlebaek, Denmark Tech Note 15, 42 pp. In: Barnes R D, Gibson G L (eds) Provenance and genetic improvement strategies in tropical forest trees. IUFRO Conf Mutare, Zimbabwe, Commonw For Inst Oxford, England, 30–69
Keiding H, Olsen H C 1965 Assessment of stem form in clones and in progenies of larch. Silvae Genet 14: 115–122
Keith C T 1961 Characteristics of annual rings in relation to wood quality for white spruce (Picea glauca). For Prod J 11: 122–126
Keith C T 1985 Defining wood quality – what's important? Wood quality consideration in tree improvement programs. Workshop Forintek Canada Corp Quebec, Canada, 21–25
Keith C T, Kellog R M 1985 Wood quality considerations in tree improvement programs. Workshop Forintek Canada Corp Quebec, Canada, 77
Keller R 1973 Caractéristiques du bois de Pin maritime. Variabilité et transmission héréditaire. Ann Sci For 30: 31–62
Keller R, Le Tacon F, Timbal J 1976 La densité du bois de Hêtre dans le Nord-Est de la France. Influence des caractéristiques du milieu et du type de sylviculture. Ann Sci For 33: 1–17
Keller R, Thoby M 1977 Liaisons entre l'état juvénile à l'état adulte pur quelques caractères technologiques et auxométriques chez le Douglasie (Pseudotsuga menziesii). Ann Sci For 34:175–203
Kellison R C 1967 A geographic variation study of yellow-poplar (Liriodendron tulipifera) within North Carolina. Tech Rept 33 N C State Univ School of For Res Raleigh, North Carolina, 41 pp
Kellison R C 1981 Characteristics affecting quality of timber from plantations, their determination and scope for modification. IUFRO Div 5 17 World Congress Kyoto, Japan, 77–88
Kellison R C, Deal E, Pearson R H 1985 More plantation wood means quality problems. For Ind, 24–26
Kellison R C, Hitchings R 1984 The Souths changing wood resource – its effect on manufacture and utility of pulp and paper. Util Changing Wood Res South U S Raleigh, North Carolina, 13 pp
Kellison R C, Hitchings R 1985 Harvesting more young southern pines will require pulpmill changes. Pulp Paper 59(7): 53–56
Kellison R C, Lea R, Frederick D J 1983 Effect of silvicultural practices on wood quality of southern hardwoods. Tappi 66(1): 67–69

Kellison R C, Zobel B J 1978 Changing resource quality – its impact in the pulp and paper industry. 32d Ann Meet For Prod Res Soc (Abstracts), 29
Kennedy R W 1961 Variation and periodicity of summerwood in some second-growth Douglas-fir. Tappi 44: 161–165
Kennedy R W 1966 Intra-increment variation and heritability of specific gravity, parallel-to-grain tensile strength, stiffness and tracheid length in clonal Norway spruce. Tappi 49: 292–295
Kennedy R W 1969 Wood formation in jack pine seedlings of various provenances. 11th Comm For Tree Breed Canada Part 2 Ottawa, Canada, 101–109
Kennedy R W 1970 Specific gravity of early- and late-flushing Douglas-fir trees. Tappi 53: 1479–1481
Kennedy R W 1971 Influence of provenance and test location on wood formation in Pinus banksiana seedlings. Silvae Genet 20: 53–100
Kennedy R W, Smith J H 1959 The effects of some genetic and environmental factors on wood quality in poplar. Pulp Pap Mag Can 60: T35–T36
Kennedy R W, Swann G W 1968 Specific gravity of Amabilis fir (Abies amabilis) determined from increment cores. Inf Rept For Prod Lab Vancouver, Brit Columbia VP-X-33, 10 pp
Khalil M A 1985 Genetics of Wood characters of black spruce (Picea mariana) in Newfoundland, Canada. Silvae Genet 34:221–230
Khurshudyan P A 1958 [Some data on the relation between annual ring width and the physical and mechanical properties of Armenian oak wood.] Selskokhozyaistvennye Nauki 11(4): 33–43
Kibblewhite R P 1984 Fibers and fines of some radiata pine corewood and slabwood thermomechanical and refinermechanical pulps. Appita 37: 650–657
Kibblewhite R P, Lloyd J A 1983 Interrelations between the wood, fibre, and kraft pulp properties of new-crop radiata pine 12-year-old thinnings. FRI N Z For Serv Bull 44, 19 pp
King J P 1967 Seed source variation in tracheid length and specific gravity of five-year-old Jack pine seedlings. 8th Lake States For Tree Improv Conf U S For Serv Res Pap NC-23, 5–9
King J P 1980 Variation in specific gravity in three-year-old coppice clones of Eucalyptus saligna growing in Hawaii. Aust For Res 10: 295–299
Kininmonth J A, Revell D H, Williams D H 1983 Utilization of New Zealand-grown eucalypts for sawn timber and veneer. N Z J For 19: 246–263
Kinloch B B 1963 Observations on the effects of fusiform rust on specific gravity and fiber length in the wood of loblolly pine. Mimeo Rep School For N C State Univ Raleigh, North Carolina, 4 pp
Kirk D B, Breeman L G, Zobel B J 1972 A pulping evaluation of young loblolly pine wood. Tappi 55: 1600–1604
Klein G S 1968 Quality of wood from fertilized forests. Tappi 51: 99A–103A
Kleinschmit J, Knigge W 1967 Durch Umwald und Erbanlagen bedingte Variation der Trockensubstanzerzeugung, Struktur and Rohdichte junger Fichten (Picea abies Karst). Allg Forst-Jagdztg 138: 189–198
Klem G G 1942 Planteavstandens inflytelse på granvedens og sulfitcellulosens kvalitet. (Effect of planting space on the quality of spruce wood and sulphite pulp.) Medd Norsk Skogforsøksves 8(28): 257–293
Klem G S 1965a Tørrvolymektsvariasjoner hos vanlig gran (Picea abies (L) Karst) i Norge. (Variations in dry density of Picea abies in Norway.) Norsk Skogind 19: 348–351
Klem G S 1965b Tørrvolymvektsvariasjoner hos fremmede bartreslag og vanlig gran fra Sør- og Vestlandet. (Variations in specific gravity of foreign softwood species and Norway spruce from South and West Norway.) Medd Norsk Skogvorsøksves 20(74): 137–169
Klem G S 1968 Quality of wood from fertilized forests. Tappi 51(11): 99A–103A
Knigge W 1961 Der Einfluß verschiedener Wuchsbedingungen auf Eigenschaften und Verwertbarkeit des Nadelholzes. Allg Forst-Jagdztg 132: 149–156
Knigge W 1962 Untersuchungen über die Abhängigkeit der mittleren Rohdichte nordamerikanischer Douglasienstämme von unterschiedlichen Wuchsbedingungen. Holz Roh- Werkst 20: 352–360
Knigge W, Koltzenburg C 1965 The influence of timber qualities and ecological conditions on the cell sizes and on the proportions of types of cell in hardwoods in the temperate zones. IUFRO Sect 41 Comm Fibre Char Melbourne, Australia, Vol 2, 51 pp
Knigge W, Schulz H 1961 Einfluß der Jahrenwitterung 1959 auf Zellartverteilung, Faserlänge und Gefäßweite verschiedener Holzarten. Holz Roh- Werkst 19: 293–303
Knigge W, Schultz N 1966 Grundriß der Forstbenutzung. Paul Parey, Hamburg, 584 pp

Knudsen M V 1958 A comparative study of some technological properties of Norway spruce in a provenance test. IUFRO 12 Congr Proc 4: 219–228
Koch C B 1967 Specific gravity as affected by rate of growth within sprout clumps of black cherry. J For 65: 200–202
Koch C B, Brauner A, Kulow D 1968 Specific gravity variation within young yellow-poplar trees. West Virginia Univ Agr Expt Sta Bull 564T, 9 pp
Koch P 1972 Utilization of the southern pines. Vol I The raw material. Vol II Processing. U S For Serv Agr Handb 420, Supt of Documents, Washington DC, 1663 pp
Koch P 1985 Utilization of hardwoods growing on southern pine sites. Agr Hand No 605 I The Raw Material, II Processing, III Products and Prospective. U S For Serv, Washington D C, 3710 pp
Koltzenburg C 1966 Die Abhängigkeit der Holzeigenschaften der Rotbuche (Fagus sylvatica L.) von Lichtgenuß, soziologischer Stellung und anderen Wuchsbedingungen. Thesis, Forst Fak Univ Göttingen, Hann Münden, 121 pp
Koltzenburg C 1967 Der Einfluß von Lichtgenuß, soziologischer Stellung und des Standortes auf Holzeigenschaften der Rotbuche (Fagus sylvatica L). Holz Roh- Werkst 25: 465–473
Koo K M, Hong S M 1967 [Tracheid length and compression strength of X Pinus rigitaeda and its back-cross hybrid.] Inst For Genet, Suwon, Korea: Res Rep 5: 85–90
Krahmer R L 1966 Variation of specific gravity in western hemlock trees. Tappi 49: 227–229
Kramer P R, Smith R L 1956 Strength properties of plantation-grown slash pine. For Prod J: 129–136
Krasnitskii A M 1959 [The microscopic structure of the wood of common ash taken from different growth conditions.] Doklady Akad Nauk SSSR 126(1–6): 125–127 (Trans Plant Sci 35(10): 24–28)
Kraus J F, Spurr S H 1961 Relationship of soil moisture to the springwood-summerwood transition in southern Michigan red pine. J For 59: 510–511
Krstińic A 1968 [Variability and heritability of height and number of branches in Salix alba populations at Bakovci and Lipovljani. Part II] Sum List 92(1/2): 27–49
Kubler H 1983 Mechanism of frost crack formation in trees – a review and synthesis. For Sci 29: 559–568
Kubler H 1987 Growth stresses in trees and related wood properties. For Abstr 48(3): 131–189, For Prod Abstr 10(3): 61–119
Kulp L 1982 The wood supply and product spectrum of the future. J Inst Wood Sci 9: 150–155
Kromhout C P, Toon R E 1978 Variation of wood properties of some tropical species in southern Africa. Workshop Genetic Impr Trop Trees Brisbane, Australia, 8–45
Laarman J G 1980 Exotic plantations and their implications for U S forest products trade: A comment. Res for Future "Issues in U S international forest products trade" Washington DC, 13 pp
Lacaze J-F, Polge H 1970 Relations phénotypiques au stade juvénile entre la densité du bois et divers caractères phénologiques et de vigeur chez Picea abies Karst. Ann Sci For 27: 231–242
Ladell J L, Carmichael A J, Thomas G H 1967 Current work in Ontario on compression wood in black spruce in relation to pulp yield and quality. 8th Lake States For Tree Improv Conf (U S For Serv Res Pap NC-23), 52–60
Ladrach W E 1984 Wood quality of Pinus patula. Carton de Colombia Cali, Colombia, Res Rep 92, 17 pp
Ladrach W E 1986a Wood quality of Pinus kesiya. Carton de Colombia Cali, Colombia (Mimeo), Res Rep 111, 11 pp
Ladrach W E 1986b Control of wood properties in plantations with exotic species. Inves For Carton de Colombia Cali, Colombia, Res Rept 106, 11 pp
Ladrach W E 1986c Control of wood properties in plantations. IUFRO Congr Ljubljana, Yugoslavia, 369–381
Ladrach W E 1987 Wood quality of Pinus oocarpa. Carton de Colombia Cali, Colombia, Res Rep 116, 7 pp
Ladrach W E, Pereira G 1987 Growth and heritability estimates for eight-year-old open-pollinated progeny of Cupressus lusitanica. Carton de Colombia Cali, Colombia, Res Rept 117, 13 pp
La Farge T 1974 Genetic differences in stem form of ponderosa pine grown in Michigan. Silvae Genet 23: 211–213
Lamb A F 1968 Fast growing timber trees of the lowland tropics No 1: Gmelina arborea. Commonw For Inst Oxford, England, 31 pp

Lamb F M, Wengert E M 1987 The drying properties of lumber containing juvenile wood. 41st Ann Meet FPRS Louisville, Kentucky, 2 pp
Land S B, Dicke S G, Tuskan G A, Patterson P E 1983 Genetic, site and within-tree variation in specific gravity and moisture content in young sycamore trees. Tappi 66: 149–155
Land S B, Lee J C 1981 Variation in sycamore wood specific gravity. Wood Sci 13: 166–170
Langlet O 1967 Regional intra-specific variousness. IUFRO Kongress München III(22): 435–458
Lantican C B, Hughes J F 1973 Variation of tracheid widths and wall thicknesses in P. caribaea from British Honduras. Trop Prov Prog Res Inter Coop Nairobi, Kenya, 528–531
Lantz C W, Hofmann J G 1969 Geographic variation in growth and wood quality of loblolly pine in North Carolina. 10th South Conf For Tree Improv Houston, Texas, 175–188
Larson P R 1957 Effect of environment on the percentage of summerwood and specific gravity of slash pine. Yale Univ Sch For Bull 63, 87 pp
Larson P R 1960 A physiological consideration of the springwood-summerwood transition in red pine. For Sci 6: 110–122
Larson P R 1962 A biological approach to wood quality. Tappi 45: 443–448
Larson P R 1963a Evaluating the environment for studies of the inheritance of wood properties. 1st World Consul For Gen Tree Improv Stockholm, 6 pp
Larson P R 1963b The indirect effect of drought on tracheid diameter in red pine. For Sci 9: 52–62
Larson P R 1967a Effects of temperature on the growth and wood formation of ten Pinus resinosa sources. Silvae Genet 16: 58–65
Larson P R 1967b Silvicultural control of the characteristics of wood used for furnish. Proc 4th TAPPI For Biol Conf New York, pp 143–150
Larson P R 1969a Incorporation of ^{14}C in the developing walls of Pinus resinosa tracheids: compression wood and opposite wood. Tappi 52: 2170–2177
Larson P R 1969b Wood formation and the concept of wood quality. Yale Univ Sch For Bull 74, 54 pp
Larson P R 1972 Evaluating the quality of fast-grown coniferous wood. Proc West For Conf Seattle, Washington, 1–7
Larson P R 1973 The physiological basis for wood specific gravity in conifers. IUFRO Div 5 Meet Brisbane, Australia 2: 672–680
Lassen L E, Okkonen E A 1969 Effect of rainfall and elevation on specific gravity of coast Douglas-fir. Wood Fiber 1: 227–235
Law K N, Koran Z 1982 Kraft pulping of white spruce bole and branch. Can J For Res 12: 299–303
Law K N, Lapointe M 1983 Chemimechanical pulping of boles and branches of white spruce, white birch and trembling aspen. Can J For Res 13: 412–417
Lawton R O 1984 Ecological constraints on wood density in a tropical montane rain forest. Am J Bot 71: 261–267
Ledig F T, Whitmore J L 1981 Heritability and genetic correlations for volume, foxtails and other characteristics of Caribbean pine in Puerto Rico. Silvae Genet 30: 88–92
Ledig F T, Zobel B J, Mathias M M 1975 Geoclimatic patterns in specific gravity and tracheid length in wood of pitch pine. Can J for Res 5: 318–329
Lee C H 1971 Trunkwood-branchwood specific gravity and tracheid length relationships in Pinus nigra. For Sci 17: 62–63
Lee C H 1974 Geographic variation of growth and wood properties in eastern white pine – 15 year results. 21st Northeast For Tree Improv Conf Univ New Brunswick Fredericton, New Brunswick, 36–41
Lee C H 1975 Geographic variation of growth and wood properties in Japanese larch in southwestern lower Michigan. 12th Lake States For Tree Improv Conf Michigan, 35–46
Lee C H 1979 Absence of growth-wood property correlation in twenty-seven black pine seed sources. Wood Fiber 11: 22–28
Lemaster R L, Troxell H E, Sampson G R 1983 Wood utilization potential of beetle-killed lodgepole pine for solid wood products. For Prod J 33(9): 64–68
Levi M 1978 Blue-flecked panneling: a new market for southern pine beetle-killed trees. South Lumberm 237(2944): 70–71
Levi M P 1981 A guide for using beetle-killed southern pine based on tree appearance. South pine beetle handbook U S For Serv Agr Handb 572, 18 pp
Lewark S 1979 Wood characteristics in Norway spruce breeding programs. IUFRO Meet Bucharest, Romania, 24 pp

Lewark S 1980 Vergleichende Untersuchung der Jahrringbreite und Rohdichte von Stecklingen und Sämlingen der Lärche. Forstarchiv 51: 47–54
Lewark S 1981a Holzqualitätsmerkmale in Fichtenzüchtungsprogrammen. Grundsätzliche Überlegungen und Literaturauswertungen aus der Sicht der Forstbenutzung. Forst Holzwirt 36: 354–357
Lewark S 1981b Variabilität der Ästigkeit und Rundholzqualität der Fichte. Holz-Zbl 107: 275–277
Lewark S 1981c [Studies in wood characteristics of young spruce (Picea abies): ring width, density and limbiness of cuttings from clones]. Dissert Georg-August Univ Göttingen, Germany, 193 pp
Lewark S 1982 Untersuchungen von Holzmerkmalen junger Fichten aus Stecklingsklonen. Forstarchiv 53: 14–21
Lewis N B 1965 Economics of pruning for long lumber in South Australia. Aust For 29: 149–160
Lima R T 1987 Stability of genotypes of Pinus oocarpa for specific gravity in three different environments in South America. M S Thesis Sch For Res N C State Univ Raleigh, North Carolina, 48 pp
Lindgren R M 1951 Deterioration of southern pine pulpwood during storage. For Prod Res Soc 5th Ann Meet Philadelphia, Pennsylvania, 12 pp
Linnartz N E, Choong E T, Underwood P 1970 Diameter growth, specific gravity and tracheid length in 4-year-old loblolly pine in response to fertilizer treatment. Louisiana State Univ Baton Rouge, Louisiana Note 90, 4 pp
Littleford T W 1961 Variation of strength properties within trees and between trees in a stand of rapid-grown Douglas-fir. Can Dept For For Prod Lab Vancouver, Canada, 20 pp
Loo J, Fowler D P, Schneider M H 1982 Geographic variation in specific gravity among Japanese larch from different provenances. Wood Fiber 14: 281–286
Loo J A, Tauer C C, McNew R W 1985 Genetic variation in the time of transition from juvenile to mature wood in loblolly pine (Pinus taeda). Silvae Genet 34: 14–19
Low A J 1964 Compression wood in conifers: A review of literature. For Abstr 25(3,4): 1–13
Lowerts G A, Kellison R C 1981 Genetically controlled resistance to discoloration and decay in wounded trees of yellow-poplar. Silvae Genet 30(2–3): 98–101
Lowery D P, Schmidt W C 1967 Effect of thinning on the specific gravity of western larch crop trees. U S For Serv Res Note INT-70, 6 pp
Lutz J 1964 How growth rate affects properties of softwood veneer. For Prod J 24: 97–102
MacDonald R G, Franklin J N 1969 Pulp and paper manufacture. 2nd Ed, Vol I: The pulping of wood. McGraw-Hill, New York, 769 pp
MacPeak M D, Burkhart L F, Weldon D 1987 A mill study of the quality, yield and mechanical properties of plywood produced from fast-grown loblolly pine. For Prod J 37(2): 51–56
Madsen T L, Moltesen P, Olesen P O 1978 [The influence of thinning degree on basic density, production of dry matter, branch thickness, and number of branches of Norway spruce.] Beretn Forst Forsøgsv Danmark 36(2): 183–203
Maeglin R R 1966 Predicting specific gravity of plantation-grown red pine. U S For Ser Res Note FPL-0149, 14 pp
Maeglin R R 1974 The effect of site quality and growth rate on the anatomy and utilization potential of northern red oak. 2d Ann Hardwood Symp Cashiers, North Carolina, 191–203
Maeglin R R, Quirk J T 1983 The effect of growth rate on tissue proportions and cell morphology for North American oaks. IUFRO Conf Div 5 Madison, Wisconsin, 39 (Abstracts)
Maeglin R R, Quirk J T 1984 Tissue proportions and cell divisions for red and white oak groups. Can J For Res 14: 101–106
Malan F S, Toon R E 1980 Natural defects in the timber of South African-grown Pinus and Eucalyptus species. Dept For Envir Conserv Pamphlet 250 South Africa, 57 pp
Manwiller F G 1972 Volumes, wood properties, and fiber dimensions of fast- and slow-grown spruce pine. Symp effect growth acceler wood prop. U S For Serv FPL Madison, Wisconsin, 1–9
Markstrom D C, Troxell H E, Boldt C E 1983 Wood properties of immature ponderosa pine after thinning. For Prod J 33(4): 33–36
Martin J W 1984 Forest management practices that will influence product characteristics of the changing wood resource. Symp util chang wood res south U S N C State Univ Raleigh, North Carolina, 9 pp
Martin R B 1961 The effect of stand density on the specific gravity of loblolly pine. M S Thesis Louisiana State Univ Baton Rouge, Louisiana

Marton R 1986 Research effort directed at funding "ultimate" chemimechanical pulp. Pulp Paper 60(6): 81–86
Marts R O 1950 Wood quality of bud-pruned longleaf pine. South Lumberm 181(2273): 197–199
Matziris D I 1979 Variation in wood density in radiata pine grown from four seed sources in two sites in Greece. Silvae Genet 22: 104–106
Matziris D I, Zobel B J 1973 Inheritance and correlation of juvenile characteristics in loblolly pine (P. taeda). Silvae Genet 22: 38–45
Mayer B K 1971 Practical experience in the use of mixed tropical hardwood for the manufacture of pulp and paper. U N Ind Grp Org Vienna, Austria, 12 pp
McAdoo J C, Murphey W K 1968 The effect of controlled nutrient environment on the tensile properties of wood. Sch For Res Penn State Univ Res Briefs 2(4): 78–82
McCutchan B G 1982 Estimated gain in form, quality and volumetric traits in American sycamore in response to the selection for tree dry weight. M S Thesis School For Res N C State Univ Raleigh, North Carolina, 82 pp
McElwee R L, Faircloth J P 1966 Variation in wood properties of water tupelo in the southeastern coastal plain. Tappi 49: 538–541
McElwee R L, Zobel B J 1962 Some wood and growth characteristics of pond pine. For Gen Workshop Soc Am For-South For Tree Improv Commit Macon, Georgia, 18–25
McGee C E 1968 Northern red oak seedlings growth varies by light intensity and seed source. U S For Ser Res Note SE-90, 4 pp
McGinnes E A, Phelps J E 1980 Color of black walnut – macroscopic and microscopic studies. IUFRO Conf Div 5 Oxford, England, 7 (Abstract)
McKee J C 1984 The impact of high volumes of juvenile wood on pulp mill operations and operating costs. Symp util chang wood res south U S Raleigh, North Carolina, 6 pp
McKimmy M D 1959 Factors related to variation in specific gravity in young-growth Douglas-fir. Oregon For Prod Res Cent Bull 8, 52 pp
McKimmy M D 1965 A variation and heritability study of wood specific gravity in 46-year-old Douglas-fir from known seed sources. 3rd TAPPI For Biol Conf Seattle, Washington, (Tappi 49: 542–549, 1966)
McKimmy M D, Campbell R K 1982 Genetic variation in the wood density and ring width trend in coastal Douglas-fir. Silvae Genet 31: 43–51
McKimmy M D, Ching K K 1968 Correlating specific gravities of branch and bole wood in young Douglas-fir. Oregon State Univ For Res Lab G-8, 8 pp
McKimmy M D, Nichols D D 1971 Genetic differences in wood traits among half-century-old families of Douglas-fir. Wood Fiber 2: 347–355
McKinley C R, Lowe W J, van Buijtenen J P 1982 Genetic improvement of wood specific gravity in loblolly pine (P. taeda) and its relation to other traits. TAPPI R & D Div Conf Asheville, North Carolina, 153–158
McKinnell F H, Rudman P 1973 Potassium fertilizer and wood density of Pinus radiata. Appita 26: 283–286
McMillin C W 1969 Ash content of loblolly pine as related to specific gravity, growth rate and distance from the pith. Wood Sci 2: 26–30
McMillin C W 1973 Fibril angle of loblolly pine wood as related to specific gravity, growth rate and distance from the pith. Wood Sci Technol 7: 251–255
McMillin C W 1978 Complete tree utilization of the southern pines. For Prod Res Soc New Orleans, Louisiana, 484 pp
McNab W H 1983 Total tree and product weight of beetle-killed loblolly pines in northeast Georgia. Georgia For Res Pap 42, 11 pp
McWilliam J R 1955 Caribbean pine (Pinus caribaea); some notes on its development and characteristics in Central America. Queensland For Serv Res Note 4, 29 pp
Megraw R A 1985 Wood quality factors in loblolly pine. TAPPI Press Atlanta, Georgia, 89 pp
Megraw R A 1986 Effect of silvicultural practices on wood quality. TAPPI Res Dev Conf Raleigh, North Carolina, pp 27–34
Megraw R A, Munk W P 1974 Effect of fertilization and thinning on wood density of young coastal Douglas-fir. TAPPI For Biol Conf Seattle, Washington, pp 85–91

Megraw R A, Nearn W T 1972 Detailed DBH density profiles of several trees from Douglas-fir fertilizer-thinning plots. Proc symp effect growth accel on wood prop. U S For Serv FPL Madison, Wisconsin, 22 pp

Meier R J, Goggans J F 1977 Heritabilities of height, diameter and specific gravity of young Virginia pine. For Sci 23: 450–456

Mell C O 1910 Determination of quality of locality by fiber length in wood. For Q 8: 419–422

Mergen F 1954 Inheritance of oleoresin yield in slash pine. Naval Stores Review (Dec Issue), 2 pp

Mergen F, Furnival G M 1960 Discriminant analysis of Pinus thunbergii×Pinus densiflora hybrids. Proc Soc Amer For, 36–40

Mergen F, Burley J, Yeatman C W 1964 Variation in growth characteristics and wood properties of Norway spruce. Tappi 48: 499–504

Mergen F, Hoekstra P E, Echols R M 1955 Genetic control of oleoresin yield and viscosity in slash pine. For Sci 1: 19–30

Meyer-Uhlenreid K H 1957 Holzanatomische Untersuchungen an der Pappel. Holzforschung 11: 150–157

Meylan B A 1968 Cause of high longitudinal shrinkage in wood. For Prod J 18: 75–78

Mikami S 1973 Preliminary studies on the genetic improvement in spiral grain in Japanese larch. Proc IUFRO Div 5, South Africa 2: 746–751

Miler Z, Miler A, Pasternak B 1979 [Wood fibre length in pine provenance samples.] Prace Komisji Nauk Rolniczychi Komisji Nauk Lesnych 48: 95–101

Mio S, Matsumoto T 1983 Wood properties of sugi (Cryptomeria japonica) cultivars. IUFRO Conf Div 5 Madison, Wisconsin, 43 (Abstracts)

Mitchell H L 1960 Development of an adequate concept of wood quality for the guidance of geneticists and forest managers. 5th World For Congr Seattle, Washington, 1341–1348

Mitchell H L 1961 Apical growth in relation to the transition from springwood to summerwood in conifers. Univ Brit Columbia For Club Res Comm 19, 71 pp

Mitchell H L 1963 Specific gravity variation in North American conifers. IUFRO Sect 41 Commit on Fibr Char Melbourne, Australia Vol 2, 15 pp

Mitchell H L 1964 Patterns of variation in specific gravity of southern pines and other coniferous species. Tappi 47: 276–283

Mitchell H L 1971 Effect of nitrogen fertilizer on the growth rate and certain wood quality characteristics of sawlog size red oak, yellow poplar, and white ash. Symp effect growth accel on wood prop U S For Serv FPL Madison, Wisconsin, 9 pp

Moehring D M 1964 Speeding up growth of the loblolly. For Farmer, 9, 13–14

Mohrdiek O 1979 Holzdichte und Faserlänge bei Populus-Leuce-Arten. Holzzucht 33(3/4): 23–26

Montagna R G, Barbosa O, Brasil M A, Veiga R A, Toledo D V 1983 Peso do matéria seca, densidade básica e dimensoes de fibras de Pinus caribaea hondurensis. Fast Growing Trees Silvicultura No 32: 837–840

Mork E 1928 Die Qualität des Fichtenholzes unter besonderer Rücksichtnahme auf Schleif- und Papierholz. Papier-Fabr 26: 741–747

Mothe F 1983 [Study of the genetic variability between and within populations of wood quality and growth in common spruce.] Centre National de Recherches Forestières Univ Nancy, 111 pp

Mottet A 1965 Considerations on the relationship between certain anatomical characteristics and density in tropical woods. IUFRO Sect 41 Melbourne, Australia, 15 pp

Mozina I 1960 Über den Zusammenhang zwischen Jahrringbreite und Raumdichte bei Douglasienholz. Holz Roh- Werkst 18: 409–413

Mullin L J, Gough D, Carter D T 1980 Provenance trials of Eucalyptus nitens. IUFRO Meet, Oxford, England, 1–5

Murphey W K, Brady S L, Behan M J 1969 Anatomy and physical properties of two-year western larch seedlings grown in mineral deficient solutions. For Prod J 19(2): 28–31

Murphey W K, Brisbin R L 1970 Influence of sewage plant effluent irrigation on crown wood and stem wood of red pine. Bul 772 Pennsylvania State Univ, 29 pp

Murthy S V 1960 Density variation in the timber of ramin, Conystylus bancanus. Empir For Rev 39(3): 324–329

Mutibaric J 1967 [Correlation between ring width and wood density in Euramerican poplars.] Sumerstvo 20(1/2): 39–46

Myers C A 1960 Estimating oven-dry weight of pulpwood in standing ponderosa pines. J For 58: 889–891

Nair K R, Mukerji H K 1957 A statistical study of variability of physical and mechanical properties of Tectona grandis (teak) grown at different localities of India and Burma and the effects of the variability on the choice of the sampling plan. Indian For Res 1(1): 49

Namkoong G, Barefoot A C, Hitchings R G 1969 Evaluating control of wood quality through breeding. Tappi 52: 1935–1938

Namkoong G, Squillace A E 1970 Problems in estimating genetic variance by Shrikhande's method. Silvae Genet 19: 74–77

Nanson A, de Jamblinne de Meux A, Istas J R, Schalk J 1975a [Inheritance of wood density in Pinus nigra "Koekelare".] Bull Soc R For Belg 82: 68–79

Nanson A, Sacré E, Fraipont L 1975b Tests précoces de la qualité du bois de provenances d'Epicéa commun. Bull Rech Agr Gembloux, Belgion 10: 527–558

Nariyoshi A H, Redko B V, Coutinho S C 1986 A study of five populations of Pinus caribaea var. hondurensis. 18th Congresso Anual da ABCP – Semano do Papel, São Paulo, (Mimeo) 18 pp

Nebgen R J, Lowe W J 1982 Inheritance of growth, branch angle and specific gravity in three American sycamore populations. Silvae Genet 31: 86–89

Nelson N D, Burk T, Isebrands J G, Ek A R 1979 Clone and spacing effects on branch characteristics of short rotation intensively cultured Populus. 1st North Central Tree Improv Conf Madison, Wisconsin, 74–85

Nelson N D, Maeglin R R, Wahlgren H E 1969 Relationship of black walnut wood color to soil properties and site. Wood Fiber 1: 29–37

Nelson P F 1977 Pulping of fertilized P. radiata. Res Note R77/132 APM Forests Victoria Australia (Mimeo Rep), 4 pp

Nelson P F, Hall M J, Hansen M W, Henderson V T, Turvey N 1980 The effect of silvicultural practices on kraft pulping properties of P. radiata. Appita 33: 368–378

Nepveu G 1984a Contrôle héréditaire de la densité et de la rétractibilité du bois de trois espèces de Chêne (Quercus petraea, Quercus robur et Quercus rubra). Silvae Genet 33: 110–115

Nepveu G 1984b Déterminisme génotypique de la structure anatomique du bois chez Quercus robur. Silvae Genet 33: 91–95

Nepveu G, Birot Y 1979 Les corrélations phénotypiques juvénile-adulte pour la densité du bois et la vigueur chez l'Epicéa. Ann Sci For 36: 125–149

Nepveu G, Keller R, Tessier du Cros E 1978 Sélection juvénile pur la qualité du bois chez certains peupliers noirs. Ann Sci For 35:69–72

Nepveu G, Tran Ngoc T 1984 Relations entre les composantes densitométriques et l'état de surface de planches rabotées chez Pseudotsuga menziesii. Ann Sci for 41: 171–194

Nepveu G, Teissier du Cros E 1976 Juvenile selection for wood quality in euramerica poplar. IUFRO Meet Bordeaux, France, 38 pp

Nepveu G, Velling P 1983a Variabilité génétique individuelle de la qualité du bois chez Betula pendula Roth. Silvae Genet 32: 37–49

Nepveu G, Velling P 1983b Rauduskoivun puuaineen laadun geneettinen vaihtelu. (Individual genetic variability of wood quality in Betula pendula.) Folia For 575, 21 pp

Nicholls J W 1967a Assessment of wood qualities for tree breeding IV. Pinus pinaster grown in western Australia. Silvae Genet 16: 21–28

Nicholls J W 1967b Preliminary observations on the change with age of the heritability of certain wood characters in Pinus radiata clones. Silvae Genet 16: 18–20

Nicholls J W 1971a The effect of environmental factors on wood characteristics. Silvae Genet 20: 26–33

Nicholls J W 1971b The effect of environmental factors on wood characteristics. II: The effect of thinning and fertilizer treatment on the wood of Pinus pinaster. Silvae Genet 20: 53–100

Nicholls J W, Brown A G 1973 The relationship between ring width and wood characteristics in double-stemmed trees of radiata pine. N Z J For Sci 4: 105–111

Nicholls J W, Brown A G 1974 The inheritance of heartwood formation in Pinus radiata. Silvae Genet 23: 138–141

Nicholls J W, Dadswell H E 1961 Structure in relation to silvicultural treatment of Pinus radiata. CSIRO Div Tech Pap For Prod, 11–12

Nicholls J W, Dadswell H E, Fielding J M 1964 The heritability of wood characteristics of Pinus radiata. Silvae Genet 13: 68–71

Nicholls J W, Dadswell H E, Perry D H 1963 Assessment of wood qualities for tree breeding. Silvae Genet 12: 105–110

Nicholls J W, Fielding J M 1964 The effect of growth rate on wood characteristics. Commonw Sci Ind Res Org 19(1): 24–30

Nicholls J W, Morris J D, Pederick L A 1980 Heritability estimates of density characteristics in juvenile radiata wood. Silvae Genet 29: 54–61

Nicholls J W, Pawsey C K, Brown A G 1976 Further studies on the ortet-ramet relationship in wood characteristics of Pinus radiata. Silvae Genet 25: 73–79

Nicholls J W, Waring H D 1977 The effect of environmental factors on wood characteristics IV: Irrigation and partial droughting of Pinus radiata. Silvae Genet 26: 107–111

Nicholls J W, Wright J P 1976 The effect of environmental factors on wood characteristics. 3. The influence of climate and site on young Pinus radiata material. Can J For Res 6: 113–121

Nikles D G 1966 Comparative variability and relationship of Caribbean pine (P. caribaea) and slash pine (P. elliottii). PhD Thesis N C State Univ Raleigh, North Carolina, 200 pp

Noskowiak A F 1963 Spiral grain in trees: a review. For Prod J 13: 266–275

Noskowiak A F, Snodgrass J D 1961 Effect of locality of growth on certain strength properties of Douglas-fir. Final Rep to Western Pine Assoc Proj RP, 547 pp

Nylinder P 1953 Volymviktsvariationer hos planterad gran. (Variations in density of planted spruce). Medd Stat Skogsforskningsinst 43(3), 43 pp

Nylinder P, Hägglund E 1954 Ståndorts och trädegenskaper inverkan på utbyte och kvalitet vid framställning av sulfitmassa av gran. (The influence of stand and tree properties on the yield and quality of sulphite pulp of Swedish spruce (Picea excelsa).) Medd Stat Skogsforskningsinst 44(11), 184 pp

Okkonen E A, Wahlgren H E, Maeglin R R 1972 Relationships of specific gravity to tree height in commercially important species. For Prod J 22(7): 37–41

Olesen P O 1973 Forest tree improvement 6. The influence of the compass direction on the basic density of Norway spruce (Picea abies L.) and its importance for sampling for estimating the genetic value of plus trees. For Tree Impr Arboretet Hørsholm, Denmark, 58 pp

Olesen P O 1976 The interrelation between basic density and ring width of Norway spruce. Forst Forsøgsv, Danmark 34(4): 341–359

Olesen P O 1977 The variation of the basic density level and tracheid width within the juvenile and mature wood of Norway spruce. For Tree Impr 12 Norway, 21 pp

Olesen P O 1982 The effect of cyclophysis on tracheid width and basic density in Norway spruce. Forest Tree Impr Arboretet Hørsholm, Denmark, 80 pp

Ollinmaa P J 1960 Eräistä ojitetuilla soilla kasvavan puun fysikaalisista ominaisuuksista. (On certain physical properties of wood growing on drained swamps.) Acta For Fenn 72.2, 24 pp

Olson J R, Arganbright D C 1977 The uniformity factor – a proposed method for expressing variations in specific gravity. Wood Fiber 9: 202–210

Orman H R 1956 Effect of thinning on wood density of conifers. Unpubl report FRI N Z For Ser, Rotorua, New Zealand

Otegbeye G O, Kellison R C 1980 Genetics of wood and bark characteristics of Eucalyptus viminalis. Silvae Genet 29: 27–31

Oteng-Amoako A, de Zeeuw C H, Meyer R W 1983 Variation of selected anatomical properties and specific gravity in the stem of naturally grown and plantation grown Triplochiton scleroxylon from Ghana. IUFRO Conf Div 5 Madison, Wisconsin, 45–46 (Abstracts)

Paddon A R 1984a Biomass energy activities. Appropr Technol 11(2): 8–10

Paddon A R 1984b An improved pit klin for charcoal production. Appropr Technol 11(2): 7–8

Palmer E R 1973 Gmelina arborea as a potential source of hardwood pulp. Trop Sci 15: 243–260

Palmer E R 1977 The potential of fast-growing tropical pines for pulp and paper. Trop Sci 19: 127–140

Palmer E R, Bin P T 1966 Pulping studies on Malayan exotic species. Pinus caribaea. Res Pam 55 Trop Prod Inst, 24 pp

Palmer E R, Ganguli S 1982 Pulping characteristics of Pinus caribaea grown in Sri Lanka. Trop Prod Inst L57 London, 21 pp

Palmer E R, Gibbs J A 1967 The pulping characteristics of Pinus caribaea from Sabah. Trop Prod Inst L12 London, 23 pp

Palmer E R, Gibbs J A 1977 Pulping characteristics of Pinus caribaea from Fiji: the effect of rate of growth. Trop Prod Inst Rep 246 London, 15 pp

Palmer E R, Gibbs J A 1974 Pulping characteristics of Gmelina arborea an Bursera simaruba from Belize. Trop Prod Inst Rep London, 5 pp

Palmer E R, Gibbs J A 1978 Pulping characteristics of six hardwoods from Guyana. Trop Prod Inst Rep London, 43 pp

Palmer E R, Gibbs J A, Dutta A P 1982 Pulping trials of wood species in plantations in Kenya. Trop Prod Inst L61 London, 58 pp

Palmer E R, Gibbs J A, Dutta A P 1983 Pulping characteristics of hardwood species growing in plantations in Fiji. Trop Prod Inst Rep London, 38 pp

Palmer E R, Greenhalgh P 1983 The production of pulp and paper on a small scale. Trop Dev Res Inst G 175 London, 53 pp

Palmer E R, Johnson J S, Ganguli S, Gibbs J A, Dutta A P 1982 Pulping trials of Pinus patula and Pinus radiata grown in plantations in Kenya. Trop Prod Inst L 63 London, 56 pp

Palmer E R, Johnson J S, Gibbs J A, Ganguli S, Dutta A P 1984 Pulping characteristics of Gmelina arborea grown in plantations in the Soloman Islands. Trop Dev Res Inst L65 London, 21 pp

Palmer E R, Tabb C B 1968 The production of pulp and paper from coniferous species grown in the tropics. Trop Sci 10(2): 79–99

Palmer E R, Tabb C B 1971 Pinus caribaea – its potential as pulpwood. Trop Prod Inst Rept London, 23–44

Palmer E R, Tabb C B 1974 Pulpwood production prospects. Trop Sci 16: 207–236

Panshin A J, de Zeeuw C H, Brown H P 1964 Textbook of wood technology. McGraw-Hill, New York, 643 pp

Parham R A, Robinson K W, Isebrands J G 1976 The effects of tension wood in kraft paper from a short-rotation hardwood (Populus tristis). For Prod Res Soc 30th Ann Meet Toronto, Canada, 23 pp

Park J C 1982 Occlusion and the defect core in pruned radiata pine. FRI Rotorua, New Zealand, Bull 2, 15 pp

Park J C, Parker C E 1982 Predicting value loss due to resin pockets in timber from radiata pine. FRI Rotorua, New Zealand, Bull 6, 14 pp

Parker M L, Hunt K, Warren W G, Kennedy R W 1976 Effect of thinning and fertilization on intra-ring characteristics and kraft pulp yield of Douglas-fir. Appl Polym Symp 28: 1095–1086

Parrot L 1960 De la variabilité génétique de la densité du bois chez l'Epicéa (Picea excelsa Link). Ann Ec Nat Eaux For 17: 265–334

Pastor Y P de Castro 1977 A provenance trial with Eucalyptus pilularis. 3rd World Consul For Tree Breed Canberra, Australia 1: 371–380

Paul B H 1927 Producing dense southern pine timber in second-growth forests. South Lumberm 128(1668): 46–47

Paul B H 1932 The relation of certain forest conditions to the quality and value of second-growth loblolly pine lumber. J For 30: 4–21

Paul B H 1941 Thinning and quality, sudden acceleration of diameter growth in vertical and leaning longleaf pine trees in relation to quality of lumber. South Lumberm, 163(2057): 203–206

Paul B H 1953 Forest genetics in relation to wood quality. Proc Lake States For Gen Conf Misc Rept 22, 55–59

Paul B H 1958 Specific gravity changes in southern pines after release. South Lumberm, 197(2465): 122–124

Paul B H 1959 The effect of environmental factors on wood quality. U S For Serv Rep FPL-2170, 50 pp

Paul B H 1960 The juvenile core in conifers. Tappi 43: 1–2

Paul B H, Marts R O 1931 Controlling the proportion of summerwood in longleaf pine. J For 29: 89–95

Paul B H, Smith D M 1950 Summary of growth in relation to quality of southern yellow pine. U S For Serv Rep FPL-1751, 19 pp

Pearson F G, Fielding H A 1961 Some properties of individual growth rings in European larch and Japanese larch and their influence upon specific gravity. Holzforschung 15: 82–89

Pearson R G, Gilmore R C 1980 Effect of fast growth rate on the mechanical properties of loblolly pine. For Prod J 30(5): 47–54
Pearson R G, Weir R J, Smith W D 1980 Utilization of pine thinnings. South For Econ Workers Conf Univ South Mississippi Long Beach, Mississippi, 27 pp
Pederick L A 1970 Variation and inheritance of stem form and bark thickness in young loblolly pine. Tech Rept 41 School For Res N C State Univ Raleigh, North Carolina, 44 pp
Perem E, Clermont L P 1961 Importance of tension wood in appraising wood quality. IUFRO 13 Congr Proc Wien 2: 41–42
Perry T O 1960 The inheritance of crooked stem form in loblolly pine (Pinus taeda). J For 58: 943–947
Perry T O, Wang C W 1958 Variation in the specific gravity of slash pine wood and its genetic and silvicultural implications. Tappi 41: 178–180
Persson A 1975 Ved och pappersmassa från gran och tall i olika förband. (Wood and pulp of Norway spruce and Scots pine at various spacings). Skogshögsk Stockholm Inst Skogsprod Rapp Upps 37, 145 pp
Persson A 1977 Kvalitetsutveckling inom yngre förbandsförsök med tall. (Quality development in young spacing trials with Scots pine.) Skogshögsk Stockholm Inst Skogsprod Rapp Upps 45, 152 pp
Persson A 1982 Contortatallen kvalitet i jämförelse med tall och gran. Massa, papper och sågad vara. (The quality of Pinus contorta in comparison to Scots pine and Norway spruce. Pulp, paper and sawn timber.) Sver Skogsvårdsförb Tidskr 80(1/2): 37–42
Peterson T A 1968 Variation in radial growth patterns and specific gravity of red pine (Pinus resinosa). For Prod J 18: 63
Petrik A W 1968 Comparison of anatomical and pulping properties of some fast- and slow-grown trees. For Prod J 18: 62
Petrini S 1959 The two oldest Swedish provenance tests in Scots pine. Med Statens Skogsforskningsinst 48(11): 1–49
Petroff G 1973 Influence of age of plantation eucalypts on their papermaking characteristics. Bois Forêts Trop 150: 63–67
Phelps J E, Isenbrands J G, Teclaw R M 1983 Raw material evaluations of short rotation intensively cultured Populus coppice. IAWA Bull 4: 10 (Summary)
Pillow M Y, Bray M W 1935 Properties and sulphate pulping characteristics of compression wood. Paper Trade J 101(26): 31–34
Pillow M Y 1952 Some characteristics of young plantation-grown red pine in relation to the properties of the wood. J For Prod Res Soc 2(1): 25–31
Pirito D D, Crews D L, Troxell H E 1974 The effects of dwarf mistletoe on the wood properties of lodgepole pine. Wood Fiber 6: 26–35
Plank M E 1984 Lumber recovery from insect-killed lodgepole pine in the northern Rocky Mountains. U S For Ser Res Paper PNW-320, 12 pp
Plumptre R A 1977 Variation of growth, stem quality, and wood properties in relation to genetic improvement of tropical forest trees. 3rd World Consul For Tree Breed Canberra, Australia, 28 pp
Plumptre R A 1978 Some wood properties of Pinus patula from Uganda and techniques developed in studying them. Commonw for Inst Oxford Univ U.K. Occasional Paper
Plumptre R A 1983 Pinus caribaea Vol II Wood properties. Trop For Pap 17 Commonw for Inst, Oxford Univ, 145 pp
Polge H 1963 [Contribution to the study of the quality of wood of the principal exotic coniferous species used in French reforestation.] Annales de l'Ecole Nationale des Eaux, de Forêts et de la Station de Recherches et Expériences 20: 403–469
Polge H 1964 Le bois juvénile des conifères. Rev For Fran 16:474–505
Polge H 1965 The use of curves of density variation for the study of environmental factors and in particular of climatic factors. Proc IUFRO Sect 41 Melbourne, Australia, 8 pp
Polge H 1968 Tests précoses de la qualité du bois sur 25 provenances d'Abies grandis. Ann Sci For 25: 3–23
Polge H 1969a [Density of planting and pruning of live branches, or why, when and how to prune.] RFF 21 Special "Silviculture" 31, 451–462
Polge H 1969b Influence de la fertilisation sur la qualité du bois de Pin maritime. Ann Sci For 26: 45–64

Polge H 1973 Facteurs écologique et qualité du bois. Ann Sci For 30:307–328
Polge H 1975 Etude préliminaire de l'influence de la fertilisation sur la qualité du bois de Chêne. Rev For Fran 27: 201–208
Polge H, Garros S 1971 Influence de défoliations sur la structure du bois de Pin maritime. Ann Sci For 28: 195–206
Polge H, Garros S, Michel P 1971 Héritabilité de la densité du bois de Sapin pectiné. Ann Sci For 28: 185–194
Polge H, Illy G 1967 Observations sur l'anisotropie du Pin maritime des Landes. Ann Sci For 24: 205–231
Polge H, Illy G 1968 Héritabilité de la densité du bois et corrélations avec la croissance étudiées a l'aide de tests non destructifs sur plants de Pins maritimes de quatre ans. Silvae Genet 17: 173–181
Polge H, Keller R 1973 Qualité du bois et largeur d'accroissements en forêt de Trançais. Ann Sci For 30: 91–125
Polge H, Keller R, Thiercelin F 1972 [Effects of soil and heredity on the growth and the anatomical characteristics of young plants of beech.] Inst Nat de la Recherche Agron Centre Nat de Recherches For Station de Recherches sur la Qualité des Bois Nancy, France, 24 pp
Polk R B 1972 Heritabilities of some first-order branching traits in Pinus banksiana. 8th Central States For Tree Improv Conf Columbia, Missouri, 33–39
Porterfield L R, Crist J B 1979 Impacts of the changing quality of timber resources. Proc Joint Meet Econ Financial Mgt Comm FPRS Atlanta, Georgia, 49 pp
Posey C E 1965 Effects of fertilization upon wood properties of loblolly pine (Pinus taeda). 8th South Conf For Tree Improv Savannah, Georgia, 126–130
Posey C E 1969 Phenotypic and genotypic variation in eastern cottonwood in the southern great plains. 10th South Conf For Tree Improv Houston, Texas, 130–135
Posey C E, Bridgwater F E, Buxton J A 1969 Natural variation in specific gravity, fiber length and growth rate of Eastern cottonwood. Tappi 52: 1508–1511
Posey C E, Bridgwater F E, Walker N 1970 Effect of seed origin on tracheid length, specific gravity, and volume of shortleaf pine in Oklahoma. For Sci 16: 66–70
Prestemon D R 1967 Variation in heart stain and density in tanoak. For Prod J 17(7): 33–41
Preston S B 1980 Characteristics of wood from a heterogeneous forest resource in relation to utilization. IUFRO Conf Div 5 Oxford, England, 157–167
Priestley J H 1930 Studies in the physiology of cambial activity. New Phytol 29: 56–75, 96–140, 316–354
Pritchard R P, Bailey I W 1916 The significance of certain variations in the anatomical structure of wood. For Q 14: 662–672
Pronin D 1971 Estimating tree specific gravity of major pulpwood species of Wisconsin. U S For Serv Res Pap FPL-161, 17 pp
Pronin D, Lassen L E 1970 Evaluating quaking aspen of Wisconsin for a site-quality to wood-quality relationship. U S For Serv Res Pap FPL-141, 8 pp
Purkayastha S K, Tandon R D, Rao K R 1973 A note on the variation in wood density in some 36-year-old teak trees from different seed origins. Indian For 99: 215–217
Radcliffe B M 1953 The influence of specific gravity and rate of growth upon the mechanical properties of sugar maple in flexure. Sta Bull 597 Purdue Univ Agr Exp Sta Lafayette, Indiana, 26 pp
Ralston R A, McGinnes E A 1964 Shortleaf pine wood density unaffected by ring growth. South Lumberman 208(2592): 17–19
Reck S 1969 Controlling growth rate and wood density within conifer species by means of genetical selection and intensive culture. 2nd World Consul For Tree Breed Washington DC, 7 pp
Reck S 1974 Growth and wood characteristics in clones of Populus trichocarpa grown in West Germany and their significance for selection. IUFRO Meeting Stockholm, Sweden
Reck S, Dieterich H 1976 Einige Wuchs- und Holzeigenschaften bei einer Kreuzung Populus alba×Populus grandidentata im Vergleich mit dem Schwarzpappelbastard "Robusta". Silvae Genet 25: 44–48
Rees L W, Brown R M 1954 Wood density and seed source in young plantation red pine. J For 52: 662–665
Reid J S 1963 Wood quality in rapidly growing conifers planted in New Zealand. IUFRO Sect 41 Comm on Fibre Char Melbourne, Australia, 1–38

Remröd J 1976 Val av tallprovenienser i Norra Sverige – analys av överlevnad, tillväxt och kvalitet i 1951 års fallproveniensförsök. (Choosing Scots pine (Pinus sylvestris L) provenances in Northern Sweden – analysis of survival, growth and quality in provenance experiments planted 1951). Skogshögsk Stockholm Inst Skogsgen Rapp Upps 19, 132 pp
Rendle B J 1955 Tension wood – a natural defect on hardwoods. Wood 20: 348–351
Rendle B J 1956 Compression wood. A natural defect of softwoods. Wood 21: 120–123
Rendle B J 1959 Fast-grown coniferous timber – some anatomical considerations. Quart J For: 1–8
Rendle B J, Phillips E W 1958 The effect of rate of growth (ring width) on the density of softwoods. Forestry 31: 112–120
Rennel J 1984 Future of paper in the telematic world. Jaakko Pöyry Review Helsinki, Finland, 231 pp
Resch H, Johnson J W, Bendtsen B A 1983 Strength and stiffness of clear Douglas-fir exposed to the Mount St. Helens volcanic eruption. For Prod J 33(7/8): 52–56
Resler P W, Gladstone W T, Marton R 1975 Effect of fertilization on paper-making properties of Douglas-fir. Tappi 58(2): 99–102
Rexrode C O, Baumgras J E 1984 Distribution of gum spots by causal agent in black cherry and effects on log and tree quality. South J App For 82(3): 22–28
Richardson C J, Koerper G J 1981 The influence of edaphic characteristics and clonal variation on quantity and quality of wood production in Populus grandidentata in the Great Lakes region of the USA. Mitteil Forstl Bundes-Versuchsanstalt Wien 142: 271–292
Richardson S D 1959 The effect of night temperature on tracheid size and wood density in conifers. Proc 9th Int Bot Congr Montreal, Canada 2: 326
Richardson S D 1964 Studies on the physiology of xylem development. III: Effects of temperature, defoliation, and stem girdling on tracheid size in conifer seedlings. J Inst Wood Sci 12: 3–11
Rink G, Thor E 1973 Inheritance and gains of three wood properties in Virginia pine. 12th South For Tree Improv Conf Baton Rouge, Louisiana, 24–30
Risi J, Zeller E 1960 Specific gravity of the wood of black spruce (Picea mariana Mill BSP) grown on a Hylocomium-Cornus site type. Laval Univ For Res Found Québec, Canada Contr 6, 70 pp
Roddy D M 1983 Jack pine wood density. Can Tree Impr Assoc Toronto, Canada, 19 pp
Rone V M 1970 [Variation in wood density and tracheid length in Picea abies progeny.] Lesoved 5: 78–82
Ross D, Buckner E, Core H, Woods F 1979 Nitrogen fertilization decreases yellow-poplar wood density. South J App For 8(3): 119–122
Rudman P 1970 The influence of genotype and environment on wood properties of juvenile Eucalyptus camaldulensis. Silvae Genet 19: 49–53
Rudman P, Higgs M, Davidson J, Malajczuk N 1969 Breeding eucalypts for wood properties. 2d World Conf For Tree Breed Washington DC, 9 pp
Rudman P, McKinnell F H 1968 Effect of various fertilizer combinations on wood density of juvenile P. radiata. Res Working Group Newsletter 1(1): 10–11
Rudman P, McKinnell F H 1970 Effects of fertilizers on wood density of young radiata pine. Aust For 34: 170–178
Rydholm S A 1965 Pulping processes. Interscience, New York, 1269 pp
Sajdak R L 1967 Variation in bark characters and wood specific gravity of sugar maple. 8th Lake States For Tree Improv Conf U S For Ser Res Pap NC-23: 10–14
Sakai K, Hatakeyama S 1963 Estimation of genetic parameters in forest trees without raising progeny. Silvae Genet 12: 152–157
Sakai H, Uegaki T 1962 On the structure of the annual ring of Japanese red pine (Pinus densiflora). Variation of annual ring width, specific gravity and tracheid length in young tree stems. Tottori Soc Agr Sci 14: 113–119
Santamour F S 1979 Inheritance of wound compartmentalization in soft maples. J Arbor 5: 220–225
Santos C F, Nogueiro I R 1971 Fiber dimensions of older age Eucalyptus saligna, at São Paulo State Anais da ESA "Luiz de Queiroz" 28: 165–175
Santos C F, Nagueiro I R 1977 The dimensions of vessels and increase in length of woody fibers in relation to the fusiform cambial initials in the annual rings of Eucalyptus saligna. Nat Congr Bot Belo Horizonte, Brazil, 1 p
Sanwo S K 1983 The influence of crown class on the specific gravity of a tropical regrowth hardwood. IUFRO Conf Div 5 Madison, Wisconsin, 48 (Abstracts)

Saucier J R 1987 Management practices and juvenile wood. FPRS 41st Ann Meet Louisville, Kentucky, 2 pp
Saucier J R, Dorman KW 1969 Intraspecific variation in growth and wood characteristics of two slash pine varieties grown in south Florida. 10th South For Tree Improv Conf Houston, Texas, 49–57
Saucier J R, Ike A J 1969 Effects of fertilization on selected wood properties of sycamore. For Prod J 19: 93–96
Saucier J R, Taras M A 1966a Wood density variation among six longleaf pine seed sources grown in Virginia. J For 64: 463–465
Saucier J R, Taras M A 1966b Specific gravity and fiber length variation within annual height increments of red maple. For Prod J 16: 33–36
Saucier J R, Taras M A 1967 Wood density and percent summerwood variation among nine loblolly pine seed sources grown in Alabama. 9th South Conf For Tree Improv Knoxville, Tennessee, pp 115–119
Saucier J R, Taras M A 1969 Regional variation in specific gravity of seven pines in the southern United States. U S For Ser Res Pap SE-45, 16 pp
Saucier J R, Taras M A 1970 Wood density surveys of the minor species of yellow pine in the Eastern United States. Part IV: pond pine (P. serotina). U S For Ser Res Pap SE-65, 12 pp
Savina A V 1956 [The physiological justification for the thinning of forests.] Trans from Russian by Israel Prog for Sci Trans For Nat Sci Found and USDA, 91 pp
Scaramuzzi G 1955 Dimensional data about fibers in Populus×euramericana (Dode) Guinier cv "I-214". 8th Sess Int Poplar Comm Madrid FAO/CIP/79-C, 20 pp
Scaramuzzi G 1957 Observaciones sobre dimensiones de las fibras leñosas en relación con las condiciones de crecimiento en algunos individuos de Eucalyptus camaldulensis. J argent Eucalipto Asoc for argent 236–200, 4 pp
Scaramuzzi G 1958 Variation of the volume proportions of wood elements within the stem in Populus×euramericana. Guinier Pap 4: 242–251
Scaramuzzi G 1960 Indagini tecnologichi sul legno di alcuni ibridi euramericani di pioppo. II. Dati anatomici. Pubbl Cent Sper Agr For Roma 3: 193–216
Scaramuzzi G 1965 The relationship of fibre wall thickness, fibre diameter and percentage of summerwood (latewood) to specific gravity. IUFRO Section 41 Melbourne, Australia, 11 pp
Scaramuzzi G, Ferrari G 1963 Relation of specific gravity to structure in poplar wood. IUFRO Sect 41 Melbourne, Australia, 6 pp
Schmidt J D, Smith W J 1961 Wood quality evaluation and improvement in Pinus caribaea. Queensland For Ser Res Note 15, 69 pp
Schmidtling R C 1973 Intensive culture increases growth without affecting wood quality of young southern pines. Can J For Res 3: 565–573
Schmidtling R C, Amburgey T L 1982 Genetic variation in decay susceptibility and its relationship to growth and specific gravity in loblolly pine. Holzforschung 36: 159–161
Schniewind A P 1961 The effect of site and other factors on specific gravity and bending strength of California red fir. For Sci 7: 106–115
Schniewind A P 1962 Horizontal specific gravity variation in tree stems in relation to their support function. For Sci 8: 111–118
Schoenike R E 1962 Natural variation in jack pine (Pinus banksiana). PhD Thesis, Univ of Minn St. Paul, Minnesota, 233 pp
Schönau A P 1973 The effect of planting espacement and pruning on growth, yield, and timber density of Eucalyptus grandis. J S Afr For 88: 16–23
Schönau A P 1982 The relationship between timber density of black wattle and fertilizers, stocking and growth rate. Rep Wattle Res Inst 35th yr Pietermaritzburg, South Africa, 114–117
Schönbach H 1960 Ergebnisse weiterer Untersuchungen über Cellulosegehalt und Rohwichte Graupaer Pappelklone. Wiss Abh Deutsch Akad Landwiss Berlin 44(4): 81–88
Schultze-Dewitz G 1958 Einfluß der soziologischen Stellung auf den Jahrringbau sowie auf die Holzelemente bei Kiefern, Spirken und Lärchen eines Naturwaldes. Holz-Zbl 84(65): 849–851
Schultze-Dewitz G 1959 Einfluß der Kronenhöhe auf den Jahrringbau sowie auf die Faserlänge und den Markstrahlanteil bei der Fichte verschiedener Altersklassen. Holzforsch Holzverwert 11: 67–74

Schultze-Dewitz G 1960 Wie wirkt sich der Einfluß der Stellung eines Baumes im Bestand auf seine Holzstruktur aus? Holzforsch Holzverwert 12: 30–33

Schultze-Dewitz G 1968 Einfluß der Vegetationszeit auf einige Strukturmerkmale bei Kiefernholz. Holzindustrie 21(2): 55–58

Schultze-Dewitz G 1969 Die Variation der Strukturmerkmale im Nadelholz. Holztechnologie 10: 185–189

Schütt P 1958 Schwankungen im Zellulose- und Ligningehalt bei einigen in Westdeutschland angebauten Pinus contorta-Herkünften. Silvae Genet 7: 65–69

Schütt P 1962 Individuelle und bestandsweise Schwankungen der Holzdichte und der Faserlänge bei Pinus contorta. Papier 16: 671–675

Scott C W, MacGregor W D 1952 Fast grown wood, its features, and value with special reference to conifer planting in the United Kingdom since 1919. 6th British Commonw For Conf Canada, 20 pp

Scott M H, Stephens R P 1947 The quality of mature Pinus patula and Pinus insignis timber grown in South Africa. Brit Emp For Conf, 23 pp

Scurfield G 1973 Reaction wood: its structure and function. Science 179: 637–656

Scurfield G, Anderson C A, Segnit E R 1974 Silica in woody stems. Aust J Bot: 211–229

Seibt G 1963 Der Einfluß der Düngung auf das Holzgewicht von Kiefer, Japanlärche und Fichte. Aus Walde 6: 51–82

Seibt G, Knigge W, Reemtsma J B, Ulrich B 1968 Untersuchungen über den Einfluß verschiedener Düngungsmaßnahmen auf Ertragsleistung, Holzeigenschaften, nadelanalytische Merkmale und Bodenzustand des Fichtenbestandes im Düngungsversuch Oerrel-Lintzel. Allg Forst-Jagdztg 139: 57–71

Sellers T 1962 Factors influencing the wood quality in plantation-grown slash pine. For Prod J 12: 443–446

Semke L K 1984 Effects of juvenile pine fibers on kraft paper properties. Symp N C State Univ Raleigh, North Carolina, P 10 (Summary)

Semke L K, Corbi J C 1974 Sources of less coarse pine fiber for southern bleached printing papers. Tappi 57(11): 113–117

Senft J F 1987 Physical and mechanical properties of juvenile wood. 41st Ann Meet FPRS Louisville, Kentucky, 2 pp

Senft J F, Bendtsen B A, Galligan W L 1983 Wood quality in plantation-grown species: practical considerations. IUFRO Conf Div 5 Madison, Wisconsin, 50

Senft J F, Bendtsen B A, Galligan W L 1985 Weak wood. J For 83: 476–485

Sesbou A 1981 [Study of the genetic variability of quality of wood and of collapse in Eucalyptus camaldulensis.] PhD Thesis Univ Nancy, France, 145 pp

Sesbou A, Nepveu G 1978 Variabilité infraspécifique du retrait avec collapse et de la densité du bois chez Eucalyptus camaldulensis. Ann Sci For 35: 237–263

Shelbourne C J 1963 Growth rates, stem form, and branching characteristics of five provenances of P. khasya grown under plantation conditions in N Rhodesia. Commonw For Rev 42: 334–336

Shelbourne C J 1969 Predicted genetic improvement from different breeding methods. 2d World Consul For Tree Breed Washington DC, 6 pp

Shelbourne C J 1971 Provenance variation in growth rate and other characterstics in 13-year-old loblolly pine (P. taeda) in New Zealand. For Res Inst Rep 54, 15 pp

Shelbourne C J, Ritchie K S 1968 Relationships between degree of compression wood development and specific gravity and tracheid characteristics in loblolly pine. Holzforschung 22: 185–190

Shelbourne C J, Stonecypher R W 1971 The inheritance of bole straightness in young loblolly pine. Silvae Genet 20: 151–156

Shelbourne C J, Thulin I J, Scott R M 1972 Variation, inheritance, and correlation amongst growth, morphological, and wood characters in radiata pine. N Z For Serv Prod For Div Gen and Tree Improvement Rep 61, 26 pp

Shelbourne C J, Zobel B J, Stonecypher R W 1969 The inheritance of compression wood and its genetic and phenotypic correlations with six other traits in 5-year-old loblolly pine. Silvae Genet 18: 43–47

Shepherd K R 1964 Some observations on the effect of drought on the growth of Pinus radiata. Aust For 28: 7–22

Shigo A L, Rogers N F, McGinnes E A, Funk D T 1978 Black walnut on Kansas strip mine soils: some observations 25 years after pruning. U S For Ser Res Pap NE-393, 15 pp

Shigo A L, Shortle W C, Garrett P W 1977 Genetic control suggested in compartmentalization of discolored wood associated with tree wounds. For Sci 23: 179

Shrikhande V J 1957 Some considerations in designing experiments on coconut trees. J Indian Soc Agr Sta 9: 82–99

Siddiqui K M 1971 Some effects of potassium fertilization on properties of wood and pulp from red pine. Tappi 54: 784

Siddiqui K M, Gladstone W T, Marton R 1972 Influence of fertilization on wood and pulp properties of Douglas-fir. Symp effect growth accel on wood prop. U S For Serv FRL Madison, Wisconsin, 16 pp

Siddiqui K M, Khan M, Akhtar S 1979 Results of 10-year-old Eucalyptus camaldulensis provenance study at Peshawar. Silvae Genet 28: 24–26

Siemon G R 1973 Effects of thinning on crown structure, stem form and wood density of radiata pine. PhD thesis Aust Nat Univ Canberra, Australia

Sinclair S A, Barnes D P 1983 Toughness of healthy and spruce budworm-killed balsam fir. For Prod J 33(7/8): 24–26

Sinclair S A, Ifju G 1977 Processing beetle-killed southern pine – an opinion survey in Virginia. South Lumberman 235: 11–14

Sinclair S A, Ifju G 1979 Lumber quality of beetle-killed southern pine in Virginia. For Prod J 29(4): 18–22

Skolman R G 1963 Wood density and growth of some conifers introduced to Hawaii. U S For Serv Res Pap PSW-12, 20 pp

Skolman R G 1972 Specific gravity variation in Eucalyptus robusta grown in Hawaii. U S For Serv Res Pap PSW-78, 7 pp

Slabaugh P E 1957 Do pruned red pine trees remain dominant? Tech Note 476 U S For Serv Lake States For Expt Stat, 2 pp

Sluder E R 1970 Variation in wood specific gravity of yellow poplar (Liriodendron tulipifera) and its relationship to environmental conditions in the Southern Appalachians. PhD Thesis North Carolina State Univ Raleigh, North Carolina, 77 pp

Sluder E R 1972 Variation in specific gravity of yellow-poplar in the southern Appalachians. Wood Sci 5: 132–138

Smith D M, Wahlgren H, Bengston G W 1971 Effect of irrigation and fertilization on the wood quality of young slash pine. Symp effect growth accel on wood prop. U S For Serv FPL Madison, Wisconsin, 8 pp

Smith DM 1962 Patterns of variation of wood structure and properties within trees of southern yellow pine and the influence of environmental factors on these patterns. Symp wood quality in relation to the southern pulp and paper industries. Raleigh, North Carolina, 14–16

Smith D M 1968 Wood quality of loblolly pine after thinning. U S For Serv Rep FPL-89, 10 pp

Smith F H 1967 Effects of balsam woolly aphids (Adelges piceae) infestation on cambial activity in Abies grandis. Am J Bot 54:1215–1223

Smith H D 1974 Wood as an energy source for the southern pulp and paper industry. Mimeo Rept School Forest Resources N C State Univ Raleigh, North Carolina, 18 pp

Smith H D 1977 Economically optimum spacing and site preparation for slash pine plantations. Tech Rep 59 School For Res N C State Univ Raleigh, North Carolina, 104 pp

Smith J H, Kennedy R W (1983) Effects of stand management on tree, log, and wood quality of Douglas-fir. IAWA Bull 4: 12

Smith J H, Heger L, Hejjas J 1966 Patterns in growth of earlywood, latewood and percentage latewood determined by complete analysis of 18 Douglas-fir trees. Can J Bot 44: 453–466

Smith J H, Ker J W 1960 Growing Douglas-fir and western hemlock trees at desired rates. Res Note 24, Fac For Univ Brit Columbia Vancouver, 32 pp

Smith J H, Ker J W, Csizmazia J 1961 Economics of reforestation of Douglas-fir, western hemlock and western red cedar in the Vancouver Forest District. Univ Brit Columbia For Bull 3, Vancouver, British Columbia, 144 pp

Smith W J 1967 The heritability of fibre characteristics and its application to wood quality improvement in forest trees. Silvae Genet 16: 41–50

Smith W J 1973 Wood yield, properties and quality in Queensland-grown Pinus caribaea. Selection and Breeding to Improve Some Tropical Conifers CFI Oxford, England, Vol 2, pp 45–69

Snellgrove T A, Ernst S 1983 Veneer recovery from live and dead lodgepole pine. For Prod J 33(6): 21–26

Snellgrove T A, Snell J K, Max T A 1983 Damage to national forest timber on Mount St. Helens. J For 81:368–371

Sohn S I, Goddard R E 1974 Genetic study of wood specific gravity in slash pine. 22nd Northeastern For Tree Improv Conf Syracuse, New York, 61–64

Son D S, Chung S B 1970 [Fibre length and specific gravity of Populus alba × glandulosa hybrid.] Inst For Genet Suwon, Korea Res Rep 8: 1–5

Son D S, Chung S B 1972 [The growth performance and specific gravity of hybrid poplar.] Inst For Genet Suwon Korea Res Rep 9: 9–16

Spurr S H, Hsuing W 1954 Growth rate and specific gravity in conifers. J For 55: 191–200

Spurr S, Hyvärinen M J 1954 Wood fiber length as related to position in tree and growth. Bot Rev 20: 561–575

Squillace A E 1978 Paraquat treatment reduces wood density after alcohol-benzene extraction. Proc Lightwood Res Coord Council Atlantic Beach, Florida, 50–53

Squillace A E, Echols R M, Dorman K W 1962 Heritability of specific gravity and summerwood percent and relation to other factors in slash pine. Tappi 45: 599–601

Staebler G R 1963 Growth along the stems of full-crowned Douglas-fir trees after pruning to specified heights. J For 61: 124–127

Stage A R 1963 Specific gravity and tree weight of single-stem samples of grand fir. U S For Serv Res Pap INT-4, 11 pp

Stairs G R 1969 Seed source and growth rate effects on wood quality in Norway spruce. 11th Meet Commit For Tree Breed in Canada, Ottawa, Canada, 231–236

Stairs G R, Marton R, Brown A F, Rizzo M, Petrik A 1966 Anatomical and pulping properties of fast and slow-grown Norway spruce. Tappi 49: 296–300

Stamm A J 1964 Wood and cellulose science. Ronald Press, New York, 58–61

Stamm A J, Sanders H T 1966 Specific gravity of the wood substance of loblolly pines as affected by chemical composition. Tappi 49: 397–400

Steinbeck K, McAlpine R G, May J T 1972 Short rotation culture of sycamore. J For 4: 210–213

Stern K 1961 Einfluß der Himmelsrichtung auf die Verteilung der Raumdichte des Holzes in Fichtenbeständen. Holz Roh- Werkst 19: 213–218

Stern K 1963 Einfluß der Höhe am Stamm auf die Verteilung der Raumdichte des Holzes in Fichtenbeständen. Holzforschung 17: 6–12

Stevens M, Lejeune J, Schalck J 1983 Variation of specific gravity and tensionwood distribution in fast growing poplar hybrids. IUFRO Conf Div 5 Madison, Wisconsin

Stonecypher R W 1966 The loblolly pine heritability study. Estimates of genetic and environmental variances and covariances in a natural population of loblolly pine (Pinus taeda). Southlands Exper. Forest Tech Bull No 5, 128 pp

Stonecypher R W, Cech F C, Zobel B J 1964 Inheritance of specific gravity in two- and three-year-old seedlings of loblolly pine. Tappi 47: 405–407

Stonecypher R W, Zobel B J 1966 Inheritance of specific gravity in 5-year-old seedlings of loblolly pine. Tappi 49: 303–305

Stonecypher R W, Zobel B J, Blair R 1973 Inheritance patterns of loblolly pines from a non-selected natural population. North Carolina Agr Exp Sta Tech Bull 200, 60 pp

Strauss S H, Critchfield W B 1982 Inheritance of β-pinene in xylem oleoresin of knobcone × monterey pine hybrids. For Sci 28: 687–696

Strickland R K 1960 Geographic variations in specific gravity and tracheid length of loblolly pine. M S Thesis Univ of Georgia Athens, Georgia. 68 pp

Stickland R K 1965 Longleaf pine specific gravity variation. Tree Improv Tips Univ Florida Gainesville, Florida, 11 pp

Strickland R K, Goddard R E 1965 a Specific gravity variation – Choctawhatchee race of sand pine. Tree Improv Tips Univ Florida, Gainesville, Florida, 12 pp

Strickland R K, Goddard R E 1965 b Inheritance of branching and crown characteristics in slash pine. 8th South Conf For Tree Improv Savannah, Georgia, 57–63

Strickland R K, Goddard R E 1966 Correlation studies of slash pine tracheid length. For Sci 12: 54–62

Sudo S 1983 Assessment for wood quality improvement of akamatsu (Pinus densiflora). IUFRO Conf Div 5 Madison, Wisconsin, 53 (Abstract)

Sunly J G, Lavers G M 1961 Variations in the strength and specific gravity of Sitka spruce grown in Great Britain. J Inst Wood Sci 7: 15–27

Sutton W R, Harris J M 1973 Effect of heavy thinning on wood density in radiata pine. N Z J For Sci 4: 112–115

Swan H S 1961 The influence of gibberellic acid on the growth and development of three hybrid poplars. Rep Pulp Pap Res Inst Canada Woodlands Res Index 124, 18 pp

Szopa P S, Tennyson L C, McGinnes E A 1976 Influence of irrigation and wood structure on white and red oaks. For Prod Res Soc Ann Meet Toronto, Canada, 26

Ta-wei N, Beng-yen Y, Wen-Chung S 1972 Variations in wood specific gravity and fiber length of Pinus taiwanensis. U S – China Coop Sci Sem For Ecol Genetics Taipai Taiwan, 26 pp

Tackle D 1962 Specific gravity of lodgepole pine in the Intermountain Region. U S For Serv Res Note I, 4 pp

Tajima T 1967 Tree growth and wood properties. Fac Agr Tokyo Univ Tokyo, 208 pp

Talbert J T, Jett J B 1981 Regional specific gravity values for plantation grown loblolly pine in the southeastern United States. For Sci 27: 801–807

Talbert J T, Jett J B, Bridgwater F E, Jahromi S T 1982 Genetic parameters of wood specific gravity in a control pollinated loblolly pine genetic test. TAPPI Res Dev Div Conf Asheville, North Carolina, pp 179–182

Tamm C O, Carbonnier C H, Hagberg N 1960 Recent results in fertilizer experiments in Swedish forests. 5th World For Congr Seattle, Washington, 558–561

Taras M A, Saucier J R 1967 Influence of extractives on specific gravity of southern pines. For Prod J 17: 97–99

Taras M A, Saucier J R 1970 Wood density surveys of the minor species of yellow pine in the Eastern United States. IV: Pond pine, Pinus serotina. U S For Serv Res Pap SE-65, 16 pp

Taylor F W 1965 A study of the natural variation of certain properties of the wood of yellow-poplar (Liriodendron tulipifera) within trees, between trees, and between geographic areas. PhD Thesis School For N C State Univ Raleigh, North Carolina, 190 pp

Taylor F W 1968 Specific gravity differences within and among yellow-poplar trees. For Prod J 18: 75–81

Taylor F W 1969a Variation in wood properties in sycamore. For Prod Utiliz Lab Mississippi State Univ Res Rep 7, 18 pp

Taylor F W 1969b Variation of wood properties in pecan. For Prod Utiliz Lab Mississippi State Univ Res Rep 8, 17 pp

Taylor F W 1969c The effect of ray tissue on the specific gravity of wood. Wood Fiber 1: 142–145

Taylor F W 1971 Variation of wood properties in sugarberry. For Prod Utiliz Lab Mississippi State Univ Res Rep 11, 17 pp

Taylor F W 1973a Differences in the wood of Eucalyptus grandis grown in different parts of South Africa. Special Rep IUFRO Meet Sect 5 Pretoria, S Afr 2: 1020–1027

Taylor F W 1973b Variation in the anatomical properties of South African grown Eucalyptus grandis. Appita 27: 171–178

Taylor F W 1977 Variation in specific gravity and fiber length of selected hardwoods throughout the Mid-South. For Sci 23: 190–194

Taylor F W 1982 The anatomical properties of fast-grown pine. TAPPI Res Dev Conf Asheville, North Carolina, 91–97

Taylor F W, Wang E I, Yanchuk A, Micko M M 1982 Specific gravity and tracheid length variation of white spruce in Alberta. Can J For Res 12: 561–566

Taylor F W, Wooten T E 1973 Wood property variation of Mississippi delta hardwoods. Wood Fiber 5: 2–13

Teissier du Cros E, Kleinschmit J, Azoeuf P, Hoslin R 1980a Drehwuchs bei Buche, Variabilität und Erblichkeit. Forstarchiv 51: 41–47

Teissier du Cros E, Kleinschmit J, Azoeuf P, Hoslin R 1980b Spiral grain in beech, variability and heredity. Silvae Genet 29: 5–13

Thiercelin F 1970 Tardivité du débourrement et densité dans une population adulte de Picea abies Karst. Ann Sci For 27: 243–254
Thomas R J 1985 The characteristics of juvenile wood. Symp Util Chang Wood Res South U S North Carolina State Univ Raleigh, North Carolina, 18 pp
Thor E 1964 Variation in Virginia pine I: Natural variation in wood properties. J For 62: 258–262
Thor E 1965 Variation in some wood properties of eastern white pine. For Sci 1: 451–455
Thor E, Bates A L 1973 Relationship of site and radial growth with wood specific gravity and extractives in Eastern white pine. J Tenn Acad Sci 48: 5–8
Thor E, Brown S J 1962 Variation among six loblolly pine provenances tested in Tennessee. J For 60: 476–480
Thor E, Core H A 1977 Fertilization, irrigation, and site factor relationship with growth and wood properties of yellow poplar. For Sci 9: 130–135
Thorbjornsen E 1961 Variation in density and fiber length in wood of yellow poplar. Tappi 44: 192–195
Tiemann M D 1944 Wood technology. Pitman Publishing Corp, New York, 328 pp
Timell T E 1969 The chemical composition of tension wood. Svensk Papperstidn 72: 173–181
Timell T E 1973a Studies on opposite wood in conifers. I: Chemical composition. Wood Sci Technol 7: 1–5
Timell T E 1973b Studies on opposite wood in conifers. II: Histology and ultrastructure. Wood Sci Technol 7: 79–91
Timell T E 1981a Recent progress in the chemistry, ultrastructure and formation of compression wood. Chemistry and morphology of wood and wood components. Int Symp Wood Pulp Chemistry. Vol I Stockholm: 99–147
Timell T E 1981b Recent progress in the chemistry and topochemistry of compression wood. Wood Sci Technol 16: 83–122
Timell T E 1983 Origin and evolution of compression wood. Holzforschung 37: 1–10
Timell T E 1986 Compression wood in gymnosperms. Springer, Heidelberg, 2150 pp
Todorovski S 1969 [Distribution of weight in the sessile oak (Quercus sessilifolia) trunk.] School Agr For Univ Skopje Terezije Belgrade, 28 pp (Trans for USDA)
Tsoumis G, Panagiotidis N 1980 Effect of growth conditions on wood quality characteristics of black pine (Pinus nigra). Wood Sci Technol 14: 301–310
Tsutsumi J, Matsumoto T, Mio S 1983 Effect of growth rate on the intrinsic wood properties of Japanes larch (Larix leptolepis). IUFRO Conf Div 5 Madison, Wisconsin, 56
Turnbull J M, du Plessis C P 1946 Some sidelights on the rate of growth bogey. J S Afr For Assoc 14: 29–36
Turnbull R F 1965 Review of current attitudes of growth stresses. IUFRO Sect 46 Melbourne, Australia, 8 pp
Uusvaara O 1974 Wood quality in plantation-grown Scots pine. Commun Inst For Fenn 80.2, 105 pp
Valentine F A 1962 Natural variation in specific gravity in Populus tremuloides in northern New York. 9th Northeastern For Tree Impr Conf Syracuse, New York, pp 17–24
van Buijtenen J P 1955 Summerwood formation in loblolly pine. PhD Thesis Texas A&M Univ Coll Station, Texas, 83 pp
van Buijtenen J P 1962 Heritability estimates of wood density in loblolly pine. Tappi 45: 602–605
van Buijtenen J P 1963 Heritability of wood properties and their relation to growth rate in Pinus taeda. 1st World Consul For Gen Tree Improv Stockholm, Sweden, 13 pp
van Buijtenen J P 1964 Anatomical factors influencing wood specific gravity of slash pines and the implications for the development of a high-quality pulpwood. Tappi 47: 401–404
van Buijtenen J P 1965 Inheritance of fiber properties in North America conifers. IUFRO Section 41 Melbourne, Australia, 12 pp
van Buijtenen J P 1969 Controlling wood properties by forest management. Tappi 52(2): 257–259
van Buijtenen J P 1978 Response of "Lost Pines" seed sources to site quality. 5th N A For Biol Worksh Gainesville, Florida, 7 pp
van Buijtenen J P 1982 Fiber for the future. Tappi 65(8): 10–12
van Buijtenen J P 1984 An integrated accelerated breeding program. Adv Gen Breed: Current Status Res Needs S-23 Meet Baton Rouge, Louisiana, 10 pp

van Buijtenen J P 1986 Computer simulation of the effect of wood specific gravity and rotation age on the production of linerboard and multiwall sack paper. TAPPI Res Dev Conf Raleigh, North Carolina, 49–53

van Buijtenen J P, Alexander S D, Einspahr D W, Ferrie A E, Hart T, Kellogg R M, Porterfield R L, Zobel B J 1975 How will tree improvement and intensive forestry affect pulp manufacture? (Digest of CA Report 54) Tappi 58(9): 129–134

van Buijtenen J P, Alexander S D, Einspahr D W, Ferrie A, Hart T, Kellogg R M, Zobel B J 1974 Quality control in the forest and in the mill. TAPPI CA Rept 54 Atlanta, Georgia, 96 pp

van Buijtenen J P, Einspahr D W, Peckham J R 1968 Micropulping loblolly pine grafts selected for extreme wood specific gravity. Silvae Genet 17: 15–19

van Buijtenen J P, Joranson P N, Einspahr D W 1958 Diploid versus triploid aspen as pulpwood sources with reference to growth, chemical, physical, and pulping differences. Tappi 41: 170–175

van der Sijde H A, Shaw M J, van Wyk G 1985 Reaction wood in Pinus taeda – a preliminary report. S Afr For J: 27–32

van der Slooten J H, Fishwick R W, Richter H G, Ferreira M, Montagna R G 1976 An examination of the wood density from plantations in Southern Brazil. PRODEPEF Tech Series 5, 47 pp

van Eck W A, Woessner R A 1964 A study of wood density in yellow poplar and red oak as related to environment. West Virginia Acad Sci 36, 8 pp

Veal M A, Blair R L, Jett J B, McKean W T 1974 Impact of fusiform rust on pulping properties of young loblolly and slash pine. Proc TAPPI For Biol Conf Seattle, Washington, 47–58

Veal M A, Marrs G R, Jackson M 1986 Control of loblolly pine wood quality for fiber products. Tappi 70(1): 51–54

Vidakovíc M, Ahsan J 1970 The inheritance of crooked bole in Shisham (Dalbergia sissoo). Silvae Genet 19: 94–98

Vital B R, Lucia R M 1980 Procedência de semintes e qualidade da madeira de Eucalyptus grandis. Revista Arvore 4(2): 170–178

Vital B R, Pereira A R, Lucia R M, De Andrade D C 1984 Effect of age of tree on density of wood of E. grandis cultivated in the Cerrado region of Minas Gerais. Bol Tec 8: 49–57

von Pechmann H 1953 Untersuchungen über die Bruchschlagarbeit von Rotbuchenholz. Holz Roh-Werkst 11: 361–367

von Pechmann H 1958 Die Auswirkung der Wuchsgeschwindigkeit auf die Holzstruktur und die Holzeigenschaften einiger Baumarten. Schweiz Z Forstwes 109: 615–647

von Pechmann H, Wutz A 1960 Haben Mineraldüngung und Lupinenanbau einen Einfluß auf die Eigenschaften von Fichten- und Kiefernholz? Forstwiss Cbl 79: 91–105

Von Wedel, Zobel B J, Shelbourne C J 1968 Prevalence and effects of knots in young loblolly pine. For Prod J 18: 97–103

Voorhies G 1969 Specific gravity studies of young-growth southwestern ponderosa pine. For Prod J 19(6): 45–46

Wahlgren H E 1965 Forest sercive western woods density survey. Phases II: (increment core processing) and III: (estimating tree specific gravity). Symp on density – A key to wood quality. Madison, Wisconsin, 21–35

Wahlgren H E, Hart A C, Maeglin R R 1966 Estimating tree specific gravity of Maine conifers. U S For Ser Res Pap FPL-61, 25 pp

Wahlgren H E, Yandle D O 1970 Development of a model for estimating tree specific gravity of loblolly pine. Wood Sci 2: 129–135

Walker L C 1960 Forest fertilization – its effect on growth and wood quality. Soil Sci Soc Am Ann Meet Chicago, Illinois, 13 pp

Walters C S, Brukman G 1965 Variation in specific gravity of cottonwood as affected by tree sex and stand location. J For 63: 182–185

Wang S, Littell R C, Rockwood D L 1984 Variation in density and moisture content of wood and bark among twenty Eucalyptus grandis progenies. Wood Sci Technol 18: 97–120

Wang I C, Micko M M 1984 Wood quality of white spruce from north central Alberta. Can J For Res 14: 181–185

Wangaard F F 1950 The mechanical properties of wood. John Wiley and Sons, New York, 377 pp

Wardrop A S 1956 The nature of reaction wood. V: The distribution and formation of tension wood in some species of Eucalyptus. Aust J Bot 4: 152–166

Wardrop A B 1961 The structure and formation of reaction wood in Angiosperms. Rec Adv Bot Vol II: 1325–1330

Wardrop A B, Dadswell H E 1950 The nature of reaction wood. II: The cell wall organization of compression wood tracheids. Aust J Sci Res 3: 1–13

Wardrop A B, Dadswell H E 1952 The nature of reaction wood. III: Cell division and cell wall formation in conifer stems. Aust J Sci Res B-5: 385–398

Wardrop A B, Dadswell H E 1955 The nature of reaction wood. IV Variation in cell wall organization of tension wood fibres. Aust J Bot 3: 177–189

Warren B R, Mirams R V 1963 A survey of basic density and fibre length variation in Pinus radiata. The Physical Properties of Pinus Radiata Rept 7, 22 pp

Warren W G 1979 The contribution of earlywood and latewood specific gravities to overall wood specific gravity. Wood Fiber 11: 127–135

Watanabe H 1965 A study of the origin of longitudinal growth stresses in tree stems. IUFRO Sect 41 Melbourne, Australia, 17 pp

Watanabe H, Tsutsumi J, Kanagawa H 1962 Properties of branchwood, especially on specific gravity, tracheid length, and appearance of compression wood. Bull Kyushu Univ For Note 35: 91–96

Watson A J 1956 Pulping characteristics of eucalypt tension wood. Appita 10: 43–59

Watson A J 1965 Fibre characteristics and wood properties; tropcial and semi-tropical hardwoods. IUFRO Sect 41 Melbourne, Australia, 7 pp

Watson A J, Dadswell H E 1957 Paper making properties of compression wood from Pinus radiata. J Aust Pulp Pap Ind Tech Assoc 11(3): 56–70

Waugh G 1972 Growth stresses – genetic and environmental influences. CSIRO For Prod Newsletter South Melbourne, Victoria, 3–4

Webb C D 1964 Natural variation in specific gravity, fibre length, and interlocked grain of sweetgum (Liquidambar styraciflua) in the South Atlantic States. PhD Thesis School For North Carolina State Univ Raleigh, North Carolina, 125 pp

Webb C D 1969 Variation of interlocked grain in sweetgum. For Prod J 19(8): 45–48

Webb C D, Belanger R P, McAlpine R G 1973 Family differences in early growth and wood specific gravity of American sycamore (Platanus occidentalis). 12th South For Tree Improv Conf Baton Rouge, Louisiana, pp 213–227

Webster N, McKechnie J L 1980 Websters new twentieth century dictionary of the English language unabridged. William Collins, 2129 pp

Weiner J, Roth L 1966 The influence of environment and genetics on pulpwood quality. Inst Paper Chem Bibliog Series 24 Appleton, Wisconsin, 176 pp

Wellwood R W, Smith J G 1962 Variation in some important qualities of wood from young Douglas-fir and hemlock trees. Res Pap 50 Univ Brit Columbia Vancouver, Canada, 15 pp

Werner R A, Elert E E, Holsten E H 1983 Evaluation of beetle-killed white spruce for pulp and paper. Can J For Res 13: 246–250

Westing A H 1965 Formation and function of compression wood in Gymnosperms I. Bot Rev 31: 381–480

Westing A H 1968 Formation and function of compression wood in Gymnosperms II. Bot Rev 34: 51–78

Wheeler E A 1987 Anatomical and biological properties of juvenile wood in conifers and hardwoods. 41st Ann Meet FPRS Louisville, Kentucky, 2 pp

Wheeler E A, Pearson R G, La Pasha C A, Zack T, Hadley W 1986 Computer-aided wood identification-reference manual. North Carolina Agr Res Ser Bull 474 Raleigh, North Carolina, 160 pp

Wheeler E Y, Zobel B J, Weeks D L 1965 Tracheid length and diameter variation in the bole of loblolly pine. Tappi 49: 484–490

White J F, Saucier J R 1966 A comparison of the specific gravity of two slash pine varieties grown in south Florida. Tappi 49: 230–232

Whitesell C D, Zobel B J, Roberds J 1966 Specific gravity and tracheid length of loblolly pine in Maryland and Delaware. Tech Rept 29 School of For Res North Carolina State Univ Raleigh, North Carolina, 11 pp

Whitmore J L 1973 Wood density in Costa Rican balsa. Wood Sci 5: 223–229

Whitemore J L, Liegel L H 1980 Spacing trials of Pinus caribaea var. hondurensis. U S For Serv Res Pap SO-162, 8 pp

Whyte A G, Adams P, McEwen S E 1980b Size and stem characteristics of foxtails compared with Pinus caribaea var hondurensis of normal habit. IUFRO Div 5 Oxford, England, 50 (Summary)

Whyte A G, Adams P M, McEven S E 1981 Foxtailing in Pinus caribaea var. hondurensis in Fiji; frequency distribution of occurrence and wood properties. For Ecol Mgt 3: 237–243

Whyte A G, Wiggins P C, Wong T W 1980a A survey of spiral grain in Pinus caribaea var. hondurensis in Fiji and its effects. IUFRO Conf Div 5 Oxford, England, 58 (Abstract)

Wicker M 1979 Le bois de tension: acquisitions récentes. Ann Biol 18: 221–254

Wilcox M D 1975 Wood brightness variation in clones of loblolly pine. Silvae Genet 24: 54–99

Wilcox M D 1977 Variation of growth, stem quality, and wood properties in relation to genetic improvement of tropical forest trees. IUFRO Workshop Brisbane, Australia, 9 pp

Wilcox M D, Smith H D 1973 Selection indices for wood quality in loblolly pine. 12th South For Tree Improv Conf Baton Rouge, Louisiana, 322–342

Wilcox W W, Pong W Y 1971 The effect of height, radial position, and wetwood on white fir wood properties. Wood Fiber 3: 47–55

Wilde S A, Paul B H 1959 Growth, specific gravity, and chemical composition of quaking aspen on different soil types. U S For Serv Rep FPL-2144, 10 pp

Wilde S A, Shaw B H, Fedkenheuer A W 1968 Weeds as a factor depressing forest growth. Weed Res 8: 196–204

Wilkes J, Abbott D 1983 Influence of rate of tree growth on anatomy of eucalypt species. Appita 37: 231–232

Williams R F, Hamilton J R 1961 The effect of fertilization on four wood properties of slash pine. J For 59: 662–665

Wilson J W 1964 Wood characteristics III: Intra-increment physical and chemical properties. Pulp Pap Res Inst Can Res Note 45, 9 pp

Wilson J W, Ifju G 1965 Wood characteristics VII: Intra-increment relationship of Douglas-fir wood density. Pulp Pap Res Inst Can Tech Rep Ser 424, 24 pp

Wiselogel A E, Tauer C G 1982 Genetic variation of wood specific gravity and tracheid length in shortleaf pine. TAPPI Res Dev Div Conf Asheville, North Carolina, 169–178

Wodzicki T 1961 Investigations on the kind of Larix polonica wood formed under various photoperiodic conditions. II Effect of different light conditions on wood formed by seedings in a greenhouse. Acta Soc Bot Polon 30: 111–131

Wodzicki T 1964 Photoperiodic control of natural growth substances and wood formation in larch (Larix decidua). J Exp Bot 15: 584–599

Wodzicki T, Witkowska L 1961 On the photoperiodic control of extension growth and wood formation in Norway spruce. Acta Soc Bot Polon 30: 755–764

Wodzicki T J, Wodzicki A B 1973 Auxin stimulation of cambial activity in Pinus silvestris. II: Dependence upon basipetal transport. Physiol Plant 29: 288–292

Worrall J 1970 Interrelationships among some phenological and wood property variables in Norway spruce. Tappi 53(1): 58–63

Worrall J 1975 Provenance and clonal variation in phenology and wood properties of Norway spruce. Silvae Genet 24: 2–5

Worrall J J, Burley J, Palmer E R, Hughes J F 1977 The properties of some Caribbean pine pulps, and their relationship to wood specific gravity variables. Wood Fiber 8: 228–234

Woods F W, Vincent L W, Moschler W W, Core H A 1979 Height, diameter, and specific gravity of "foxtail" trees of Pinus caribaea. For Prod J 29(5): 43–44

Wooten T E 1968 The identification and properties of compression wood. Mississippi For Prod Util Lab Inf Ser 7, 7 pp

Wright J A 1987 Results of micropulping wood samples of seven pine hybrid families in Zululand. 19th South For Tree Impr Conf College Station, Texas, 399–406

Wright J P 1970 Site and silvicultural influences on log quality in radiata pine. Pinus radiata Symp Aust Nat Univ Canberra, Australia

Wright J W 1976 Introduction to forest genetics. Academic Press, New York, 463 pp

Wright J W, Read R A, Barrett W H 1970 Use of geographic variation in the improvement of plantations in Oregon pinus in the Republic of Argentina. IDIA – Supplemento Forestal, 11 pp

Wylie R E 1976 Radiata pine – a basic for selection of trees for pruning and thinning. N Z For Ser 4th Ed, 24 pp

Yanchuck A D, Dancik B P, Micko M M 1983a Intraclonal variation in wood density of trembling aspen in Alberta. Wood Fiber Sci 15: 8 pp

Yanchuk A D, Dancik B P, Micko M M 1983b Variation and heritability of wood density and fiber length of trembling aspen in Alberta, Canada. Silvae Genet 32: 65–74

Yang K C 1983 Vertical variation of sapwood thickness in Pinus banksiana and Larix laricina. IUFRO Conf Div 5 Madison, Wisconsin, 60 (Abstract)

Yazawa K, Fukazawa K 1959 Studies on the relation between physical properties and growth conditions for planted sugi (Cryptomeria japonica D. Don) in central district of Japan. V: On the specific gravity, fiber saturation point and volumetric shrinkage of the spring- and summerwood. Bull Coll Exp For Fac Agr Hokkaido Univ 20: 93–117

Yiannos P N 1964 The apparent cell-wall density of wood and pulp fibers. Tappi 47: 468–471

Young H E, Kramer P J 1952 The effect of pruning on the height and diameter growth of loblolly pine. J For 50: 474–476

Youngs R L, Erickson J R 1984 Utilization of our changing wood resource. Symp util chang wood res south United States. Symp North Carolina State Univ Raleigh, North Carolina, 3 (Summary)

Zahner R 1955 Soil moisture and tree growth. Gulf States Sect Ann Meet SAF Biloxi, Mississippi, 25–30

Zahner R 1968 Site quality and wood quality in upland hardwoods: theoretical considerations of wood density. For Soils Conf Raleigh, North Carolina, 43 pp

Zahner R, Lotan J E, Baughman W D 1964 Earlywood-latewood features of red pine grown under simulated drought and irrigation. For Sci 10: 361–370

Zahner R, Oliver W W 1962 The influence of thinning and pruning on the rate of summerwood initiation in red and jack pines. For Sci 8: 51–63

Zahner R, Whitemore F W 1960 Early growth of radically thinned loblolly pine. J For 58: 628–633

Zavarin E, Mirov N T, Colling E N, Snajberk K, Costello K 1968 Chemical composition of turpentine from Pinus khasya. For Sci 14: 55–61

Zerges R V, Newman R D 1980 Kraft pulp and paper properties of Populus clones under short-rotation intensive culture. Tappi 63(7): 91–94

Zimmermann M H (ed) 1964 The formation of wood in forest trees. Academic Press, New York, 562 pp

Zobel B J 1956 Genetic, growth, and environmental factors affecting specific gravity of loblolly pine. For Prod J 6: 442–477

Zobel B J 1961 Inheritance of wood properties in conifers. Silvae Genet 10: 65–70

Zobel B J 1963 Improving wood qualities through use of seed from selected parents and the effect of different environmental and growth conditions. Symp wood quality North Carolina State Univ Raleigh, North Carolina, 10 pp

Zobel B J 1964 Breeding for wood properties in forest trees. Unasylva 18: 89–103

Zobel B J 1965a Inheritance of fiber characteristics and specific gravity in hardwoods – A review. IUFRO Section 41 Meet Melbourne, Australia, 13 pp

Zobel B J 1965b Inheritance of spiral grain. IUFRO Section 41 Melbourne, Australia, 9 pp

Zobel B J 1970 Developing trees with wood properties more desirable for paper. Tappi 53: 2320–2325

Zobel B J 1971 Genetic manipulation of wood of the southern pines including chemical characteristics. Wood Sci Technol 54: 255–271

Zobel B J 1972 Genetic effects on wood qualities. For Biol Comm Meet TAPPI Appleton, Wisconsin, 9 pp (mimeo)

Zobel B J 1973 Wood of fast growing exotic pines. Selection and Breeding to Improve some Tropical Conifers Dept For Queensland, Australia, Vol II, 1–3

Zobel B J 1974 Improving yields from forest land. Proc Fiber Conser Util Seminar Pulp Pap: 17-1–17-9

Zobel B J 1975a Using the juvenile wood concept in the southern pines. South Pulp and Paper Mfg, 14–16

Zobel B J 1975b Our changing wood resource – its effect on the pulp industry. Appl Polym Symp 28: 47–54

Zobel B J 1976 Wood properties as affected by changes in the wood supply of southern pines. Tappi 59(4): 126–128

Zobel B J 1979a Timber supply trends in South America. CICEPLA Meet Cartagena, Colombia (mimeo), 15 pp

Zobel B J 1979b The southeast timber supply, how will it be affected by changing forestry in South America, Canada and the Northeast? South J. Appl For 3: 37–42
Zobel B J 1980a Genetic improvement of forest trees for biomass production. Progress in biomass conversion, C. Academic Press, New York, 37–58 (Eds Sarkenen K V and Tillman D A)
Zobel B J 1980b Inherent differences affecting wood quality in fast-grown plantations. IUFRO Conf Div 5 Oxford, England, 169–188
Zobel B J 1981 Wood quality from fast-grown plantations. Tappi 64(1): 71–74
Zobel B J 1984 The changing quality of the world wood supply. Wood Sci Technol 18: 1–17
Zobel B J 1985 Juvenile wood in tropical forest plantations; its characteristics and effect on the final product. CAMCORE Bull Trop For 2, 29 pp
Zobel B J 1986 Advances in forestry in South America and their effect on competition with the southern United States. Tappi 69(10): 34–36
Zobel B J, Blair R 1976 Wood and pulp properties of juvenile wood and topwood of the southern pines. Appl Polym Symp 28: 421–433
Zobel B J, Campinhos E, Ikemori Y 1983 Selecting and breeding for desirable wood. Tappi 66(1): 70–74
Zobel B J, Cole D, Stonecypher R 1962 Wood properties of clones of slash pine. For Genet Workshop Macon, Georgia, 32–39
Zobel B J, Goggans J F, Maki T E 1961a Wood properties affected by fertilizers. For Farmer 20(9): 6, 7, 18
Zobel B J, Haught A E 1962 Effect of bole straightness on compression wood of loblolly pine. Tech Rep 15 North Carolina State Univ School For Raleigh, North Carolina, 4 pp
Zobel B J, Henson F, Webb C 1960a Estimation of certain wood properties of loblolly and slash pine trees from breast height sampling. For Sci 6: 155–162
Zobel B J, Jett J B, Hutto R 1978 Improving wood density of short rotation Southern pine. Tappi 61(3): 41–44
Zobel B J, Kellison R C 1972 Short-rotation forestry in the Southeast. Tappi 55: 1205–1208
Zobel B J, Kellison R C 1973 Should wood be included in a pine tree improvement program? Selection and breeding to improve some tropical conifers. II: 4–12
Zobel B J, Kellison R C 1984 Wood – Where will it come from, where will it go? (A comparison of the southern United States with South America). Utiliz Changing Wood Res South U S Raleigh, North Carolina, 12 pp
Zobel B J, Kellison R C, Kirk D G 1973 Wood properties of young loblolly and slash pine. Effect of growth acceleration on wood properties. For Prod Lab U S For Serv Madison, Wisconsin, 22 pp
Zobel B J, Kellison R C, Mathias M 1969a Genetic improvement in forest trees – growth rate and wood characteristics in young loblolly pine. 10th South Conf For Tree Improv Houston, Texas, 57–75
Zobel B J, Kellison R C, Mathias M F, Hatcher A V 1972 Wood density of the southern pines. Tech Bull 208 North Carolina Agr Exp Sta Raleigh, North Carolina, 56 pp
Zobel B J, Mathias M, Kellison R C, Roberds J H 1968 Moisture content of southern pine trees. Tech Rep 37 School For Res North Carolina State Univ Raleigh, North Carolina, 44 pp
Zobel B J, McElwee R L, Browne C 1961b Interrelationship of wood properties of loblolly pine. 6th South Conf For Tree Improv Gainesville, Florida, 142–163
Zobel B J, Ralston J, Roberds J H 1965 Wood yields from loblolly pine stands of different age, site and stand density. Tech Rep 26, North Carolina State Univ School For Raleigh, North Carolina, 23 pp
Zobel B J, Ralston J J, Roberds J H 1969b Dry wood weight yields of loblolly pine. J For 69: 822–824
Zobel B J, Stonecypher R L, Browne C 1968 Inheritance of spiral grain in young loblolly pine. For Sci 14: 376–379
Zobel B J, Stonecypher R, Browne C, Kellison R C 1966 Variation and inheritance of cellulose in southern pines. Tappi 49: 383–387
Zobel B J, Talbert J 1984 Applied forest tree improvement. Wiley, New York, 511 pp
Zobel B J, Thorbjornsen E, Henson E 1960c Geographic, site, and individual tree variation in the wood properties of loblolly pine. Silvae Genet 9: 149–158
Zobel B J, van Wyk G, Stahl P 1987 Growing exotic forests. Wiley, New York, 505 pp
Zoliver R 1968 Site quality and wood quality in upland hardwoods: theoretical considerations of wood density. For Soils Conf Raleigh, North Carolina (Mimeo Rept)

Subject Index

Species Index

Conifers

Hardwoods

MIX
Papier aus verantwortungsvollen Quellen
Paper from responsible sources
FSC® C105338

If you have any concerns about our products,
you can contact us on
ProductSafety@springernature.com

In case Publisher is established outside the EU,
the EU authorized representative is:
Springer Nature Customer Service Center GmbH
Europaplatz 3, 69115 Heidelberg, Germany

Printed by Libri Plureos GmbH
in Hamburg, Germany